Crystallization
as a Separations Process

A C S   S Y M P O S I U M   S E R I E S **438**

Crystallization as a Separations Process

Allan S. Myerson, EDITOR
Polytechnic University

Ken Toyokura, EDITOR
Waseda University

Developed from a symposium sponsored
by the International Chemical Congress
of Pacific Basin Societies,
Honolulu, Hawaii,
December 17–22, 1989

American Chemical Society, Washington, DC 1990

Library of Congress Cataloging-in-Publication Data

Crystallization as a separations process/Allan S. Meyerson, editor. Ken Toyokura, editor.

   p.   cm.—(ACS symposium series; 438)
"Developed from a symposium sponsored by the International Chemical Congress of Pacific Basin Societies, Honolulu, Hawaii, December 17–22, 1989."

   Includes bibliographical references and indexes.

   ISBN 0–8412–1864–1

   1. Crystallization—Congresses. 2. Separation (Technology)—Congresses.

   I. Meyerson, Allan S., 1952–   . II. Toyokura, Ken, 1933– .
III. International Chemical Congress of Pacific Basin Societies (1989: Honolulu, Hawaii) IV. Series.

QD921.C774   1990
548′.5—dc20   90–1162
                                        CIP

The paper used in this publication meets the minimum requirements of American National Standard for Information Sciences—Permanence of Paper for Printed Library Materials, ANSI Z39.48–1984. ∞

Copyright © 1990

American Chemical Society

PRINTED IN THE UNITED STATES OF AMERICA

ACS Symposium Series

M. Joan Comstock, *Series Editor*

1990 ACS Books Advisory Board

Foreword

THE ACS SYMPOSIUM SERIES was founded in 1974 to provide a medium for publishing symposia quickly in book form. The format of the Series parallels that of the continuing ADVANCES IN CHEMISTRY SERIES except that, in order to save time, the papers are not typeset, but are reproduced as they are submitted by the authors in camera-ready form. Papers are reviewed under the supervision of the editors with the assistance of the Advisory Board and are selected to maintain the integrity of the symposia. Both reviews and reports of research are acceptable, because symposia may embrace both types of presentation. However, verbatim reproductions of previously published papers are not accepted.

Contents

Preface

CRYSTALLIZATION IS AN IMPORTANT separation and purification technique used in a wide variety of industries. Interest in crystallization as a separations process has increased in recent years; as a result of this interest, important and significant research is being done in a number of countries. The international crystallization community has traditionally met every three years at the industrial crystallization meetings sponsored by the European Federation of Chemical Engineering. The most recent of these meetings was in Prague in September 1987; the next meeting is scheduled for September 1990 in the Federal Republic of Germany.

With the increased level of research and interest in the field, an additional international meeting held between the two European meetings seemed desirable. With this in mind, a symposium was organized as part of the International Chemical Congress of Pacific Basin Societies. The symposium included oral presentations and a poster session, with a total of 50 contributions. The contributions came from nine countries: 11 from the United States, 20 from Japan, 3 from Australia, 1 from the United Kingdom, 2 from Italy, 4 from the Federal Republic of Germany, 5 from the Netherlands, 3 from Sweden, and 1 from Czechoslovakia.

The presentations made at the symposium covered the broad spectrum of current research in crystallization. The contributions covered the major areas of current interest and research in crystallization.

This volume contains a selection of chapters presented at the symposium organized into four areas: basic studies, crystallizer operation and control, crystallization of organic molecules and biomolecules, and crystallization and precipitation of inorganic compounds. In addition, an overview chapter is included, which reviews important areas in the field.

We would like to thank the symposium contributors and participants, the contributors to this volume, and the organizing committee of the International Chemical Congress of Pacific Basin Societies for allowing us the opportunity to organize and conduct the symposium.

ALLAN S. MYERSON
Polytechnic University
Brooklyn, NY 11201

KEN TOYOKURA
Waseda University
Tokyo, Japan

June 27, 1990

Chapter 1

Crystallization Research in the 1990s
An Overview

Allan S. Myerson

Department of Chemical Engineering, Polytechnic University, 333 Jay Street, Brooklyn, NY 11201

Crystallization from solution is an important separation and purification process in a wide variety of industries. These range from basic materials such as sucrose, sodium chloride and fertilizer chemicals to pharmaceuticals, catalysts and specialty chemicals. The major purpose of crystallization processes is the production of a pure product. In practice however, a number of additional product specifications are often made. They may include such properties as the crystal size distribution (or average size), bulk density, filterability, slurry viscosity, and dry solids flow properties. These properties depend on the crystal size distribution and crystal shape. The goal of crystallization research therefore, is to develop theories and techniques to allow control of purity, size distribution and shape of crystals.

The past 30 years have seen great advances in our understanding of the fundamentals of crystallization and has resulted in improved crystallizer design and operation. A dominant theme during this period was the analysis and prediction of crystal size distributions in realistic industrial crystallizers. This led to the development and refinement of the population balance technique which has become a routine tool of the crystallization community. This area is best described in the book of Randolph and Larson (1) which has been an indispensable reference and guide through two editions.

Another important development which altered our view of crystallization processes was the realization of the importance of secondary nucleation due to contact between crystals and the impeller and vessel. Secondary nucleation of this type has been shown (2-6) to often have a dominant role in determining crystallizer performance. Our understanding of crystal growth, nucleation, fluid mechanics and mixing have all greatly improved. A number of review (7-10) have appeared in recent years which describe the advances in these and

0097–6156/90/0438–0001$06.00/0
© 1990 American Chemical Society

other areas. It is the purpose of this overview chapter to discuss areas which, in my opinion, will be of greatest interest and potential use in the 1990's. The list is by no means complete, however I believe that in the next decade much progress will be made in these areas and they will have a profound impact on industrial crystallization processes.

Fundamental Research

Metastable Solution Structure

Classical nucleation theory has long postulated the existence of molecular clusters in supersaturated solutions (11, 12). This is the result of the size dependence of the free energy change between a result of the size dependence of the free energy change between a small particle and the solution. A critical size exists above which a cluster of molecules will grow and below which the cluster will dissolve. The critical size is the maximum in the free energy curve where $d\Delta G/dr = 0$. Nucleation theory tells us that supersaturated solutions have a population of subcritical sized molecular clusters which form and dissolve. Nucleation occurs when one of these clusters reaches the critical size. Based on this concept the following questions arise.

1. Do these clusters actually exist?
2. What are the size distribution and/or average size of the clusters?
3. How do the clusters affect solution properties?
4. How do the clusters participate in the crystal growth process?

The last five years have seen an increasing level of research activity aimed at answering these questions which is likely to continue during the 1990's.

Indirect evidence of the existence of molecular clusters was reported by Mullin and Leci (13) who demonstrated that isothermal columns of aqueous supersaturated solutions of citric acid developed concentration gradients with higher concentrations in the bottom of the column, than at the top. No gradients were reported in saturated or undersaturated columns. These experiments were repeated by Larson and Garside (14) for a number of solutes with similar results. Larson and Garside (15) attempted to explain the gradient formation using a quasi-equilibrium theory which estimated an average cluster size of 1000 - 10,000 molecules. This is clearly far too large since clusters of this size would be easily detectable by light scattering and have not been detected in these types of experiments.

A number of investigators have attempted to obtain more direct evidence of cluster formation in supersaturated solutions by using spectroscopic techniques (16-22). Studies in bulk solutions and in highly supersaturated droplets using Raman and FTIR spectroscopy resulted in reports of features present in the supersaturated solution spectrum which were not present in the undersaturated solution spectrum or in the crystalline material. Rusli et al. (19) used Raman spectroscopy to study $NaNO_3$ solutions. They reported that the clusters do not have a crystalline structure but are solvated consisting of clusters of solute and solvent molecules. They also reported that the clusters occur to a small degree in the undersaturated solutions.

While providing important evidence concerning the existence of the clusters and their structure, the spectroscopy work has not yet provided quantitative data on the cluster size and evolution.

Additional indirect evidence of cluster formation has been obtained through observed changes in the properties of supersaturated solutions. Myerson and coworkers (23-28) have demonstrated that diffusion coefficients of a variety of substances (urea, glycine, KCl, NaCl, valine) decline rapidly with increasing concentration in the supersaturated region. It has also been shown (27, 28) that the diffusion coefficients are a weak function of the solution 'age', declining with increasing age.

Employing experimental supersaturated solution diffusion coefficient data and the cluster diffusion theory of Cussler (29), Myerson and Lo (27) attempted to estimate the average cluster size in supersaturated glycine solutions. They estimated an average cluster size on the order of two molecules. Their calculations indicated that while the average cluster size was small, large clusters of hundreds of molecules existed, only there were very few of them. Most of the molecular association was in the form of dimers and trimers.

The current status of research in supersaturated solution structure centers on the direct determination of the cluster size and size distribution and the role of the clusters in crystal growth and nucleation. This will likely be accomplished through sophisticated spectroscopic and light scattering techniques using bulk solutions and using very highly supersaturated droplets suspended without a container using the electrodynamic balance technique (30, 31). Hopefully this will yield quantitative data which will answer some of the important questions relating to cluster size, form and evolution.

Growth Rate Dispersion

The development and refinement of population balance techniques for the description of the behavior of laboratory and industrial crystallizers led to the belief that with accurate values for the crystal growth and nucleation kinetics, a simple MSMPR type crystallizer could be accurately modelled in terms of its CSD. Unfortunately, accurate measurement of the CSD with laser light scattering particle size analyzers (especially of the small particles) has revealed that this is not true. In many cases the CSD data obtained from steady state operation of a MSMPR crystallizer is not a straight line as expected but curves upward (1, 32, 33). This indicates more small particles than predicted by the population balance and could be the result of crystal breakage, size dependent growth or some other growth phenomenon.

A number of investigators developed empirical growth rate expressions that included a size dependence. These models were summarized by Randolph (33, 34) who showed that they all produced a concave upward semi-log population density plot; thus are useful for empirical fits of non-linear MSMPR CSD data. These models however, supply no information on what is actually happening to cause the non-linear CSD.

The explanation which has emerged after a number of theoretical (35-40) and experimental studies (41-46) is known as growth rate dispersion. This unexpected concept says that crystals of the same size and material exposed to identical conditions of supersaturation, temperature and hydrodynamics do not necessarily grow at the same rate. This is not the same as size dependent growth (47) in which crystals of different sizes display differences in growth rate. The concept of growth rate dispersion was first used by White and Wright (41) to explain an observed widening of the size range during the batch growth of sucrose.

Studies of growth rate dispersion have resulted in two distinctly different mechanisms being proposed. In one mechanism it is assumed that crystals have a distribution of growth rates but each individual crystal grows at a constant rate (at a fixed set of conditions). This implies that nuclei are born with a distribution of growth rates and the observation of two nuclei in a crystallizer at steady state conditions will reveal each nuclei growing at a constant but different rate. This type of behavior has been demonstrated experimentally for crystal fragments produced by attrition (49), for secondary nuclei and (43, 44) for single crystal produced by primary nucleation (45).

The second mechanism for growth dispersion says that while all crystals have the same time averaged growth rate, the growth rates of individual crystals can fluctuate significantly with time. This mechanism, therefore, implies that the observation of two different crystals growing under identical conditions might be different at any time but that the time average growth rate over a long time period would be identical. Experimental evidence has also appeared (42-43) which supports this mechanism.

Explanations of growth rate dispersion which have appeared in the literature employ the surface integration step of the Burton Cabrera Frank (BCF) theory of crystal growth (50) as the primary cause of the observed phenomena. The BCF theory indicates that the growth rate of a crystal is dependent on screw dislocations which are present on the surface. Experimental work has shown that changes in the location or density of screw dislocations can cause large changes in the crystal growth rate. Collisions of the crystals with the impeller, walls and each other can result in damage to the dislocations and therefore changes in the crystal growth rate. This is especially true with secondary nuclei which could display very different dislocation densities. In addition, the imperfect nature of the crystal growth process could lead to changes in the dislocations of the crystal faces. Zumstein and Rousseau (40) developed a model that includes both growth rate dispersion mechanisms. They found that both growth rate distribution and growth rate fluctuation could be observed in batch crystallizers experimentally by monitoring the increase in the variance of the CSD during growth periods. In continuous crystallization, however, only the growth rate dispersion mechanism could be observed through the upward curvature of the CSD plot.

Both experimental and theoretical work has demonstrated that growth rate dispersion exists, and has a measurable effect on the CSD in both batch and continuous crystallization processes. Further understanding of this phenomenon on a fundamental level will be required to develop methods to make use of or control growth rate dispersion and make it a tool in control of particle size and shape.

Impurity–Crystal Interactions and Crystal Morphology

The interaction of impurities (including the solvent) with growing crystals has long been of interest. A tremendous amount of qualitative data exists which shows the effect of many ionic and organic impurities on crystal habit, crystal growth rate and nucleation rate (51, 52). Industrially, impurities are used as scale inhibitors and habit modifiers in a number of processes. These additives, however, were developed through experimental programs which are essentially trial and error. There has not been a technique available to predict, in advance, the effect of impurities on the crystal morphology or growth rate. There is also no reliable method to choose an appropriate solvent to obtain desired solid properties. In recent years, however, progress has been made in these areas so that the concept of tailoring a solvent system and/or impurities to achieve a desired crystal morphology and size distribution seems a realistic goal.

A group at the Weitzmann Institute in Israel has for a number of years been investigating the concept of crystal growth and morphology control through the use of tailor made additives (53-57). A tailor made additive is a molecule very similar to the crystallizing species in its structure but different in some specific way. The difference in the structural characteristics of the additive are such that once incorporated in the structure, the additive will disrupt the bonding sequences in the crystal and interfere with the growth process at the crystal surface. The Weitzmann group found that these additives could dramatically affect crystal growth and habit. The mechanism for the additive's effectiveness is based on two steps. The additive first adsorbs preferentially on specific crystal faces where the modified part of the additive points away from crystal interior. Once bound, the additive disrupts the layer growth of the particular face thus lowering the growth rate of the face. This results in an increased surface area of the faces in which the impurity is adsorbed. By careful design of additives it is thus possible to design a crystal morphology. ·Examples of systems studies using this technique include the crystallization of benzamide from ethanol using benzoic acid and o and p toluamide as the additives (54), crystallization of racemic mixtures of amino acids using chiral additives which cause a kinetic resolution of the isomers by adsorbing only onto surfaces of similar chirality (55) and the use of 1-glutamic acid on 1-asparagine (58).

The use of tailor made additives holds great promise in the area of crystal growth and morphology control. The routine selection and use of these type of additives will require a fundamental understanding of the mechanism by which the additives work on a molecular basis. At the same time, the effect of solvent molecules on the crystal growth process is another related and important problem. In both instances, the relationship between internal crystal structure, crystal growth rate, solvent and impurities are needed to predict the habit of a crystal and thus allow selection of the proper conditions and components required to obtain a desired habit.

The early investigation of crystals led to interest in the correlation of crystal morphology (shape) with internal structure. A simple correlation was noticed by Donnay and Harker (59) in 1937 between the interplanar spacing of a crystallographic plane, d_i, and its area on an average crystal. Similar correlation holds between d_i and the frequency with which the plane (hkl) appears in an ensemble of crystals. Since the area of a plane is roughly proportional to the inverse of its linear growth velocity, R, the Donnay-Harker law is equivalent to stating that $R_i \sim 1/d_i$.

Hartman and Perdok (60-62) in 1955 developed a theory which related crystal morphology to its internal structure on an energy basis. They concluded that the morphology of a crystal is governed by a chain of strong bonds (called periodic bond chains (PBC)), which run through the structure. The period of these strong bond chains is called the PBC vector. In addition, Hartman and Perdok divided the crystal face into three types. These types are:

1. F-faces (flat) each of which is paralled to at least two PBC vectors.
2. S-faces (stepped) parallel to at least one PBC vector
3. K-faces (kinked) not parallel to any PBC vector.

If the sum of the energy within a slice (E slice) of each type of face is compared, the F face will be the largest (followed by S and then K). the attachment energy (E^{att}), which is the difference between the crystallization energy and the slice energy will, therefore, be the smallest. A small attachment energy means a low growth velocity. The slow growth velocity of F faces means they will be the large faces on a grown crystal. The higher velocities of S and K faces mean they rarely (S) and almost never (K) develop.

The result of Hartman Perdok theory is to allow prediction of the growth shape of crystals from the slice energy of different F faces.

Hartman and Bennema (63) looked at attachment energy as a habit controlling factor. They found that the relative growth velocity always increases with increasing E^{att}, however, the relationships between the two depends on the mechanism of crystal growth and variables such as supersaturation, temperature and solid-fluid interactions. They demonstrated at low supersaturation, however, the relative growth velocity of a face is directly proportional to the attachment energy of that face. Hartman (64, 65) employed this assumption to calculate the habit of naphthalene and sulfur and had good agreement with observed forms.

A number of investigators (66, 67) have recently employed the critical Ising temperature (transition temperature from smooth to rough interface) to determine the relative importance of F faces. In general, results obtained by this method are quite similar to those obtained from attachment energy calculations.

A major weakness in the calculations described above is that they can only be used to represent vapor grown crystals. In crystals grown from solution, the solvent can greatly influence the crystal habit as can small amounts of impurities. Several investigators (68, 69) accounted for discrepancies between observed crystal habit and those obtained using attachment energies by assuming preferential solvent (or impurity) adsorption on crystal faces.

In a recent paper (70) Hartman studies the effect of surface relaxation on the habit of corundum and hematite. The habits observed on natural and synthetic crystals of these systems did not agree with calculated relaxed equilibrium or growth habits. Hartman concluded that these observations could be understood by invoking specific solvent adsorption on (111) faces.

The work discussed in the previous paragraphs provides the framework for the prediction of crystal habit from internal structure. The challenge is to add realistic methods for the calculation of solvent and impurities effects on the attachment energies (hence the crystal habits) to allow this method to provide prediction of crystal habit. Initial attempts of including solvent effects have been recently described (71, 72). The combination of prediction of crystal habit from attachment energies (including solvent and impurity effects) and the development of tailor made additives (based on structural properties) hold promise that practical routine control and prediction of crystal habit in realistic industrial situations could eventually become a reality.

Crystallizer Design and Operations

The design and operation of industrial crystallizers is where developments in the laboratory are confirmed and their practical significance determined. In recent years, crystallization processes involving specialty chemicals and pharmaceuticals have increased. This has led increased interest in batch crystallization operation, optimization and design. At the same time, the advent of powerful computers and their routine availability has stimulated interest in the area of on-line control of crystallization process (both batch and continuous). Progress in batch crystallization is summarized in a number of recent papers and reviews (73-80). In this section I will discuss two areas which I think will have an impact in the next decade.

Crystal Growth and Nucleation Kinetics from Batch Experiments

The measurement of crystal growth and nucleation kinetics is a fundamental problem which confronts everyone interested in crystallizer process design and development. Application of the population balance model to analyze and/or predict the CSD requires knowledge of the kinetic constants. Crystallizer design also requires some knowledge of kinetic parameters, particularly the ratio of the exponents of nucleation to crystal growth. The standard method for the determination of growth and nucleation kinetics is the use of the CSD obtained from a steady state MSMPR crystallizer. At steady state, assuming size independent growth and clear feed, the population balance equation can be written as:

$$G\tau \frac{dn}{dL} + n = o \qquad (1)$$

where G is the crystal growth, τ the residence time, n the population density and L the crystal size. The solution to equation 1 is

$$n = n^o \exp (-L/G\tau) \qquad (2)$$

A plot of the ln n versus L yields a straight line where the slope is $- 1/G \tau$ and the intercept is n_o (the population density of nuclei). Since the MSMPR is at steady state, the supersaturation is known. This experiment can then be repeated at a number of different supersaturations and fit to growth and nucleation expressions of the form below:

$$G = k_g \Delta C^g \qquad (3)$$

$$B = k_b \Delta C^b \text{ or } k_b^1 M_T^j \Delta C^b \qquad (4)$$

where Δ C is the supersaturation, M_T the slurry density, k_g, k_b and k_b^1 the kinetic constants and g, b and j the kinetic orders.

This is the standard method for obtaining growth and nucleation kinetics. While not difficult (for those with experience in crystallization) the technique is very time consuming, uses up significant amounts of material and data analysis often presents problems (because of a concave upward CSD). Because of these factors, industrial investigators often do not attempt to obtain growth and nucleation kinetics in crystallization process development and design. This had led to interest in the development of methods to employ batch crystallization to obtain the needed data. The simplest method which can be used to obtain kinetic data from a batch crystallizer, involves measurement of the desupersaturation curve in a seeded isothermal experiment. If a small supersaturation is employed it is assumed that nucleation is negligible. Garside et al. (81) developed a method to employ the initial derivatives of the desupersaturation curve to obtain growth and dissolution kinetics. The experimental technique and analysis required for this technique are simple however, as in most techniques that employ derivatives, there is a large variability in the results. The technique is useful in estimating growth and dissolution kinetics with small amounts of material and is relatively quick to do.

Tavare and Garside (82) developed a method to employ the time evolution of the CSD in a seeded isothermal batch crystallizer to estimate both growth and nucleation kinetics. In this method, a distinction is made between the seed (S) crystals and those which have nucleated (N crystals). The moment transformation of the population balance model is used to represent the N crystals. A supersaturation balance is written in terms of both the N and S crystals. Experimental size distribution data is used along with a parameter estimation technique to obtain the kinetic constants. The parameter estimation involves a Laplace transform of the experimentally determined size distribution data followed by a linear least square analysis. Depending on the form of the nucleation equation employed four, six or eight parameters will be estimated. A nonlinear method of parameter estimation employing desupersaturation curve data has been developed by Witkowki et al (85).

The availability of computers and sophisticated parameter estimation techniques have led to interest in other methods for the estimation of growth and nucleation kinetics. These methods include measurement of the desupersaturation curve and/or size distribution in a seeded batch crystallizer undergoing controlled cooling and in a non-seeded batch crystallizer undergoing controlled cooling (83-84). The parameter estimation techniques being proposed to analyze these systems are quite complex and non-linear. With the advances in computing and the availability of advanced numerical programs, it should be possible to do these calculations on a PC.

The determination of kinetics from batch experiments and parameter estimation is experimentally easy but involves difficult and sophisticated calculations. The accuracy of the results, however, is not significantly worse than obtained in MSMPR experiments in many systems. Development of standard calculation techniques and software should make this method of kinetics determination routine in the next decade.

On-Line Control of Crystallizers

The development of the population balance approach has done much to further our understanding of CSD transients and instabilities which are observed in the plant and the laboratory. Advances in computing and control technology which have occurred in the chemical process industry together with the recognition that on-line CSD control might be achievable has led to an increased level of interest and investigation of this area.

Until recently most work on CSD control was theoretical with not attempt at implementation, since measurement of the necessary variables employed in these theoretical studies was usually not possible. Descriptions of many of the theoretical studies are summarized in several places (1, 86, 87).

One approach which has resulted in experimental implementation is that of Randolph and co-workers (88-92). Using a simulation (91) Randolph and Beckman demonstrated that in a complex RTD crystallizer, the estimation of nuclei density could be used to eliminate cycling or reduce transients in the CSD. Randolph and Low (88) experimentally attempted feedback control by manipulation of the fines dissolver flow rate and temperature in response to the estimated nuclei density. They found that manipulation of fines flow rate upset the fines measurement indicating that changes in the manipulated variable disturbed the measured variable. Partial fines dissolution resulting from manipulation of the fines dissolver temperature appeared to reduce CSD transients which were imposed upsets in the nucleation rate. In a continuation of this work Randolph et. al. (92) used proportional control of inferred nuclei density to control an 18 liter KC1 crystallizer. Nuclei density in the fines loop was estimated by light scattering. This technique was shown to be effective in minimizing the effect of disturbances in nucleation rate of the product CSD.

Another experimental approach (93, 94) to CSD control involved the use of on line solution and slurry density measurements as measured variables with jacket temperature and residence time as manipulated variables. This was attempted in a 10 liter MSMPR crystallizer employing K-alum. Results showed that this scheme was able to return the crystallizer to a steady state after the introduction of an upset and that on-line density measurements of solution and slurry could be obtained and used.

A group at Delft University (95-97) has for several years been involved in experimental and theoretical studies of on-line control of crystallization processes. Their control scheme involves the on-line measurement of the CSD by light scattering or by Fraunhofer diffraction. The main experimental implementation problem is that neither of these techniques work with concentrated slurries so that dilution is necessary. The Delft group has designed an automatic dilution unit which employs saturated mother liquor as the dilution liquid which is separated from the product using a hydrocyclone and filters. A semibatch dilution mode has been employed, however complete implementation of this system with control results has not yet been reported.

The future of on-line control of crystallization should see the use of parameter estimation for estimation and correction of model parameters along with higher level nonlinear control schemes. The major challenge continues to be realistic measurement of the necessary variables such as the CSD or its moments.

References

1.   A.D. Randolph and M.A. Larson "Theory of Particulate Processes" 2nd Edition Academic Press (San Diego, CA, 1988).
2.   D.C. Timm, and M.A. Larson, AIChE J., 14, 448 (1968).
3   C.Y. Sung, J. Estrin and G.R. Youngquist AIChE J., 19, 957 (1973).
4.   R.F. Strickland-Constable, and R.E.A. Mason Trans Faraday Soc; 518, 455 (1966).
5.   D.P. Lal, R.E.A. Mason, and R.F. Strickland-Constable, J. Cryst. Growth, 12, 53 (1972).
6.   N.A. Clontz and W.L. McCabe Chem. Eng. Prog. Symp. Ser., 110, 67 (1971).
7.   J. Garside, Chem. Eng. Sci., 40, 3 (1985).
8.   J.W. Mullin, in the "Kirk Othmer Encyclopedia of Chemical Technology", Vol. 7, p. 243, Wiley, New York (1979).
9.   M.A. Larson, AIChE Symp. Ser. 240, 39 (1984).
10. J. Garside, AIChE Symp. Ser. 240, 23 (1984)
11. A.C. Zettlemoyer, (Ed.), "Nucleation", Dekker New York (1969).
12. B. Lewis, "Nucleation and Growth Theory" in Crystal Growth, B.R. Pamplin, Ed., Pergamon Press, Oxford (1980).
13. J.W. Mullin and C. Leci, Philos. Mag., 19, 1075 (1969).
14. Larson, M.A. and J. Garside, Chem. Eng. Sci., 41, 1285 (1986).
15. Larson, M.A. and J. Garside, J. Cryst. Growth, 76, 88 (1986).
16. K.H. Fung, and I.N. Tang, Chemical Physics Letters, 147, 509 (1988).
17. G.A. Hussmann, M.A. Larson, and K.A. Berglund in "Industrial Crystallization 84", S.J. Jancic and E.J. de Jong, eds., Elsevier, Amsterdam, p. 21 (1984)
18. P.M. McMahon, K.A. Berglund, and M.A. Larson, in "Industrial Crystallization 84", S.J. Jancic and E.J. de Jong, eds., Elsevier, Amsterdam, p. 229 (1984).

19. I.T. Rusli, G.L. Schrader and M.A. Larson, J. Cryst. Growth, 97, 345 (1989).
20. M.K. Cerreta and K.A. Berglund in "Industrial Crystallization 84", p. 233, S.J. Jancic and E.J. de Jong, eds., Elsevier, Amsterdam, (1984).
21. G.S. Grader, R.C. Flagan, J.H. Seinfeld and S. Arnold Rev. Sci. Inst., 58, 584 (1987).
22. G.S. Grader, S. Arnold, R.L. Flagan and J.H. Seinfeld, J. Chem. Phys., 86, 5897 (1987).
23. L. Sorell and A.S. Myerson, AIChE Journal, 28, 772, (1982).
24. Y.C. Chang and A.S. Myerson, AIChE Journal, 30, 820 (1984).
25. Y.C. Chang and A.S. Myerson, AIChE Journal, 32, 1567 (1986)
26. Y.C. Chang and A.S. Myerson, AIChE Journal, 33, 697, (1987).
27. P.Y. Lo and A.S. Myerson, J. Cryst. Growth 99, 1048 (1980).
28. P.Y. Lo and A.S. Myerson, J. Cryst. Growth (in press).
29. E.L. Cussler, AIChE J., 26, 43 (1980).
30. S. Arnold, E.K. Murphy, and G. Sageev, Applied Optics, 24, 1048 (1985).
31. S. Arnold, M. Newman and A.B. Plachino, Optics Letters, 9, 13, (1984).
32. J. Nyvit, O. Sohnel, M. Matuchova and M. Brout "The Kinetics of Industrial Crystallization" Elsevier (Amsterdam 1985).
33. A.D. Randolph, AIChE Symp. Ser., 240, 14 (1984).
34. A.D. Randolph in "Industrial Crystallization 78", E.J. De Jong and S.J. Jancic eds., Norton Holland (Amsterdam 1979).
35. A.D. Randolph and E.T. White, Chem. Eng. Sci., 32, 1067 (1977).
36. K.A. Berglund and M.A. Larson, AIChE Symp. Ser., 215, 9 (1982).
37. K.A. Berglund and M.A. Larson AIChE J., 30, 280 (1984).
38. K.A. Ramanarayanan, K. Athreya, and M.A. Larson, AIChE Symp. Ser., 80, 76 (1984).
39. K.A. Ramanarayanan, K.A. Berglund, and M.A. Larson, Chem. Eng. Sci., 40, 1604 (1985).
40. R.C. Zumstein and R.W. Rousseau, AIChE J., 33, 122 (1987).
41. E.T. White and A.G. Wright Chem. Eng. Prog. Symp. Ser., 110, 18 (1971).
42. H.J. Human, W.J.P. Van Enckerork and P. Bennema in "Industrial Crystallization 81", S.J. Jancic and E.J. de Jong editors, North Holland Amsterdam, p. 387 (1982).
43. K.A. Berglund, E.L. Kaufman and M.A. Larson, AIChE J., 25, 876 (1983).
44. K.A. Berglund and M.A. Larson AIChE Symp. Ser., 78, 9 (1982).
45. J. Garside and R.I. Ristic, J Cryst. Growth, 61, 215 (1983).
46. J. Garside and M.A. Larson, J. Cryst. Growth, 43, 694 (1978).
47. C. Herndon and D.J. Kirwan, AIChE Symp. Ser., 78, 192 (1978).
48. J. Garside and R.J. Davey, Chem. Eng. Comm. 4, 393 (1980).
49. C.M. Van't Land and B.G. Wunk in "Industrial Crystallization", J.W. Mullin Ed. Plenum Press, N.Y. (1976), p. 5.
50. W.K. Burton, N. Cabrera and F.C. Frank Phil. Trans. Royal Soc., 243, 299 (1951).

51. H.E. Buckley, "Crystal Growth", (Wiley, New York 1951).
52. J.W. Mullin "Crystallization" (Butterworths London, 1971).
53. L. Addadi, Z. Berkovitch-Yellin, N. Domb, E. Gati, M. Lahav and L. Leiserowitz, Nature, 296, 21 (1982).
54. Z. Berkovitch Yellin, L. Addadi, M. Idelseon, M. Lahav and L. Leiserowitz, Agnew Chem. Suppl. 1336 (1982).
55. L. Addadi, S. Weinstein, E. Gabi, I. Weissbach and M. Lahav, J. Am. Chem. Soc., 104, 4610 (1982).
56. L. Addadi, Z. Berkovitch-Yellin, I. Weissbuch, M. Lavar, L. Leiserowitz, Mol Cryst. Liq. Cryst., 96, (1983).
57. Z. Berkowitch-Yellin, S. Ariel and L. Leiserowitz, J. Am. Chem. Soc., 105, 765, 1985.
58. S.N. Black, R.J. Davey and M. Halcrow, J. Cryst. Growth, 79, 765 (1986).
59. J.D.H. Donnay and D. Harker, Am. Mineralogist, 22, 446 (1937)
60. P. Hartman and W.G. Perdok, Acta Crystallogr. 8, 49 (1955).
61. P. Hartman and W.G. Perdok, Acta Crystallogr., 8, 521 (1955).
62. P. Hartman and W.G. Perdok, Acta Crystallogr. 8, 525 (1955).
63. P. Hartman and P. Bennema J. Cryst. Growth 49, 145 (1980).
64. P. Hartman J. Cryst. Growth 49, 157 (1980).
65. P. Hartman, J Cryst. Growth 59, 166 (1980).
66. M.H.J. Hottehuis, J.G.E. Gardiniers, L. Jetten and P. Bennema, J. Cryst. Growth 92, 171 (1988).
67. R.A. Terpstra, J.J.M. Rizpkema and P. Bennema, J. Cryst. Growth 76, 494 (1986).
68. M. Saska and A.S. Myerson, J. Cryst. Growth, 61, 546 (1983).
69. S.N. Black, R.J. Davey J. Cryst. Growth, 99, 136 (1988).
70. P. Hartman, J. Cryst. Growth, 96, 667 (1989).
71. Z. Berkowitz Yellin, J. Am. Chem. Soc., 107, 8539 (1985).
72. A.S. Myerson and M. Saska (This volume).
73. N.S. Tavare, Chem. Eng. Comm., 1987.
74. N.S. Tavare, J. Garside and M. Chivate Ind. Eng. Chem. Process Des. Dev., 19, 673 (1980).
75. M. Ulrich, Ger. Chem. Eng. 4, 195 (1979).
76. A.G. Jones, J. Budz and J.W. Mullin Chemical Eng. Sci., 42, 619 (1981).
77. B.G. Palwe, M.D. Chivate and N.S. Tavare, Ind. Eng. Chem. Proc. Des. Dev., 24, 914 (1985).
78. A.G. Jones, Chem. Eng. Sci., 29, 1075 (1974).
79. A.J. Jones and J.W. Mullin, Chem. Eng. Sci., 29, 105 (1975).
80. N.S. Tavare and M. Chivate, J. Chem. Eng. Japan, 13, 371 (1980).
81. J. Garside, I.G. Gibilaro and N.S. Tavare, Chem. Eng. Sci., 37, 1625 (1982).
82. N.S. Tavare and J. Garside, Chem. Eng. Des. & Design 64, 129 (1986).
83. A. Rasmusson, Personal Communication.
84. R. Farrell, Personal Communication.
85. W.R. Witkowski, S.M. Miller and J.B. Rawlings (This volume).
86. N.S. Tavare AIChE J. 32, 705 (1986).

87.   A.D. Randolph, AIChE Symp. Ser., 76, 193 (1980).
88.   A.D. Randolph, C.C. Low and E.T. White, Ind. Eng. Chem. Process Prod. Dev., 20, 496 (1981).
89.   A.D. Randolph and R.D. Rovang, AIChE Sym. Ser., No. 193 76, 18 (1980).
90.   A.D. Randolph, J.R. Beckman and Z. Kraljvich, AIChE J 23, 500 (1977).
91.   A.D. Randolph and J.R. Beckman, AIChE J 23, 510 (1977).
92.   A.D. Randolph, L. Chen and A. Tavang, AIChE J, 33, 583 (1987).
93.   J.L. Johnson, F.J. Schork and A.S. Myerson, Separation Process Eng. (Japan) 17, 86 (1987).
94.   S. Rush and A.S. Myerson, unpublished data.
95.   J. Jager, S. deWolf, W. Klapwijk and E.J. de Jong in "Proceedings of the 10th Symposium on Industrial Crystallization" J. Nyvlt and S. Zacek ed (Elsevier, N.Y.) 1987 p. 415.
96.   E.J. De Jong "Proceedings of the 10th Symposium on Industrial Crystallization", J. Nyvlt and S. Zacek ed (Elsevier, N.Y. 1987) p. 419.
97.   J. Jager, S. De Wolf, H.J.M. Kramer, E.J. de Jong, This volume.

RECEIVED May 30, 1990

BASIC STUDIES

Chapter 2

Integral Equation Analysis of Homogeneous Nucleation

Günther H. Peters, John Eggebrecht, and Maurice A. Larson

Department of Chemical Engineering and Center for Interfacial Materials
and Crystallization Technology, Iowa State University, Ames, IA 50011

The assumptions of homogeneous nucleation theory are
examined for the liquid-vapor transition of a Lennard-
Jones fluid. Approximate solutions of the first Yvon-
Born-Green integro-differential equation in a
spherically symmetric and finite volume provide the
dependence of the density profile of a small droplet on
temperature and supersaturation. The structure
provides a mechanical approach via the pressure tensor
to the interfacial properties. Classical thermodynamic
and statistical mechanical expressions for the surface
tension are compared. This approach allows the
calculation of the free energy of formation of a
droplet from a metastable vapor, avoiding most of the
usual assumptions of homogeneous nucleation theory.
These theoretical results, which are tested by comparison with molecular dynamics simulations, indicate that
the droplet size dependence of the interfacial free
energy is sufficiently strong that, for the state
points considered, the free energy barrier prior to the
nucleation is absent. The atomic kinetics of condensation are examined visually using molecular dynamics
temperature quenching experiments, providing insight
into the kinetic hindrance of the nucleation process.
Recent theoretical and simulation studies of cluster
structure and solvation in ionic solutions are
discussed.

When a phase transition occurs from a pure single state and in the
absence of wettable surfaces the embryogenesis of the new phase is
referred to as homogeneous nucleation. What is commonly referred to
as classical nucleation theory is based on the following physical
picture. Density fluctuations in the pre-transitional state result
in local domains with characteristics of the new phases. If these
fluctuations produce an embryo which exceeds a critical size then
this embryo will not be dissipated but will grow to macroscopic size
in an open system. The concept is applied to very diverse phenomena:

vapor condensation, thin film growth, polymer precipitation, lattice defect formation, and of particular interest here, to crystallization.

The applicability of the concept to all of these processes relies upon several postulates:

1. Thermodynamics remains relevant as phase dimensions are reduced to a molecular scale.
2. Embryos assume the symmetry of lowest specific surface area.
3. Embryos are sufficiently dilute that interactions between them may be neglected.
4. Equilibrium thermodynamic functions can be used to represent non-equilibrium states.

The implementation of the concept usually involves additional approximations:

a. The phases are ideal and/or incompressible.
b. The phase boundary is a density discontinuity.
c. The surface tension is independent of curvature.
d. The system is in contact with a material reservoir which instantaneously replenishes molecules depleted by the growing embryo.

It is difficult to judge the limitations of the postulates while imposing these additional assumptions. Our interest lies in the development of a molecular treatment of crystallization. However, it is useful as a preliminary to that effort to consider the effects, when these commonly used assumptions are lifted, upon the basic physical picture of the process. For this purpose we examine the condensing vapor, upon which applications to other transitions are based by analogy.

If we accept the first three postulates, we can lift each of these approximations using statistical mechanics and the companion techniques of computer simulation. But to do so we must consider a material for which complete thermodynamic and the necessary structural information is available. We, therefore, consider the Lennard-Jones fluid in most of the following discussion.

Ambiguities associated with rates of mass transport are removed by investigating a finite system. Finite rates of mass transport will result in a depletion zone surrounding a growing embryo and if these are not dilute in the transforming system then these embryos will compete for material. At some stage in the process, perhaps near completion, interactions of growth centers cannot be neglected. The general thermodynamic formulation of the model (1) is not restricted to a particular ensemble and conclusions reached should be transferable to other ensembles.

The approximation of vapor ideality is easily removed. Retaining the approximations of an incompressible liquid phase, a discontinuous density profile and curvature independent surface tension the conditions are those studied by Rao, Berne and Kalos (2). The essential physics was unchanged from the usual treatment in an open system, except that a minimum in the free energy of formation is found which corresponds to the unique equilibrium phase separated state whose symmetry, in the absence of an external field, is spherical.

The remaining approximations of liquid phase incompressibility and a discontinuous mass distribution can be removed through the use of the Yvon-Born-Green (YBG) equation (3), which is simply a

molecular statement of the conservation of linear momentum. We have
recently obtained solutions of this equation in a finite system by
appending the constraint of mass conservation (4). The method is
summarized below. For a planar liquid-vapor interface the density
profile and surface tension have been shown to be accurately
predicted by solutions of the YBG equation through comparisons with
molecular dynamics computer simulations (5,6). Such comparisons are
essential in the development of molecular theories of the structure
of matter since simulations can provide the exact result, for a given
model, within limits of precision imposed by finite sampling. Here
we use recent simulation studies (7) to test our solutions of the YBG
equation for the Lennard-Jones droplet.

The YBG equation is a two point boundary value problem requiring
the equilibrium liquid and vapor densities which in the canonical
ensemble are uniquely defined by the number of atoms, N, volume, V,
and temperature, T. If we accept the applicability of macroscopic
thermodynamics to droplets of molecular dimensions, then these
densities are dependent upon the interfacial contribution to the free
energy, through the condition of mechanical stability, and
consequently, the droplet size dependence of the surface tension must
be obtained.

We must also determine the location of the surface of tension,
R_s. Although expressions for this parameter exist, they are derived
by a hybrid of molecular mechanical and thermodynamic arguments which
are not at present known to be consistent as droplet size decreases
(8). An analysis of the size limitation of the validity of these
arguments has, to our knowledge, never been attempted. Here we
evaluate these expressions and others which are thought to be only
asymptotically correct. We conclude, from the consistency of these
apparently independent approaches, that the surface of tension, and,
therefore, the surface tension, can be defined with sufficient
certainty in the size regime of the critical embryo of classical
nucleation theory.

Having described the equilibrium structure and thermodynamics of
the vapor condensate we then re-examine homogeneous nucleation
theory. This combination of thermodynamics and rate kinetics, in
which the free energy of formation is treated as an activation energy
in a monomer addition reaction, contains the assumption that
equilibrium thermodynamic functions can be applied to a continuum of
non-equilibrium states. For the purpose of elucidating the effects
of the removal of the usual approximations, we retain this assumption
and calculate a radially dependent free energy of formation. We
find, that by removing the conventional assumptions, the presumed
thermodynamic barrier to nucleation is absent.

Finally we consider recent investigations of electrolyte
solutions. Larson and Garside (9) have proposed the existence of
stable clusters in the pretransitional solution as an explanation of
the observed rapidity of crystallization. Ionic clusters in sub-
saturated solutions have long been thought to exist (10,11), though
direct experimental observation is difficult. Their existence has
provided useful models for extensions of the dilute solution limiting
behavior of activity and conductance to finite concentrations. Monte
Carlo computer simulation results are presented which characterize
ionic cluster size distributions in sub-saturated solutions and which
examine the morphology of large ionic clusters in super-saturated

solutions and solvation structure as the saturation line is approached ($\underline{12}$).

The Droplet Density Profile

In this section we consider the manner in which the radially dependent density distribution, $\rho(r)$, of the equilibrium droplet can be obtained from the Yvon-Born-Green equation ($\underline{3}$)

$$\vec{\nabla}_1 \; \rho(\vec{r}_1) = -\beta \; \rho(\vec{r}_1) \int d\vec{r}_2 \; \rho(\vec{r}_2) \; \vec{\nabla}_1 \; u(r_{12}) \; g(\vec{r}_1, \vec{r}_2) \tag{1}$$

which is the first in a hierarchy of equations relating n-particle density distributions to n+1-particle distributions. Here r_i is the position vector of particle i (i=1,2); β is 1/kT where k is the Boltzmann constant and T is the absolute temperature; and $g(\vec{r}_1, \vec{r}_2)$ is the pair correlation function. The statement of the molecular model to which this equation is applied is contained in the intermolecular potential, $u(r_{12})$. Here we consider only the pair-wise additive Lennard-Jones potential

$$u(r_{12}) = 4 \; \varepsilon \left[\left(\frac{\sigma}{r_{12}} \right)^{12} - \left(\frac{\sigma}{r_{12}} \right)^{6} \right] \tag{2}$$

where ε is the depth of the potential well and σ is the collision diameter. The pair correlation function of the inhomogeneous fluid is unknown and must be approximated. We use a local density weighted linear interpolation of bulk liquid and vapor phase pair correlation functions ($\underline{13}$), g_L and g_V, which are well known ($\underline{14}$).

$$g(\vec{r}_1, \vec{r}_2) = w_L \left(\rho(\vec{r}_1), \rho(\vec{r}_2) \right) g_L(r_{12}) + w_V \left(\rho(\vec{r}_1), \rho(\vec{r}_2) \right) g_V(r_{12}) \tag{3}$$

where the weights are simple functions of the local densities

$$w_L \left(\rho(\vec{r}_1), \rho(\vec{r}_2) \right) = \frac{1}{2(\rho_L - \rho_V)} \left[\rho(\vec{r}_1) + \rho(\vec{r}_2) - 2\rho_V \right] \tag{4a}$$

and

$$w_V \left(\rho(\vec{r}_1), \rho(\vec{r}_2) \right) = \frac{1}{2(\rho_L - \rho_V)} \left[2\rho_L - \rho(\vec{r}_1) - \rho(\vec{r}_2) \right] \tag{4b}$$

The constraint of a finite volume requires that the one-particle density distribution satisfy the condition

$$N = 4\pi \int_0^{R_L} dr \; r^2 \; \rho(r) \tag{5}$$

where N is the total number of atoms contained in the system whose volume has a radial dimension R_L. Details of our numerical method of solving Equations 1 through 5 are given in reference 4.

In Figure 1 we compare our numerical solutions with the molecular dynamics computer simulations of Thompson, et al. (7). In this comparison we use liquid and vapor densities obtained from the simulation studies. In the next section we obtain the required boundary values by approximate evaluation of vapor-liquid equilibrium for a small system.

Droplet Thermodynamics

In a finite, one component, system defined by the number of molecules, N, the volume, V, and the temperature, T, the Helmholtz free energy of formation, ΔA_f, of a stable droplet is written

$$\Delta A_f = \mu_L N_L - p_L V_L + \mu_V N_V - p_V V_V + 4\pi R_s^2 \gamma(R_s) - \bar{\mu} N + \bar{p} V \tag{6}$$

when the conditions of thermal, chemical and mechanical equilibrium are satisfied:

$$T_L = T_V \tag{7}$$

$$\mu_L = \mu_V \tag{8}$$

and

$$p_L - p_V = \frac{2\gamma(R_s)}{R_s} \tag{9}$$

The latter is referred to as the Laplace-Young equation. μ_i, p_i, and V_i are the chemical potential, pressure and volume of phase i, i=(L,V). In a purely thermodynamic formulation we must regard these as properties of a macroscopic fluid at the temperature and vapor and liquid densities, ρ_V and ρ_L, of the microscopic phases. $\bar{\mu}$ and $\bar{p}$ are the chemical potential and pressure of the homogeneous metastable, or unstable vapor, prior to condensation.

The Laplace-Young equation refers to a spherical phase boundary known as the surface of tension which is located a distance R_s from the center of the drop. Here the surface tension is a minimum and additional, curvature dependent, terms vanish (15). The molecular origin of the difficulties, discussed in the introduction, associated with R_s can be seen in the definition of the local pressure. The pressure tensor of a spherically symmetric inhomogeneous fluid may be computed through an integration of the one and two particle density distributions.

$$\beta \, p(r) = \rho(r) \, I - \left[\frac{\beta}{2} \int d\vec{r}_{12} \, \frac{\vec{r}_{12}\vec{r}_{12}}{r_{12}} \, \frac{\partial u(r_{12})}{\partial r_{12}} \right.$$

$$\int_0^1 d\alpha \; \rho(r-\alpha r_{12}) \; \rho(r+(1-\alpha)r_{12}) \; g(r-\alpha r_{12},r+(1-\alpha)r_{12}) \Bigg]$$ (10)

where I is the unit matrix. The pressure tensor in a spherical symmetry has two independent components, $p_N(r)$ and $p_T(r)$,

$$p(r) = p_N(r) \; \hat{r}\hat{r} + p_T(r) \hat{\theta}\hat{\theta}$$ (11)

where r and θ are unit vectors in spherical polar coordinates. The α-integral provides a localization of the intermolecular forces which contribute to the pressure at r. In the following we adopt the convention of Irving and Kirkwood in which a contribution to the local pressure is obtained on every differential area at r intersected by the intermolecular vector, r_{12} (16). Other conventions are thought to be equally valid. When this differential area lies within a distance of the interface for which contributions to the integrand of Equation 10 are non-negligible the value of p at r will depend upon the localization function. This distance is approximately the range of the intermolecular forces and contains the regime of interest in classical nucleation theory.

Both the surface tension and the surface of tension may be expressed in terms of integrals of radial moments of the components of the pressure tensor (17-20)

$$\gamma(R_s) = \int_0^\infty dr \; \left(\frac{r}{R_s}\right)^{-1} [p_N(r) - p_T(r)]$$ (12)

$$\gamma(R_s) = \int_0^\infty dr \; \left(\frac{r}{R_s}\right)^{2} [p_N(r) - p_T(r)]$$ (13)

Equations 12 and 13 are two different expressions for the surface tension (21) which at the present time are not known to be consistent. Although Equation 12, and Equation 9, are independent of the form of the localization of the pressure tensor, Equation 13 is not (22). With the assumption of the equivalence of Equations 12 and 13 an expression for the surface of tension is obtained

$$R_s^{(p)} = \left\{\int_0^\infty dr \; r^2 \; [p_N(r) - p_T(r)] \; / \int_0^\infty dr \; r^{-1} \; [p_N(r) - p_T(r)] \right\}^{1/3}$$ (14)

The superscript (p) will be used to distinguish the surface of tension defined by Equation 14 from other definitions introduced below.

Equations 12 through 14 may be derived from thermodynamic and mechanical expressions for the transverse force and torque acting on a surface which intersects the interfacial region (21). Moments of an isotropic pressure force and a surface tension acting at R_s are equated with moments of the transverse component of the pressure tensor. However, when the radial dimension of a droplet is smaller than the range of the intermolecular force the pressure near the center of the drop departs significantly from the thermodynamic limit. Replacement of the uniform pressure of the Gibbs model with the normal component of the pressure tensor leads to the following additional expressions for the surface of tension,

$$R_s^{(0)} = \int_0^\infty dr \ [p_N(r) - p_T(r)] \ / \int_0^\infty dr \ r^{-1} \ [p_N(r) - p_T(r)] \tag{15}$$

$$R_s^{(1)} = \int_0^\infty dr \ r \ [p_N(r) - p_T(r)] \ / \int_0^\infty dr \ [p_N(r) - p_T(r)] \tag{16}$$

$$R_s^{(2)} = \int_0^\infty dr \ r^2 \ [p_N(r) - p_T(r)] \ / \int_0^\infty dr \ r \ [p_N(r) - p_T(r)] \tag{17}$$

so that $R_s^{(p)} = (R_s^{(0)} R_s^{(1)} R_s^{(2)})^{1/3}$. Equations 15 through 17 may be regarded as consistent with Gibbs' thermodynamic model only to terms of order $1/R_s$ and so, within that model, are only asymptotically correct as the droplet surface approaches the planar limit.

We have obtained numerical solutions of Equations 1-5, with boundary values which satisfy Equations 7-9 for a wide range of state points. At each state the pressure tensor was computed from Equation 10 and used to evaluate Equations 12 and 13 and Equations 14-17. Each pair of expressions for the surface tension and surface of tension was then used, through Equation 9, to obtain the liquid and vapor densities. This process was repeated until the surface tension was stationary. While values of the surface tension and surface of tension are sensitive to variations in the phase densities, these densities are insensitive to variations in γ and R_s. Consequently, with initial estimates of the surface tension taken from the planar limit, convergence was obtained in seven, or fewer, iterations.

Bulk phase fluid structure was obtained by solution of the Percus-Yevick equation (23) which is highly accurate for the Lennard-Jones model and is not expected to introduce significant error. This allows the pressure tensors to return bulk phase pressures, computed from the virial route to the equation of state, at the center of a drop of sufficiently large size. Further numerical details are provided in reference 4.

In Figure 2 we compare surface tensions computed from Equation 12 using the surface of tension expressions indicated. In Figure 3 Equation 13 has been used. We observe that all methods produce very similar results down to surfaces of tension of approximately 4 atomic diameters. Equation 17 and, therefore, Equation 14, yields non-physical negative values of R_s for small drops and the iteration sequence diverges. Surface tensions obtained through the use of Equations 17 and 14 in Figures 2 and 3 are truncated at the last convergent state. This failure may either signal a breakdown of the purely thermodynamic prescription for R_s or be due to numerical instability. A discussion of possible physical origins are provided

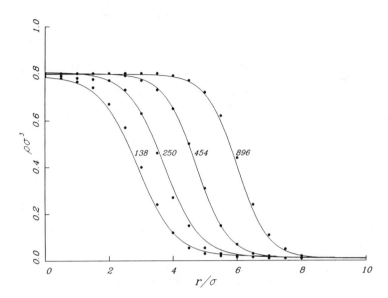

Figure 1. Density profiles for a droplet confined to a finite volume as predicted by YBG theory (solid line) are compared with the results of molecular dynamics simulations (7) (•). The reduced temperature, $kT/\epsilon = 0.71$, is near the triple point. The total number of atoms in each of the systems is indicated.

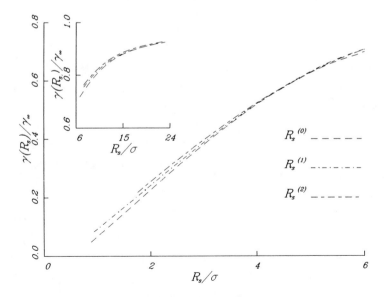

Figure 2. Radially dependent surface tensions computed from equation 12 are shown for the indicated expressions for R_s.

in reference 4. Our analysis could not be extended to droplets whose equimolar radius, R_e, was less than approximately four atomic diameters. For such small droplets, R_s approaches a single atomic diameter.

That the degree of consistency observed for the various routes to the determination of the surface tension is not the result of a cancellation of errors is demonstrated by the agreement displayed by the surfaces of tension in Figure 4. Also shown is the limiting form whose slope is one and whose intercept defines the Tolman parameter (24)

$$\delta = \lim_{\frac{1}{R_e} \to 0} (R_e - R_s) \qquad (18)$$

The surface tension is very insensitive to changes in supersaturation. While the planar surface tension is, of course, temperature-dependent, the ratio of the small system surface tension to the planar limit is nearly independent of temperature. This simplicity in the functional dependence of the radially dependent surface tension has been anticipated by a number of workers in terms of the parameter defined by Equation 18. Tolman obtained the following differential equation

$$\frac{1}{\gamma(R_s)} \frac{d\ \gamma(R_s)}{d\ R_s} = \frac{\frac{2\delta}{R_s^2}\left[1 + \frac{\delta}{R_s} + \frac{1}{3}\left(\frac{\delta^2}{R_s^2}\right)\right]}{1 + \left(\frac{2\delta}{R_s}\right)\left[1 + \left(\frac{\delta}{R_s}\right) + \frac{1}{3}\left(\frac{\delta^2}{R_s^2}\right)\right]} \qquad (19)$$

which he then evaluated to terms of order $1/R_s$ to obtain

$$\frac{\gamma(R_s)}{\gamma_\infty} = \frac{1}{1 + \frac{2\delta}{R_s}} \qquad (20)$$

Another approximate evaluation of Equation 19 has been given by Larson and Garside (9)

$$\frac{\gamma(R_s)}{\gamma_\infty} = \exp\left(\frac{-1.3\delta}{R_s}\right) \qquad (21)$$

Falls, et al. (25) obtained an analytic integration of Equation 19.

$$\ln \frac{\gamma(R_s)}{\gamma_\infty} = 1.3927 - 0.4425 \ln\left(\frac{\delta}{R_s} + 1.7937\right)$$

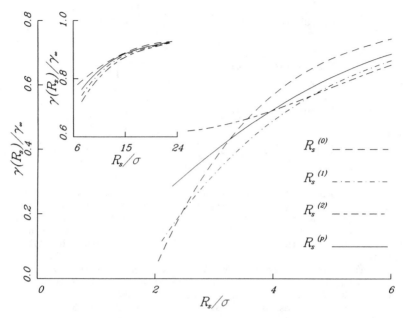

Figure 3. Radially dependent surface tensions computed from equation (13) are shown for the indicated expressions for R_s. The reduced temperature is 0.76.

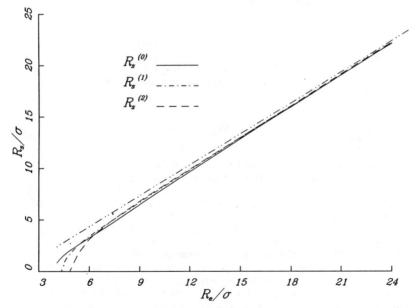

Figure 4. The dependence of the surface of tension on droplet size and the limiting form given by equation (18) (-..-..-) are shown. The reduced temperature is 0.76.

$$- 0.2788 \ln \left[\left(\frac{\delta}{R_s} \right)^2 + 1.2063 \left(\frac{\delta}{R_s} \right) + 0.8363 \right] \tag{22}$$

$$- 1.6439 \tan^{-1} \left[1.4548 \left(\frac{\delta}{R_s} \right) + 0.8775 \right]$$

Rasmussen (26) proposed the following expression

$$\frac{\gamma(R_s)}{\gamma_\infty} = \left(1 - \frac{\delta}{R_s} \right)^2 \tag{23}$$

γ_∞ is the surface tension of the planar interface. Each of these expressions gives a qualitatively similar representation. To compare these with the results of our integral equation theory, as shown in Figure 5, we have obtained solutions of the YBG equation for a planar interface, as described in reference (5), to determine the Tolman parameter.

Finally, we compare the results of our calculations with the computer simulation results of Thompson et al. (7) in Figure 6. In these studies the surface tension was determined by a direct evaluation of the pressure tensor. Equations 12 and 13 were combined, replacing R_s with Equation 9 to obtain

$$\gamma^3(R_s) = -\frac{1}{8} (p_L - p_V)^2 \int_0^\infty dr \, r^3 \frac{dp_N(r)}{dr} \tag{24}$$

The statistical quality of simulation estimates of the pressure tensor was poor near the origin resulting in large uncertainties in the estimate of the surface tension through Equation 24. By combining a truncated Taylor series expansion of Equation 20 with Equation 9 they obtained

$$\gamma(R_s) = \frac{3}{2} \gamma_\infty - \frac{1}{2} [9\gamma_\infty^2 - 4\gamma_\infty R_e(p_L - p_V)]^{\frac{1}{2}} \tag{25}$$

Again this equation is sensitive to uncertainties in the pressure drop. The truncated series expansion of Tolman's equation may be expected to introduce significant error for small drops.

The potential model used in these simulations was truncated at 2.5 atomic diameters, while in our calculations the potential was truncated at 8.0 diameters. The effect of this difference in the model may be significant, particularly for small droplets. However, given the considerable difficulties in the evaluation of the surface tension by either Equation 24 or 25, the qualitative agreement with simulation reinforces the observation which is essential to an analysis of nucleation theory: the radial dependence of the surface tension is much stronger than previously thought.

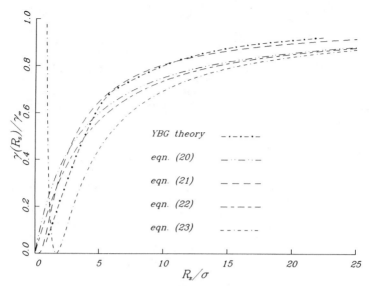

Figure 5. Radially dependent surface tensions computed from the YBG theory are compared with representations in the Tolman parameter, which for this state is 1.668σ, using the Irving-Kirkwood localization function. The reduced temperature is 0.76.

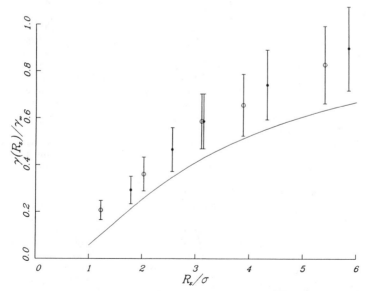

Figure 6. The radially dependent surface tension predicted by YBG theory are compared with the results of computer simulation using equations (24) (o) and (25) (●). The estimated error bars are taken from reference 7.

Nucleation

Having developed a self-consistent treatment of the structure and
interfacial energy of the Lennard-Jones droplet in a finite volume
which avoids the usual assumptions by which homogeneous nucleation
theory is implemented, we return to original question and examine how
this new information alters the physical picture of nucleation. It
is common to apply the analog of Equation 6 for an isothermal-
isobaric ensemble,

$$\Delta G_f = \mu_L N_L + \mu_V N_V + 4\pi R_s^2 \gamma(R_s) - \bar{\mu} N \tag{26}$$

to a continuum of thermodynamic states by postulating the validity of
this expression for droplets of all sizes, r. We retain this
approach for the purpose of examining the effects of the radial
dependence of the surface tension from our analysis on the classical
picture of homogeneous nucleation. We evaluate the usual expression
(27),

$$\Delta G_f(r) = 4\pi r^2 \gamma + V_L(p_V - p_\infty) - N_L kT \ln \left(\frac{p_V}{p_\infty} \right) \tag{27}$$

using the planar limit, the correction given by Equation 20 and the
results of the YBG theory for the surface tension in Figure 7. Here
p_V is the pressure of the supersaturated material reservior and p_∞ is
the vapor pressure of the planar interface. We note the absence of a
thermodynamic barrier to nucleation.

   We see in Figure 7 that Tolman's representation of the radially
dependent surface tension also leads to a vanishing thermodynamic
barrier, at high but metastable supersaturations, when a value of δ
computed from solutions of the YBG equation on the planar interface
is used. This value of the Tolman parameter is consistent with
values obtained from simulation studies of the planar Lennard-Jones
surface (28,29). It is apparent that the physical picture of
nucleation is highly dependent upon the assumed radial dependence of
the surface tension.

   These calculations suggest that hindrances other than a thermodynamic barrier are required to account for the observation that a
metastable state may exist without undergoing phase transition. We
believe these barriers are kinetic and are using non-equilibrium
molecular dynamics to elucidate them. A detailed description of the
computer simulation technique is provided elsewhere (4). We quench a
uniform vapor into a metastable state and determine, among other
things, the cluster size distribution following the quench. The
precise form of the trace of this function is dependent upon the
initial condition: the set of atomic positions, velocities and so on,
at the time of the quench. A typical trace is displayed in Figure 8.
Oscillations in the cluster size distribution are followed by
collapse to the stable drop as smaller clusters are scavenged from
the surrounding vapor.

   In another simulation, whose cluster size trace is shown in
Figure 9, a more gradual transition is apparent. We have produced a

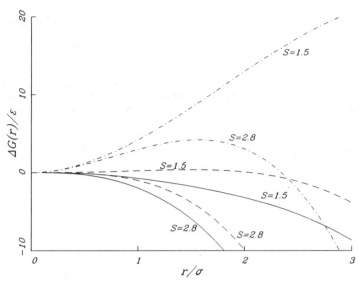

Figure 7.   The Gibbs free energy of formation for small values of r
using: classical nucleation theory (-.-.-), classical
nucleation theory with Tolman's representation of the
radial dependence of γ(r) (- - - -) and classical
nucleation theory using the YBG theory for γ(r) (————).
Supersaturations, defined by $S=p_v/p_\infty$, are shown.   The
temperature is near the triple point.

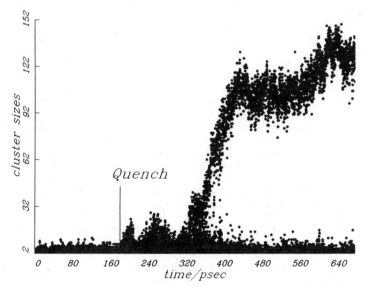

Figure 8.   The evolution of the cluster size distribution in a
molecular dynamics temperature quenching experiment.   This
system is composed of 216 Lennard-Jones atoms at a reduced
temperature of 0.8 and a supersaturation of 4.2.

computer visualization of this simulation. The two droplets, each
composed of a few hundred atoms, seen in the cluster size distribution trace between 100 and 300 picoseconds, are observed to interact
strongly but resist coalescence over this relatively long period
before collapsing to form the final droplet. It is apparent from
this visualization that rate determining factors in the nucleation
process involve cluster-cluster interactions and that growth is not
predominated by monomer addition. We will present a quantitative
analysis of nucleation dynamics elsewhere.

One interpretation of the process displayed in Figure 8 might be
the formation of a critical embryo at roughly 360 picoseconds. In
fact, using the arguments of conventional nucleation theory a
critical embryo size of approximately 25 atoms is predicted. The
dependence of completion time on supersaturation for this state, at a
supersaturation ratio of four, and other quenching experiments at
lower supersaturations is, also, in rough agreement with conventional
nucleation theory. A challenge for future work lies in derivations,
without reference to a thermodynamic barrier, of expressions similar
to those upon which the classical theory is based. Such work may
uncover elements which account for the limited success of the conventional theory in predicting experimental nucleation rates.

Electrolyte Solutions

We now discuss recent investigations of ionic solutions as they
relate to molecular mechanisms of crystallization. Monte Carlo
simulations have been performed employing a very simple solution
model: a mixture of charged and multipolar hard spheres (12). Though
these models are not highly realistic representations of an aqueous
electrolyte solution, we take the view that a molecular treatment of
crystallization will be preceded by a quantitative description of
less complex systems. With that objective an accurate equation of
state has been developed (30) based on the λ-expansion thermodynamic
perturbation theory (31). The novel feature of this approach is an
explicit treatment of ion-solvent interactions which yet retains
great simplicity. The free energy is a sum of three rational
polynomials of low degree in the temperature and density and contains
no free parameters. We demonstrate the accuracy of the resulting
expression for the compressibility factor by comparison with Monte
Carlo in Figure 10. Thermodynamic perturbation theory for the
continuum solvent model (32), which does not take ion-solvent
interactions into account but which significantly improves upon the
Debye-Hückel theory, is also shown.

At higher densities in Figure 10 are two simulated state points
which are super-saturated. The solubility limit for this solution
model is approximately at 1.2 molar. We have used Monte Carlo
simulation to investigate the size distribution and geometry of ionic
clusters as the saturation line is approached in concentration. In
Figure 11 the ion cluster size distribution is displayed for four
sub-saturated states. As ionic charge or concentration is increased
the concentrations of these cluster species increase. A 17-mer of
roughly the same fraction as the 7-mer was observed for state B whose
concentration is near saturation. In super-saturated states ion
cluster size distributions are found to be dominated by very large
clusters whose morphologies are charged chains of alternating sign.

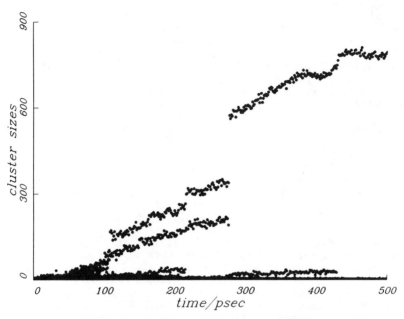

Figure 9. The evolution of the cluster size distribution in a molecular dynamics temperature quenching experiment. This system is composed of 1000 Lennard-Jones atoms at a reduced temperature of 0.76 and a supersaturation of 5.0.

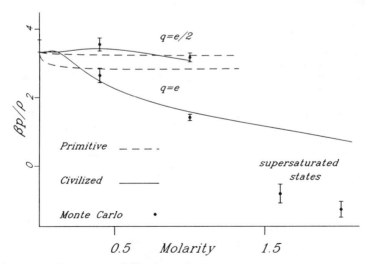

Figure 10. The compressibility factor for a charged and dipolar hard sphere mixture predicted by perturbation theory is compared with the results of Monte Carlo simulation. The elementary electronic charge is denoted e.

A central assertion of homogeneous nucleation theory is that interfacial free energy costs induce a spherical symmetry in the phase embryo. However, these simulation studies indicate that inter-molecular interactions may not permit the development of spherical symmetry when these interactions are strong and highly asymmetric. The morphologies of the large ionic clusters observed in these simulations rather suggest free chain end folding to produce rudimentary lattice structure as a possible pre-transitional mechanism.

An important consideration about which little is currently known is the dehydration which must accompany crystallization. In our simulations we observe that solvent species adjacent to an ion are strongly bound and orientationally ordered due the to the intense local electric field at ionic concentrations both above and below the saturation line. In Figure 12 the variance in the order parameter $\cos\Theta$ is close to zero near contact for the 0.4 and 2.0 molar states. At a concentration approaching the solubility limit, from the sub-saturation side, solvent molecules in the primary solvation shell develop an enhanced orientational mobility which is greater than the free rotor limit of 1/3. We interpret this loss of orientational rigidity as a cancellation of local ionic field contributions as ionic clustering occurs. Solvent orientation near an ion dimer is frustrated by the presence of two energetic minima which are separated by a barrier whose height is defined by the orientations of other adjacent solvent species. When the ion dimer shown in the figure is joined by a third ion of the appropriate sign the height of the barrier separating local energy minima is increased. Hence, at super-saturated states a variance reduction results on interior segments of a charged ionic chain while dehydration is permitted at free chain ends.

Conclusions

Integral equation theory and computer simulation have been used to examine current assumptions through which the physical picture of homogeneous phase nucleation has been constructed. We conclude that the importance of interfacial free energy effects have been over-stated in the classical theory, that strong anisotropy in inter-molecular forces may play a central role in pre-transitional cluster structure and dynamics and that growth is not monomeric and cluster-cluster interactions must be taken into account. Basic research, in areas suggested by these methods, is required before a firm understanding of the molecular mechanics of phase transitions, and techniques for the control of these mechanisms, can be developed.

Acknowledgments

Support for that portion of this work dealing with electrolyte solution theory has been provided to J.E. by the National Science Foundation (CBT-8811789) and by a grant of Cray X-MP time at the National Center for Supercomputing Applications. The authors wish to express appreciation for assistance provided by Mr. John Potter with the development of computer software used in the production of a movie of a nucleation event shown at this symposium.

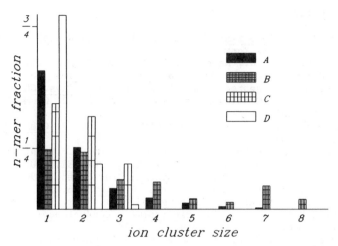

Figure 11. The ion cluster size distribution obtained from computer
simulations of the charged and dipolar hard sphere
mixture at several states: half charge, 1 Molar (A);
fully charged, 1 Molar (B); fully charged, 0.4 Molar (C);
and half charge, 0.4 Molar (D).

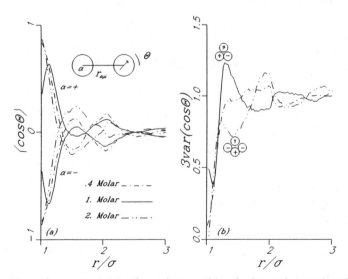

Figure 12. The mean (a) and variance (b) of the orientational
ordering of a dipolar solvent molecule as a function of
distance from an ion for the indicated solution states.
Configurations which enhance and reduce orientational
mobility are displayed.

Literature Cited

1. Gibbs, J.W. The Scientific Papers of J.Willard Gibbs; Dover Publications, New York, 1961.
2. Rao M.; Berne, B.J.; Kalos, M.H. J. Chem. Phys. 1978, 68, 1325.
3. Born, M.; Green, H.S. Proc. Roy. Soc. 1946 A 188, 10.
4. Peters, G.H.; Eggebrecht, J. Submitted to J. Phys. Chem.
5. Eggebrecht, J.; Gubbins, K.E.; Thompson, S.M. J. Chem. Phys. 1987, 86, 2286.
6. Eggebrecht, J.; Thompson, S.M.; Gubbins, K.E. J. Chem. Phys. 1987, 86, 2299.
7. Thompson, S.M.; Gubbins, K.E.; Walton, J.P.R.B.; Chantry, R.A.R.; Rowlinson, J.S. J. Chem. Phys. 1984, 81, 530.
8. Henderson, J.R. In Fluid Interfaces; Croxton ed., 1987.
9. Larson, M.A.; Garside, J. J. Crys. Growth 1986, 76, 88.
10. Friedman, H.L.; Larsen, B. Pure and Appl. Chem. 1979, 51, 2147.
11. Fuoss, R.M.; Accasina, F. Electrolytic Conductance; Interscience: New York, 1959; Chap. 18.
12. Eggebrecht, J.; Ozler, P. Accepted by J. Chem. Phys.
13. Toxvaerd, S. Mol. Phys. 1973, 26, 91.
14. Hansen, J.P.; MacDonald, I.R. Theory of Simple Liquids; Academic Press: New York, 1976.
15. Tolman, R.C. J. Chem. Phys. 1948, 16, 758.
16. Irving, J.H.; Kirkwood, J.G. J. Chem. Phys. 1950, 18, 817.
17. Schofield, P.; Henderson, J.R. Proc. Roy. Soc. 1982, A 379, 231.
18. Buff, F. J. Chem. Phys. 1955, 23, 419.
19. Bakker, G. Wien-Harms' Handbuch der Experimentalphysik, Band VI; Akademische Verlagsgesellschaft, Leipzig, 1928.
20. Ono, S.; Kondo, S. Handbuch der Physik, Band X; Springer Verlag, Berlin, 1960.
21. Rowlinson, J.S.; Widom, B. Molecular Theory of Capillarity; Claredon Press, Oxford, 1982, Chap. 4.
22. Hemingway, S.J.; Henderson, J.R.; Rowlinson. Faraday Sympos. Chem. Soc. 1981, 16, 33.
23. Percus, J.K.; Yevick, G.T. Phys. Rev. 1958, 110, 1.
24. Tolman, R.C. J. Chem. Phys. 1949, 17, 333.
25. Falls, A.H.; Scriven, L.E.; Davis, H.T. J. Chem. Phys. 1981, 75, 3986.
26. Rasmussen, D.H.; Sivaramakrishnan, M.; Leedom, G.L. AIChE Symposium Series, Crystallization Process Engineering, 1982, 78(215), 1.
27. Abraham, F.F. Homogeneous Nucleation Theory; Academic Press, New York, 1974, Chap. 2.
28. Walton, J.P.R.B. Doctoral Dissertation, University of Oxford, 1984.
29. Rao, M.; Berne, B.J. Mol. Phys. 1979, 37, 455.
30. Eggebrecht, J.; Ozler, P. To be submitted.
31. Zwanzig, R. J. Chem. Phys. 1954, 22, 1420;
    Pople, J.A. Proc. Roy. Soc. 1954, A 221, 498;
    Stell, G.; Rasaiah, J.C.; Narang, H. Mol. Phys. 1972, 23, 393 (1972);
    Rushbrooke, G.S.; Stell, G.; Høye, J.S. Mol. Phys. 1973, 26, 1199.
32. Stell, G.; Lebowitz, J.L. J. Chem. Phys. 1968, 49, 3706.

RECEIVED May 12, 1990

Chapter 3

Solubilities in Multicomponent Systems

Jaroslav Nývlt and Jiří Stávek

Institute of Inorganic Chemistry, Czechoslovak Academy of Sciences, Majakovského 24, 16600 Prague 6, Czechoslovakia

The basic problem in determining phase equilibria in multicomponent systems is the existence of a large number of variables, necessitating extensive experimental work. If ten measurements are considered satisfactory for acceptable characterization of the solubility in a two-component system in a particular temperature range, then the attainment of the same reliability with a three-component system requires as many as one hundred measurements. Therefore, a reliable correlation method permitting a decrease in the number of measurements would be extremely useful. Two different methods - the first of them based on geometrical considerations, and the second on thermodynamic condition of phase equilibria - are presented and their use is demonstrated on worked examples.

The basic problem in determining phase equilibria in multicomponent systems is the existence of a large number of variables, necessitating extensive experimental work. If ten measurements are considered satisfactory for acceptable characterization of the solubility in a two-component system in a particular temperature range, then the attainment of the same reliability with a three-component system requires as many as one hundred measurements;with more complicated systems, the necessary number of measurements will be several orders of magnitude higher. Therefore, a reliable correlation method permitting decrease in the number of measurements would be extremely useful.

Methods that more or less comply with these requirements can be classified into two groups : purely empirical methods based on certain geometrical concepts, and methods derived from thermodynamic descriptions of phase equilibria, which replace unknown quantities by an empirical function. Two typical methods will be introduced.

0097–6156/90/0438–0035$06.00/0
© 1990 American Chemical Society

Clinogonial Projection

Clinogonial projection belongs to the empirical methods enabling us
to obtain the solubility in a three component system from binary
data. It is based on the assumption that the plane corresponding to
the solubility of one of the components can be obtained by geometrical projection of the binary solubility (or melting) curve for the
mixture with the other component from the apex corresponding to the
solvent. The principle of the method is explained in"Figure 1".
   A three-component system A + B + C, represented in"Figure 1".
is considered. The quantity monitored, e.g. the saturation temperature in a given point of the system, generally has t_{AB}, t_{AC} and t_{BC}
values in the corresponding binary systems.
A number of planes are constructed at right angles to triangle ABC,
so that they pass through the same coordinate in binary systems AC
and BC, i.e. $t_{AC} = t_{BC}$. Another system of planes perpendicular to
triangle ABC   passes through point C. The required value
of the saturation temperature at point S, corresponding to t_S is
given by the t_1 value corrected for projection of the difference
appropriate t_{AB} value from linear dependence in the binary system
AB, i.e. :

where
$$t_S = t_1 + \Delta t_{AB} \; 1 - x_C(S)] \tag{1}$$

$$\Delta t_{AB} = t_{AB} - t^*_{AB} \tag{2}$$

$$t^*_{AB} = t_{OB} + \frac{x_A}{x_A + x_B} \; (\; t_{OA} - t_{OB} \;) \; . \tag{3}$$

On substitution into "Equation 1", the resulting relationship is obtained :   (4)

$$t_S = t_1 + [\; 1 - x_C(S)] \; [\; t_{AB} - \frac{x_A}{x_A + x_B} (t_{OA} - t_{OB}) - t_{OB}]$$

The calculation according to "Equation 4" is very simple and the
whole procedure can be mechanized by assembling a suitable table :
$t_1 \quad x_C(S) \quad t_{AB} \quad x_A/(x_A + x_B) \quad t_{OA} - t_{OB}$ useful products t_s.

   Very frequently, one isotherm in a ternary diagram is known
rather than the shape of function t_{AB}; thus the shape of curve t_0
in "Figure 1" is known and the shape of another isotherm has to be
determined. On the basis of "Equation 1" and an analogous "Equation 5"

$$t_T = t_2 + \Delta t_{AB} \; [1 - x_C(T)] \tag{5}$$

written for an arbitrary point T located on the same straight line
with $x_A/(x_A + x_B)$, the relation can be written :

$$\frac{t_s - t_1}{1 - x(S)} = \frac{t_T - t_2}{1 - x_C(T)} = \Delta t_{AB} \quad \text{for } x_A/(x_A + x_B) = \text{const.} \tag{6}$$

The "Equation 6" expresses a linear dependence between t_2 and $x_C(T)$. The other dependence between t and x_C is obtained directly from ternary diagram. For example, the following corresponding pairs of values can be read directly from "Figure 1" : $x_C(R)$ ... t_0 ; $x_C(S)$ ... t_1 ; $x_C(T)$ ... t_2 etc. Both the dependencies are depicted schematically in "Figure 2". Firstly, the $t(x_C)$ values corresponding to the intercepts of various connecting lines $t_{AC} = t_{BC}$ with the straight line $x_A/(x_A + x_B)$ are found from the triangular diagram. Further, the t_C value is plotted on the graph on the vertical line constructed through the point $x_C = 1$. The line connecting points t_C with point t_1 , located on the vertical line passing through point $x_C = x_C(S)$, determines the length of segment KL = t_{AB}. An arbitrary t_T value is selected and the appropriate segment is plotted on the vertical line passing through point $x_C = 1$, a line is drawn parallel to straight line $t_S - L$ and the intercept of this line with curve $t(x_C)$ yields the $x_C(T)$ value, corresponding to the isotherm in "Figure 1".

As an example, the calculation is shown for the system ammonium sulphite - ammonium sulphate - water. The isotherm of the ammonium sulphite solubility curve at 20° C is drawn in "Figure 3" and the points representing aqueous solutions of ammonium sulphite and ammonium sulphate saturated at 0, 20, 50 and 70° C are also given. For an arbitrary chosen straight line "a", the x_{H_2O} values corresponding to the intercepts of this line with H_2O dashed connecting lines for identical temperatures in the binary mixtures are read from this diagram. The procedure described above leads to the construction of diagram in "Figure 4". Open circles represent the dependence $t(H_2O)$ read from "Figure 3", calculated points represented by full circles are redrawn back into "Figure 3". The whole procedure repeated for other arbitrary lines "a" would give another set of points corresponding to individual isotherms.

The method described is very rapid and yields satisfactory results unless the extrapolations are carried out over too wide a range.

Expansion of Relative Activity Coefficients

A three component system consisting of a solvent (0) and two further components (1 and 2) can be considered. The phase equilibrium between the solid (s) and liquid (1) phases is characterized by equality of the chemical potentials of a given component in the two phases. Supposing that the component are completely immiscible in the solid phase we obtain from the condition of equality of chemical potentials :

$$\log x_i = -\log \Gamma_i = \Phi_i \qquad [T, P, \text{sat}] \tag{7}$$

where x_i is the relative molality of component i and Γ_i is its relative activity coefficient :

$$x_i = m_i/m_{0i} \qquad \Gamma_i = f_i/f_{0i} \quad . \tag{8}$$

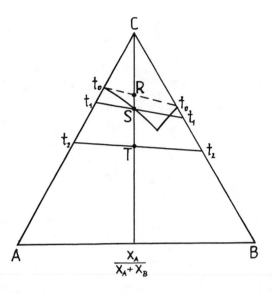

Figure 1. Principle of clinogonial projection. (Reproduced with permission from ref. 1. Copyright 1977 Academia.)

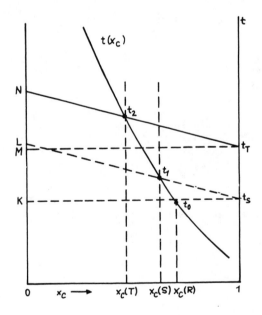

Figure 2. Graphical solution by the clinogonial projection method. (Reproduced with permission from ref. 1. Copyright 1977 Academia.)

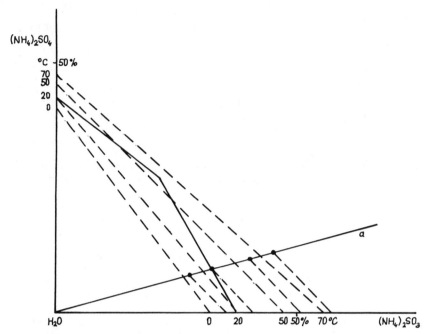

Figure 3. Example of construction by the clinogonial projection method. (Reproduced with permission from ref. 1. Copyright 1977 Academia.)

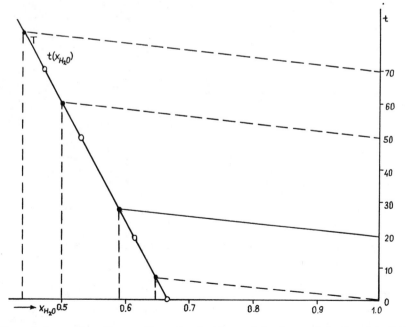

Figure 4. An auxiliary diagram. (Reproduced with permission from ref. 1. Copyright 1977 Academia.)

In these equations, subscript i denotes the substance which is in equilibrium and subscript $0i$ denotes its saturated solution in pure solvent 0 at given temperature.

If the compound 1 is dissociated and ions M and X in solution do not originate from any other component, i.e. :

$$M_{n+} X_{n-} = n_+ M^{z+} + n_- X^{z-} \qquad (9)$$

"Equation 7" can be rewritten to give :

$$\log (S_1/S_{10})^{1/n_1} = \log x_1 = \Phi_1 \qquad (10)$$

where S_1 and S_{10} are the analytical solubility products of component 1 in the multicomponent system and in the pure solvent, respectively, and $n_1 = n_+ + n_-$ is the total number of particles dissociated from one molecule of the substance.

If the presence in solution of two electrolytes with a common cation is assumed, then holds :

$$x_{i+} = n_{i+} x_1 + n_{i+} x_2 m_{20}/m_{10} \qquad (11)$$

$$x_{i-} = n_{i-} x_i \qquad (12)$$

and from the basic equation we obtain :

$$\log [x_1^{n_1-/n_1} . (\sum_{i=1}^{k} \frac{n_{i+} m_{i0}}{n_{1+} m_{10}} x_i)^{n_{1+}/n_1}] = \Phi_1 \qquad (13)$$

and a similar equation holds also for a system containing two electrolytes with a common anion (only the + and - signs in subscripts are interchanged). For all possible combinations of mono- and divalent ions, a common equation is obtained :

$$\Phi_1 = \frac{1}{\alpha + \beta} \log [x_1^\alpha . (x_1 + F B)^\beta] \qquad (14)$$

where

$$B = x_2 . m_{20}/m_{10} \qquad F = n_{2+-}/n_{1-+} \qquad (15)$$

$$\alpha = n_{1+-} \qquad \beta = n_{1-+}$$

(the first sign in subscripts is valid for a common anion and the second sign holds for a common cation). For example, in the case of solubility of $CuSO_4$ in the system $CuSO_4 - K_2SO_4 - H_2O$ it holds $\alpha = 1$; $\beta = 2$ and $F = 1$.

On the left-hand side of "Equation 14" is, of course, still an unknown function involving the relatively activity coefficient. Its value must, in general, depend on the composition, temperature and pressure of the system :

$$\Phi_1 = \Phi_1 (m_1, m_2, m_3, \ldots m_k, T, P) . \qquad (16)$$

As, however, the composition of the condensed two-phase system is unambigously determined by $(k - 1)$ concentration values, this general functional dependence can be rewritten in the form :

$$\Phi_1 = \Phi_1 \, (m_2, \, m_3, \, \ldots \, m_k) \quad (T, \, P = const.). \tag{17}$$

This form has the advantage of not containing concentration value m_1 and thus permits the explicit expression of x_1 from the basic equation. The expansion of the general function given by "Equation 17" into the MacLaurin series with respects to molalities $m_{i>1}$ yields the equation :

$$\Phi_1 = \sum_{i=2}^{k} Q_{1i} m_i + \sum_{i=2}^{k} \sum_{j=2}^{k} Q_{1ij} m_i m_j + \ldots \tag{18}$$

or, for a three-component system :

$$\Phi_1 = Q_{12} m_2 + Q_{122} m_2^2 + \ldots \tag{19}$$

The adjustable interaction constants Q can be evaluated from the experimental data for three-component systems; these constants can then be employed for concentration of temperature interpolations and also for calculation of phase equilibria in multicomponent systems. Moreover, the constants Q usually depend very little on temperature, as the relative molalities, related to the solubility of the substance in the pure solvent, are employed; hence calculations of other isotherms can be carried out easily.

For example, values of Q_{12} for the system $NaNO_3$ - NaCl - H_2O in the temperature range 0 to 100° C are (higher[3] terms are zero) :

t [° C]	Q_1
0	-0.025
25	-0.020
30	-0.022
40	-0.019
50	-0.017
75	-0.015
91	-0.008
100	-0.008

As an example, solubility in the system KCl - NaCl - H_2O at 20° C has been calculated. Necessary data are given as follows :

$M_{KCl} = 74.551 \quad M_{NaCl} = 58.443 \quad \alpha = \beta = F = 1$

Solubility of KCl ;

$$Q_{12} = -0.015 \; ; \quad m_{10} = 4.586$$

Solubility of NaCl :

$$Q_{21} = 0.000 \; ; \; Q_{211} = 0.0007 \; ; \; m_{20} = 6.135.$$

The result of calculation is visualized in "Figure 5" ; dotted lines correspond to ideal system data (all Q = 0).

It can be seen that even small value of the interaction constants Q can significantly move the resulting equilibrium curve. As the interaction constants are usually rather small, a rough information can be drawn even from the supposition of all Q = 0, i.e. without any previous knowledge of the ternary system.

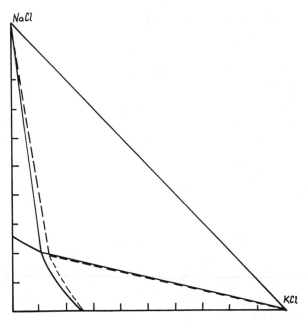

Figure 5. Comparison of real and ideal behavior in idalized system $NaCl–KCl–H_2O$. (Reproduced with permission from ref. 1. Copyright 1977 Academia.)

Conclusions

Two different methods have been presented in this contribution for correlation and/or prediction of phase equilibria in ternary or multicomponent systems. The first method, the clinogonial projection, has one disadvantage : it is not based on concrete concepts of the system but assumes, to a certain extent, additivity of the properties of individual components and attempts to express deviations from additivity of the properties of individual components and attempts to express deviations from additivity by using geometrical constructions. Hence this method, although simple and quick, needs not necessarily yield correct results in all the cases. For this reason, the other method based on the thermodynamic description of phase equilibria, reliably describes the behaviour of the system. Of cource, the theory of concentrated ionic solutions does not permit a priori calculation of the behaviour of the system from the thermodynamic properties of pure components; however, if a satisfactory equation is obtained from the theory and is modified to express concrete systems by using few adjustable parameters, the results thus obtained are still substantially more reliable than results correlated merely on the basis of geometric similarity. Both of the methods shown here can be easily adapted for the description of multicomponent systems.

Literature Cited

1.   Nývlt, J. Solid - Liquid Equilibria; Academia, Prague, 1977

RECEIVED May 12, 1990

Chapter 4

Hardness of Salts Used in Industrial Crystallization

J. Ulrich and M. Kruse

Verfahrenstechnik, FB 4, Universität Bremen, Postfach 330 440, D-2800 Bremen 33, Federal Republic of Germany

The ultra-micro-hardness of inorganic and organic salts has been measured for 15 substances. These are products usually produced in industrial crystallization. The hardness-force-dependency was examined and data are compared to those from literature. In the case of potassium nitrate a strong direction dependency of the hardness was observed. Also effects of impurities in the crystal lattice were analysed. In the end an attempt has been introduced to calculate the hardness of crystals from a physical model.

The design of the crystallizers for the crystallization of aqueous solutions is depending on the usual informations for plant design and on the kinetic data of the material to crystallize. The material fixes the physical properties, the crystal growth and the nucleation behaviour. Special difficulties exist in finding the nucleation data. Here a special focus is on the secondary nucleation, because this is in most cases the dominating nucleation mechanism. From this secondary nucleation process, it is again the part which is originated by mechanically introduced energies which is of most interest. In other words the questions are: How many abrasion particles (secondary nuclei) are produced in a crystallizer? Which rules and laws can be used to calculate the number of the nuclei? Finally how is it possible to do a scale up operation, and how is the effect of a different products to be considered?

There is a parameter necessary to describe the abrasion resistance of each crystalline substance. This abrasion resistance must be correlated with parameters of the power input devices such as pump or stirrer diameter or the stirrer tip speed. The abrasion resistance of the crystalline particles produced by secondary nucleation in industrial crystallization processes is therefore a physical property of the substance. So far there is no physical property known containing all information about this abrasion resis-

0097-6156/90/0438-0043$06.00/0
© 1990 American Chemical Society

tance and the influence of the fracture mechanics on the crystals. According to (1-8) the Vickers hardness could be a first step in defining such a material property.

Up to today it has been impossible to measure reproducible microhardness data of non-metallic crystals. This was due to the too high indentation forces of commercial available hardness testing devices. Today with the new ultra-microhardness devices, which allow to reduce the indentation force down to $5 \cdot 10^{-4}$ N, it is no longer the limiting problem to measure the Vickers hardness of such brittle crystals.

Results of the Vickers hardness of 15 inorganic and organic salts will be presented. The hardness-force dependency, and the effects of direction dependency were examined. The measured values of the Vickers hardnesses were taken for an attempt to prove a model to calculate the hardness. This model describes the hardness purely by physical properties of the substances. The use of such a model may be an approach for the description of the abrasion resistance of salts. Data describing the abrasion resistance could help in the understanding and interpretation of secondary nucleation phenomena.

Experimental Set-Up and Measurements

The ultra-micro-hardness of the salts are measured according to Vickers. The Vickers hardness is calculated from the length of the diagonals of a plastic indentation in the surface of the material to be tested by a diamond pyramid. If there exists no direction-dependency the hardness indentation has a quadratic shape. The Vickers hardness for a single indentation is calculated from the average value of the two diagonals. Equation 1 shows how to calculate the Vickers hardness H_V from the indentation force F and the average of the two diagonal lengths d of the indentation. The constant 1854 is valid for the case that the force F has the unit pond and the length d the unit millimeters.

$$H_V = 1854 \cdot F/ d^2 \tag{1}$$

All experiments were carried out with an ultra-microhardness measuring device (Vickers-hardness) produced by Anton-Paar-Company. It is attached to a microscope with an adapted photo camera. The force of indentation, the time of indentation and the time-gradient of the indentation force can be selected and adjusted separatly.

For most substances used in the experiments an indentation force greater than 0.1 N leads to cracks or even lift-offs at the edges or the ends of the diagonals of the indentation produced by the Vickers pyramid. Therefore the values of the measured hardnesses have a large spread and usually are not reproducible. Most of the literature data were produced at a time where ultra-micro-hardness devices were not available, so that forces larger $5 \cdot 10^{-2}$ N have been used.

Results

A comparison of the results of our experiments with data from
literature are shown in Table 1. Our experimental data were found
with an indentation force of $2 \cdot 10^{-2}$ N, a time of indentation of 10
s, and a gradient of 10^{-2} N/s. The shown data represent an average
of at least 10 single measurements for all the different substances.
In most cases of the literature data no information about the experimental conditions are available. Therefore an exact comparison
is unfortunately not possible.

Table 1: Comparison of the Vickers Hardness Measurements
and the Data from Literature

Substance	Vickers Hardness H_V	
	Own Results	Literature
RbJ	6.3	5.7 (9)
KJ	9.7	6.1 (9)
KCl	10.6	9.9 (9)
		9.7 (10)
		10.0 (11)
KBr	11.0	7.2 (9)
NaCl	22.3	17 (9)
		24 (11)
		18 (12)
KNO_3	47.1	13 (9)
CsBr	51.6	17 (9)
Citric acid	52.6	
$MgSO_4 \cdot 7\ H_2O$	66.8	
$NH_4Al(SO_4)_2 \cdot 12\ H_2O$	74.9	56 (9)
$KAl(SO_4)_2 \cdot 12\ H_2O$	75.7	56 (9)
K_2SO_4	112.6	102 (9)
		130 (13)
$CuSO_4 \cdot 5\ H_2O$	118.7	
LiF	122.7	102 (9)
$NaClO_3$	134.6	118 (9)

The force-dependency of the Vickers hardness was measured for
several salts. This effect is shown in Figure 1 e.g. for sodium
chlorate. With an increase of the indentation force the Vickers
hardness decreases, e.g. up to 40 % for sodium chlorate. The photos
of indentations by the Vickers pyramid in a sodium chlorate crystal
are presented in Figure 2 and give an idea of resulting cracks and
lift-offs. This cracks are due to too high indentation forces (the
indentation force to be seen in the Figures are: 2a, $2 \cdot 10^{-2}$ N; 2b,
$4 \cdot 10^{-2}$ N; 2c, 0.25 N; 2d, 0.6 N).

   In the case of potassium nitrate the Vickers hardness indentations have not a quadratic shape. This is due to a direction dependent hardness in the crystal lattice of this material. The direction
dependency can be explained with an anisotropic effect in the lat-

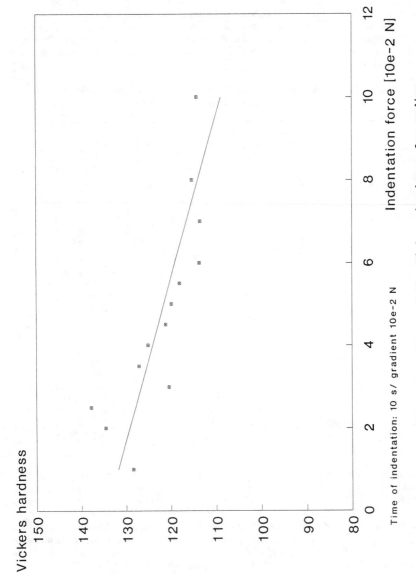

Figure 1. Force-dependency of the Vickers hardness for sodium chlorate

tice. The force-dependency of the Vickers hardness isshown in the plot in Figure 3. In the diagram there are two curves of the Vickers hardness resulting always from the same indentation image. The hardness values differ by a factor of two. The one curve results from the data of the longer and the other from the shorter diagonal of each intendation. This direction dependency was so far only observed for the case of potassium nitrate.

Also the effect of impurities in a crystal on the Vickers hardness was analysed. In Figure 4 are shown the force dependency curves of the Vickers hardnesses of a pure sodium chloride crystal and a sodium chloride crystal grown in a solution with an impurity of 10 % urea. The hardness of the pure sodium chloride crystal is up to 25 % higher than the hardness of the impure crystal.

Modelling of the Hardness

The mathematical modelling of hardness from physical properties was not the object of research work for quite a time, since there had been no possibility to verify the theoretical results with the experimental data.

First attempts (14-16) described only the hardness of simple cubic crystals of the type AB as a function of their interatomic distances and the valencies of the atoms. This is shown in Equation 2:

$$H = s \cdot [Z_1 \cdot Z_2] / r^m \qquad (2)$$

Here is H the hardness, s and m are empirical constants, Z_i are the valencies of the atoms, and r is the interatomic distance.

Another possibility to calculate the hardness from physical properties (17) is given in Equation 3:

$$H = A \cdot U \cdot r^{-3} \qquad (3)$$

Where the hardness H is combined with the lattice energy of the AB-type crystal U, the interatomic distance r and an empirical constant A.

Equation 3 was modified by the authors (18,19) so that it results in Equation 4:

$$H = K \cdot T_D \cdot r^{-3} \qquad (4)$$

Here the hardness H is connected with a constant K, the Debye temperature T_D and the interatomic distance r.

All these models for the calculation of the hardness need empirical parameters which cannot be explained by physical properties of the crystals. This problem is solved if the hardness of crystals is described as a volumetric cohensive energy as suggested by Plendl et. al.(20). Then the hardness can be written as the ratio of the cohensive energy U to the molecular volume V, as to be seen in Equation 5.

$$H = U / V \qquad (5)$$

Figure 2a. Photo of an indentation in a sodium chlorate crystal
with an indentation force of $2 \cdot 10^{-2}$ N

Figure 2b. Photo of an indentation in a sodium chlorate crystal
with an indentation force of $4 \cdot 10^{-2}$ N

Figure 2c.   Photo of an indentation in a sodium chlorate crystal
with an indentation force of 0.25 N

Figure 2d.   Photo of an indentation in a sodium chlorate crystal
with an indentation force of 0.6 N

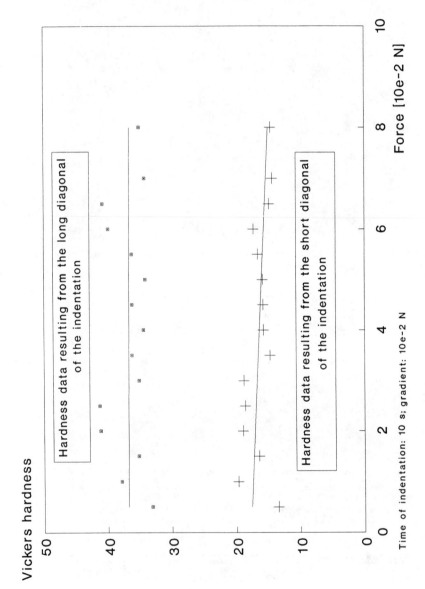

Figure 3. Force-dependency of the Vickers hardness for potassium nitrate.

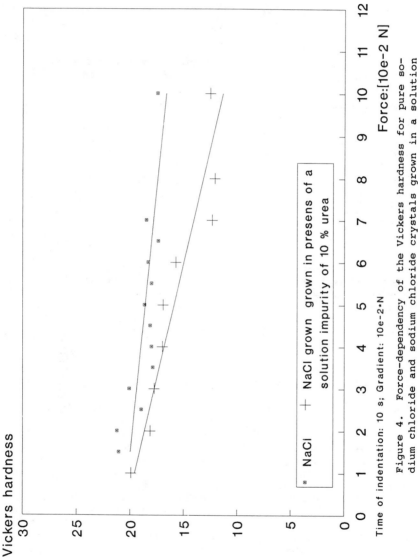

Figure 4. Force-dependency of the Vickers hardness for pure sodium chloride and sodium chloride crystals grown in a solution with an impurity of 10 % urea

An analyses of the measured microhardness of non-metallic substances
leads to new atomic subdivisions namely to:

- the 'hard-core' substances, e.g. LiF, CaO, SiO_2, and
- the 'soft-core' substances, e.g. NaCl, CaF_2, Si, Ge.

In Table 2 the results are shown for the calculated volumetric
cohensive energies of crystals with NaCl-type structure and their
measured Vickers hardness.

Table 2: Comparison of the Measured Vickers Hardness and
the Calculated Volumetric Cohensive Energy for
NaCl-Type Crystals

Substance	Vickers Hardness	Volumetric Cohensive Energy ([20]) [kcal / cm^3]
KJ	9.7	2.9
KCl	10.6	4.4
KBr	11.0	3.8
NaCl	22.3	6.7
LiF	122.7	21.0

This last model of the hardness is modified and given in Equation 6.
The hardness H is according to ([21]) a function of several physical
properties. $C_{str.}$ is the constant defined in Equation 7, Z is the
largest common valency of the atoms, f is the force constant per
unit charge and r_0 represents the interatomic distance.

$$H = C_{str.} \cdot Z \cdot f / r_0 \qquad\qquad (6)$$

$$C_{str.} = N \cdot \rho \cdot r_0^3 / 2 \cdot M_r \qquad\qquad (7)$$

Here N is the Avogadro number, ρ is the density of the crystal and
M_r the the reduced mass of the atoms.

In Figure 5 a comparison is presented for the calculated Vickers hardness (Equation 6) and the measured Vickers hardnesses for
several ionic bonded crystals with NaCl-structure.

The comparison of the measured and calculated Vickers
hardnesses is of acceptable quality. With this physical model of the
hardness it should also be possible to compute the Vickers
hardnesses of salts with a more complicated lattice structure than
the NaCl-structure. This may be an approach for a description of the
abrasion resistance of salts. Such a description of the abrasion
resistance could be useful in calculating the secondary nucleation
rates.

Measured Vickers hardness

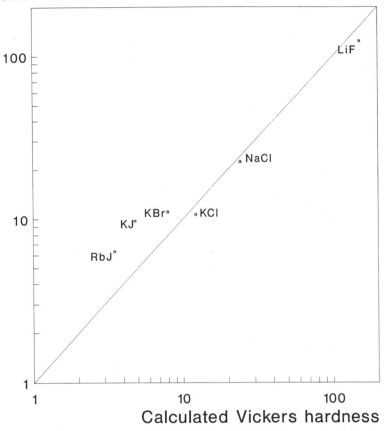

Calculated Vickers hardness

Figure 5. Comparison of the calculated and measured Vickers hardness for NaCl-type crystals

Acknowledgments

The support of this project by the Deutsche Forschungsgemeinschaft (DFG) is gratefully acknowledged.

Legend of Symbols

A	constant
d	length of diagonals
F	indentation force
f	force constant per unit charge
H	hardness
H_V	Vickers hardness
K	constant
M_r	reduced mass

m   constant
N   Avogadro's number
r, r_0 interatomic distance
s   structure factor
T_D Debye temperature
U   lattice energy or cohensive energy
V   molecular volume
z_i valency
ρ density

Literature Cited

1.   Tai, C. Y.; McCabe, W. L.; Rousseau, R. W. AIChE Journal
      1975, 21, 351.
2.   Offermann, H.; Ulrich, J. Verfahrenstechnik 1980,
      14, 815.
3.   Ulrich, J. Ph.D. Thesis, RWTH Aachen, Aachen FRG, 1981.
4.   Offermann, H.; Ulrich, J. In Industrial Crystallization
      81; Jancić, S. J.; de Jong, E. J., Ed.; Elsevier:
      Amsterdam, 1981, p 313.
5.   Heffels, S. K. Ph.D. Thesis, TH Delft, Delft NL, 1986.
6.   Pohlisch, R. J. Ph.D. Thesis, TU München, München FRG,
      1987.
7.   Offermann, H. In Industrial Crystallization 87, Nývlt,
      J.; Zácek, S., Ed.; Elsevier: Amsterdam, 1989, p 121.
8.   Mersmann, A.; Sangl, R.; Kind, M.; Pohlisch, J.
      Chem. Eng. Tech. 1988, 11, 80.
9.   Engelhardt, W. v.; Haussühl, S. Fortschr. Miner. 1965,
      42, 5.
10.   Balaramaiah, A.; Pratap, K. J.; Hami, Babu, V. Cryst.
      Res. Technol. 1987, 22, 1205.
11.   Plendl, J. N.; Gielisse, P. J.; Mansur, L. C.; Mitra, S.
      S.; Smakula, A.; Tarte, P. C. Applied Optics 1971,
      10, 1129.
12.   Chin, G. Y. Trans. Amerc. Crystall. Asso. 1975, 11, 1.
13.   Ridgway In Crystallization; Mullin, J. W.; Butterworths:
      London, 1972
14.   Goldschmidt, V. M. In Geochemische Verteilungsgesetze;
      Oslo, 1927
15.   Reis, A.; Zimmermann, L. Z. physik. Chem. 1922,
      102, 298.
16.   Friedrich, E. Fortschr. Chem. Phys. phys. Chem. 1926,
      18, 1.
17.   Wolff, G. A.; Toman, L.; Field, N. J.; Clark, J. C.
      In Halbleiter und Phosphore; Schön, M.; Walker, H.,
      Eds., Braunschweig, 1958, p 463.
18.   Ulrich, J.; Kruse, M. Cryst. Res. Technol. 1989,
      24, 181.
19.   Ulrich, J.; Kruse, M. Chem.-Ing.-Tech. 1989, 61, 962
20.   Plendl, J. N.; Gielisse P. J. Phys. Rev. 1962, 125, 828.
21.   Plendl, J. N.; Gielisse P. J. Z. Kristall. 1963,
      118, 404

RECEIVED April 12, 1990

Chapter 5

Calculation of Crystal Habit and Solvent-Accessible Areas of Sucrose and Adipic Acid Crystals

Allan S. Myerson[1] and Michael Saska[2]

[1]Department of Chemical Engineering, Polytechnic University, 333 Jay Street, Brooklyn, NY 11201
[2]Audubon Sugar Institute, Agricultural Center, Louisiana State University, Baton Rouge, LA 70803

The crystal habit of sucrose and adipic acid crystals were calculated from their internal structure and from the attachment energies of the various crystal faces. As a first attempt to include the role of the solvent on the crystal habit, the solvent accessible areas of the faces of sucrose and adipic acid and were calculated for spherical solvent probes of different sizes. In the sucrose system the results show that this type of calculation can qualitatively account for differences in solvent (water) adsorption hence fast growing and slow growing faces. In the adipic acid system results show the presence of solvent sized 'receptacles' that might enhance solvent interactions on various fares. The quantitative use of this type of data in crystal shape calculations could prove to be a reasonable method for incorporation of solvent effects on calculated crystal shapes.

Crystallization from solution is an important industrial separation and purification process for a variety of materials ranging from bulk commodity chemical to pharmaceuticals. Besides purity, specifications on the product crystals often includes such properties as crystal size distribution (or average size), bulk density, filterability, slurry viscosity and flow properties of the dry powder. All of these properties depend entirely on the crystal size distribution and the crystal shape.

Crystals are ordered atoms (or molecules) in a crystal lattice. This internal structure is accessible by x-ray diffraction and is available for most common solids. Unless a crystal displays polymorphism, the crystals internal structure will not vary with the conditions of its growth. This, however, is not the case for the external crystal habit. Crystal habit can vary with the conditions of growth, solvent used and the presence of impurities.

0097–6156/90/0438–0055$06.00/0

Hartman and Perdok(1-3) in 1955 developed a theory which related crystal morphology to its internal structure on an energy basis. They concluded that the morphology of a crystal is governed by a chain of strong bonds (called periodic bond chains)(PBC), which run through the structure.

Hartman and Bennema (4) looked at attachment energy as a habit controlling factor. They found that the relative growth velocity always increases with increasing E^{att} however, the relationships between the two depends on the mechanism of crystal growth and variables such as supersaturation, temperature and solid-fluid interactions. They demonstrated at low supersaturation, however, the relative growth velocity of a face is directly proportional to the attachment energy of that face. Hartman (5,6) employed this assumption to calculate the habit of naphthalene and sulfur and had good agreement with observed forms.

A major weakness in the calculations described above is that they can only be used to represent vapor grown crystals. In crystals grown from solution, the solvent can greatly influence the crystal habit as can small amounts of impurities. Several investigators (7-9) accounted for discrepancies between observed crystal habit and those obtained using attachment energies by assuming preferential solvent (or impurity) adsorption on crystal faces.

It is the purpose of this work to employ calculations of the solvent accessible area (SAA) of sucrose and adipic acid crystals as a first attempt to quantify the effect of solvent on the crystal morphology.

Calculation of Crystal Shape

It is a fundamental problem to predict the shape that a crystal will adopt when growing from a submicroscopic nucleus to its macroscopic form. Generally, both the intrinsic properties of the crystallizing matter and the external conditions (supersaturation, temperature, etc) will effect the shape.

A simple correlation was noticed by Donnay and Harker, (10) between the interplanar spacing of a crystallographic plane, d_i, and its area on an average crystal. A similar correlation holds between d_i and the frequency with which the plane (hk1) appears in an ensemble of crystals. Since an area of a plane is roughly proportional to the inverse of its (linear) growth velocity R, the Donnay-Harker law is equivalent to stating that $R_i \sim 1/d_i$.

Equilibrium thermodynamics require a minimum free energy, F, for a stable system. Splitting F into contributions of the crystal volume and its surface gives

$$F - F_{vol} + F_{surf} \tag{1}$$

For a perfect crystal, F_{vol} (kJ/mole) is independent of both the crystal size and shape, and

$$dF = dF_{surf} \tag{2}$$

With $F_{surf} = \sum_i A_i \gamma_i$, equation (2) at equilibrium becomes

$$\sum_i A_i d\gamma_i = 0 \tag{3}$$

Where, γ_i and A_i are the surface energy and area of a face i, and i runs over all crystal faces.

It can be proven that an equivalent form to equation (3) in terms of displacements h_i of all faces from the center of the crystal is

$$\gamma_i / h_i = constant \tag{4}$$

for all possible crystallographic orientations i = 1, 2, . . . n. Since $h_i \sim R_i$, a crystal complies with the Wulff (equilibrium) condition given by equation (4) when $R_i \sim \gamma_i$.

An equivalent method was used by Kitaigorodski and Ahmed (11) to predict the equilibrium habit of anthracene. Starting from a nucleus of just one molecule, the growth was simulated by adding one molecule at a time at positions such that the free energy minimum condition was always satisfied. A simple 6-exp two-parameter potential function was used to estimate F_{surf}.

All real crystals deviate more or less from their equilibrium habits since all grow at finite velocities R_i. Hartman and Bennema (4) and Hartman (5, 6) showed how the empirical law of Donnay-Harker can be explained on the basis of current molecular theories of crystal growth. The energy required to split a crystal along the plane A--B parallel to the plane (hk1) is the sum

$$\sum_i iE_i \tag{5}$$

where E_i is the interaction of the crystal slices (parallel to (hk1)) i-l lattice spacings apart and i runs from 1 to infinity. However, in molecular crystals, the interaction energy drops off rapidly with distance and the leading term in (5), E_i, predominates. It was shown that if a crystal grows with the spiral (BCF) mechanism, $R_i \sim E^{att}$, is a good approximation for a wide range of growth conditions. E^{att}, the attachment energy of the plane i, was defined as

$$\sum_i E_i \tag{6}$$

Since E_i decays rapidly with growing i (distance),

$$E_i^{att} = \sum_i E_i \cong \sum_i iE_i \tag{7}$$

and the surface energy γ_1 can be calculated as

$$\gamma_1 = Zd_i E_i^{att}/2V \tag{8}$$

Here, Z is the number of molecules per unit cell and V is the unit cell volume.

Using interaction potential functions, E^{att} can be obtained by summing the interactions of individual atoms. A computer program has been developed by the authors (7, 12, 13) to obtain these attachment energies. The program CENER is a Fortran program which calculates the lattice energy of molecular crystals. All lattice parameters can be varied and the lattice energy minimized as their function. The atoms coordinates are calculated relative to the new cell so that the relative atomic positions remain unchanged. The lattice energy is calculated by summing up the interatomic interactions of the control molecule with the neighboring ones. A description of the methods used can be found in Saska (7).

To obtain the crystal habit, the attachment energy is calculated for a large number or possible low index faces (F faces) which are then used to calculate relative growth velocities since $R^{rel} \sim E^{att}$. R_i^{rel} is the growth velocity, R_i, normalized with respect to R_{max}. Knowing the set R_i^{rel} and the crystallographic structure (unit cell), the position of each crystallographic plane relative to a fixed origin is determined. The crystal form is then given by intersections of the planes enclosing the smallest volume. A program (C-shape) was developed by the authors to project 3-dimensional crystal forms on an arbitrary plane. The program performs these operations:

1. Set up the analytical equations of all planes specified.
2. Calculate the position of points resulting from intersections of all planes.
3. Discard the points lying outside the crystal boundary. The criterion used was the same as proposed recently by Strom (14). If the dot product

$$(xi + yj + zk) \cdot (p_{hkl}i + q_{hkl}J + r_{hkl}k) < h_{hkl} \tag{9}$$

For any of the planes present, the point is discarded. Here x, y and z are the spatial coordinates of the point tested p_{hkl}, q_{hkl}, and r_{hkl} are the coordinates of the point lying on a normal to the plane (hk1) at a unit distance from the origin.

4. Decide whether a plane is visible in the required rotation.
5. Connect the corners of the crystal.
6. Calculate relative areas of the crystal faces.

Examples of the results of this type of calculation are shown for sucrose (7) and terephthalic acid (3) in Figures 1 and 2.

Incorporation of Solvent Effects in Crystal Habit Calculations

Interactions between solute and solvent molecules can have a significant effect on the shape of a crystal. This can be accounted for by specific adsorption of the solvent molecule on crystal faces. Current crystal growth theories indicate that when interactions between solute and solvent are strong the solute molecules are solvated and a solvation layer exists at the crystal-liquid interface which likely can vary as a function of crystal face. Crystal growth requires desolvation of the solute molecule and desolvation of the surface site on the crystal. The molecule then surface diffuses until it reaches an incorporation (kink) site.

When the desolvation energy is large (strong solvent interactions) this can become the rate determining step in the growth of a face. The more energy required for desolvation, the slower the face growth rate, hence the more important that face will be in a grown crystal. The challenge is to quantify the interaction of solvents with the crystal faces and to predict a 'true' solvent grown crystal shape and predict how a change of solvent will effect crystal shape.

Solvent Accessibility

As an attempt to quantify the effect of solvent on individual faces of sucrose crystal we have employed a solvent accessibility of Van der Waals sized surface atoms for a spherical probe representing the solvent molecule (15). The smooth surface generated by rolling such a probe along the crystal surface consists of

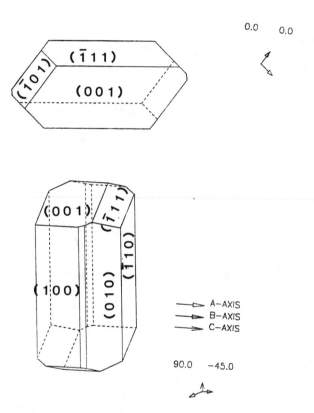

Figure 1.   Calculated growth form of the TPA polymorph 1 crystal,
Hartman condition. Kitaigorodski potential function.

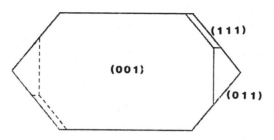

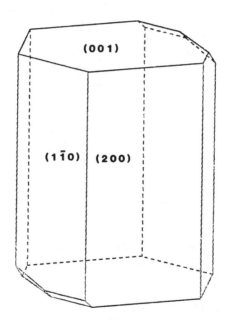

Figure 2.   Growth form of a sucrose crystal (R ˜ E) from the Kitaigorodski potential function.

solvent-accessible parts (the probe touches only one atom at a time) and reentrant parts (where the probe touches two or three atoms simultaneously) that bound regions where the probe (solvent) is excluded from direct contact with the surface atoms. This approach was previously used (15) in characterizing surfaces of proteins, studies of enzyme-substrate complexes, location of crevices in the molecular structures, etc.

Solvent - Accessible Areas of the Sucrose Surfaces

The carbon backbone of the sucrose structure (Figure 3) is almost completely shielded by the oxygen and hydrogen atoms and is expected to play only a minor role, if any, in solvent interactions. The relative contact areas of oxygens and hydrogens are orientation dependent and may impart a specific character to the different faces of the sucrose crystal.

The Connolly's MS program (16) was slightly modified in order to categorize the contact areas for the individual atoms of the sucrose structure (i.e. carbon, oxygen, and hydrogen) and run on the LSU IBM 3090 "supercomputer". The input, consisting of the atomic radii (RC = 1.6, RO = 1.4, RH = 1.2 A) and the positions of the atoms belonging to a multi-molecular crystal slice (such as for example the one in Figure 4) parallel to the desired crystal surface (i.e. a, c, d, p and p') was generated with our modified CENER program used previously (7) (a = (100), c = (001), d = (101), p = ($\bar{1}$10), p' = (110))in calculating the attachment energies of the sucrose surfaces. The probe was rolled over the 90 atoms of the unit cell located in the surface while the rest of the atoms served as blockers only to provide solid neighbors. The "buried" function was used to distinguish between the surface atoms exposed to the solvent and those "buried" in the interior of the crystal.

The size of the probe (the solvent molecule) is of prime importance for the (relative) solvent accessible areas - SAA (Figures 5-8). While the surfaces of the five most common sucrose crystal faces tend towards identical behavior in terms of their relative SAA's for small probes, their characters become distinguishable at larger solvent radii, 1.5 and 3 angstroms (1.5 A is the approximate radius of the water molecule). As expected the SAA of the carbon atoms becomes negligible and carbon is not expected to play a role in solvation properties of sucrose. For a water-sized solvent, the a face exposes about equal areas of oxygens and hydrogens while c, d, and p surfaces expose primarily hydrogens.

Although an alternate mechanism was also proposed (17), it is likely that the difference in growth rates between the right and left pole faces, most notably p' and p is due to a much stronger growth rate inhibition through solvent (water) adsorption at the left pole (p face) (18). A detailed breakdown of the solvent

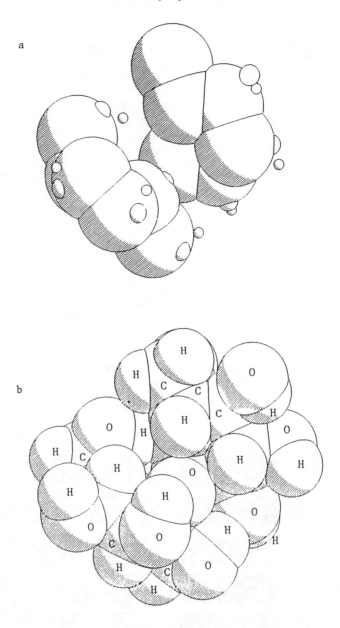

Figure 3.   a) Space-filling model of the sucrose molecule with undersized oxygens and hydrogens to expose the carbon backbone. b) Space-filling model of the sucrose molecule (same orientation as in a) with van der Waals - sized atoms. RC = 1.6 A, RO = 1.4A, RH = 1.2 A. The carbon backbone is almost completely obscured by the oxygens and hydrogens and thus shielded from the solvent molecules.

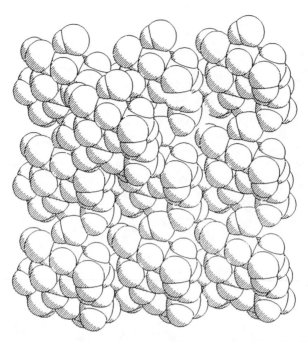

Figure 4.   A nine-molecule space-filling model of the (100) sucrose crystal. Atoms of one of the molecules are labeled. The sub-layer of the screw-axis related molecules is completely shielded from possible solvent effects.

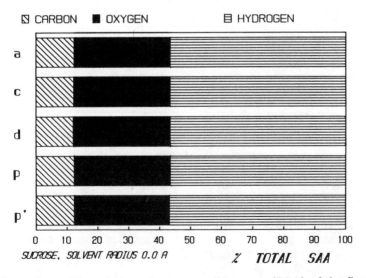

Figure 5.   The relative solvent accessible areas (SAA) of the five most common faces on the sucrose crystal. Solvent radius 0 A.

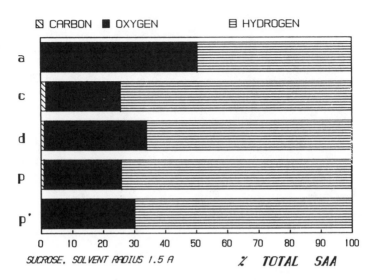

Figure 6.   The relative solvent accessible areas (SAA) of the five most common faces on the sucrose crystal. Solvent radius 1.5 A corresponds to the approximate radius of the water molecule.

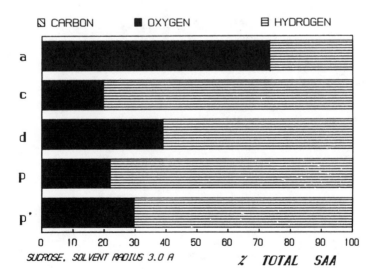

Figure 7.   The relative solvent accessible areas (SAA) of the five most common faces on the sucrose crystal. Solvent radius 3 A.

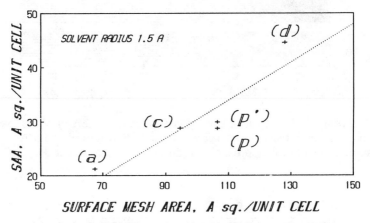

Figure 8.   The solvent accessible areas (SAA) of five sucrose crystal faces in angstroms per unit cell and their mesh areas (same units). Solvent radius 1.5 A.

contacts on the two respective faces reveals the well known fact that mostly fructose atoms are exposed on the p face while glucose atoms are favored on p'. On p only 3 out of 19 contacts (at R = 1.5A) are from glucose, while on p' 4 contacts out of 17 are with fructose.

Even though still in a preliminary stage, it is hoped that this approach will result in a better solvent - effect corrector to the attachment energy calculations (18) than the broken hydrogen bond model and a better fit of the predicted sucrose crystal habits with the observed ones. It is already clear that the present model can, at least qualitatively, distinguish between the fast growing right pole of the crystal and its slow left pole.

Adipic Acid - Solvent Accessibility

The attachment energies of (100, (010), (001), (110), and (011) faces of adipic acid were calculated using the 6-12-1 potential functions of Lifson, Hagler and Dauber (19) with the exception of hydrogen-bonded atoms. The hydrogen bond energy was obtained from the difference of the calculated crystal energy (from summations of interactions of non-bonded atoms and excepted HB bonded atoms) and the experimental sublimation energy, 32.1 kcal/mole (19). The obtained hydrogen bond energy (EHB) is 5.45 kcal/mole which is of the expected order to magnitude. Each molecule of adipic acid is involved in four hydrogen bonds similar as in terephthalic acid and other dicarboxylic acids.

Results are shown in Table 1 and Figure 9. The corresponding growth habit (R ⁻ E_{att}) is platelike with (100, (011) and (001) planes. This seems in apparent disagreement with a report (20) on the habit of ethyl acetate grown adipic acid (b-axis elongated with (001), (100), (110) and (011). The corresponding relative areas of the faces are:

(100)   25.3%
(001)   14.4%
(011)   60.0%

Three planes have been analyzed using solvent accessibility analysis and a wide range of solvent radii. Results are shown in Figures 10 and 11. Of interest for further work are the solvent accessible area (SAA) vs solvent radius (R) curves. As shown in Figure 11, there are discontinuities on some of the curves, e.g. hydrogen on (100) (triangles in figure) indicating the existence of solvent-sized "receptacles" (formed by surface hydrogens) on that particular surface. On the other hand, where the slope dSAA/dR is a smooth function of R the surface arrangement is dissimilar in size to the solvent molecule.

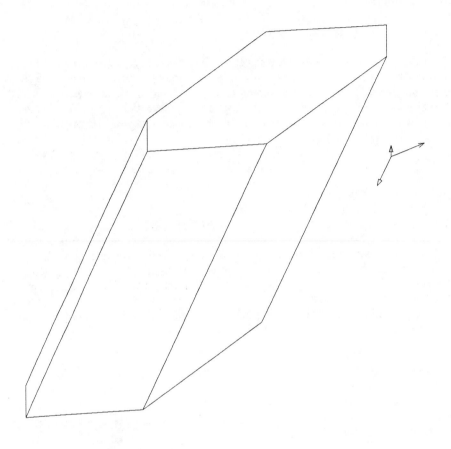

-65.0 55.0
→▷ A-AXIS
→▷ B-AXIS
─◄ C-AXIS

Figure 9.   Adipic Acid Growth Form R ~ E_{att}.

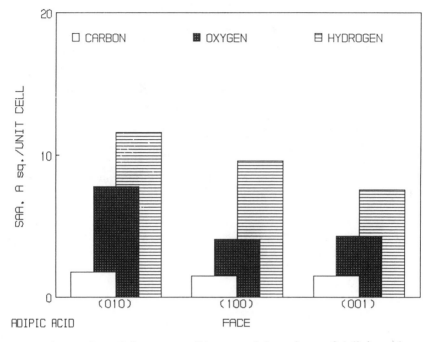

Figure 10.   Solvent accessible area of three faces of Adipic acid solvent radius 1.5 A.

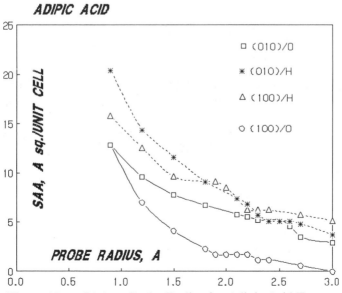

Figure 11.   SAA vs. Probe Radius for Adipic Acid Faces.

Table 1
Adipic Acid Attachment Energies

face	E_{att}
(100)	19.1 kcal/mole (2 HB intersected)
(001)	15.3
(010)	18.8
(011)	12.8
(110)	31.0 (2 HB intersected)

There are at least two features of the results of the surface analysis that yield a measure of surface-solvent interactions:

1. The relative SAA's of the surface atoms that may later be used as weighing factors in estimations of the surface-solvent interactions.
2. Recognition of solvent sized cavities or receptacles on the surface from the calculated SAA vs R. curves. These might further enhance solvent interactions on specific faces (similar as in Aquilano et al (18) for water adsorption on (101) of sucrose).

Discussion

The work presented demonstrates that the solvent effects can be seen through a comparison of solvent accessible areas of crystal faces for different size solvent molecules. The next step in this research will involve the incorporation of this effect in the attachment energy (hence crystal shape) calculation. One possible method is the calculation of a potential 'map' of the crystal faces which would be used along with polar solvent molecules to calculate and electrostatic interactions between solvent and crystal surfaces. This energy will be used to modify the attachment energy using the solvent accessible area as a 'weighing' factor. Other approaches might involve calculation of Van der Waals and hydrogen bond energy interactions between the solvent and the crystal faces, or the estimation of solvent molecule locations on the crystal faces and their inclusion in the surface energy calculation. More realistic solvent accessible areas based on the 'shape' of the solvent molecules will also be attempted.

Nomenclature

A_i	Face area
d_{hkl}	Interplanar spacing parallel to (hkl)
E^{att}	Attachment energy of a face
F	Free energy
h_{hkl}	Central distance of face hkl
R_i	Linear growth velocity

V   Unit cell volume
Z   Number of molecules per unit cell

γ_i Surface energy of face i

References

1. Hartman, P., and Perdok, W.G., Acta Crystallogr 8, 49 (1955).
2. Hartman, P., and Perdok, W.G., Acta Crystallogr 8, 521 (1955).
3. Hartman, P., and Perdok, W.G., Acta Crystallogr 8, 525 (1955).
4. Hartman, P., and Bennema, P., J. Cryst. Growth 49, 145 (1980).
5. Hartman, P., J. Crystal Growth 49, 157 (1980).
6. Hartman, P., J. Crystal Growth 49, 166 (1980).
7. Saska, M., and Myerson, A.S., J. Cryst. Growth 61, 546 (1983).
8. Black, S.N., Davey, R.J. and Halcrow, M., J. Cryst. Growth, 79 (1986).
9. Hartman, P., J. Cryst. Growth, 96, 667 (1989).
10. Donnay, J.D.H., and D. Harker, Am. Mineralogist, 22, 446 (1937).
11. Kitaigorodski, A.I., and Ahmed, N.A., Acta Cryst., A28 207 (1982).
12. Saska, M., Ph.D. Dissertation, Georgia Institute of Technology, Atlanta, GA (1985).
13. Myerson, A.S., and Saska, M. AIChE Journal, 33, 848 (1987).
14. Strom, C.S., J. Cryst. Growth, 46, (187) (1979).
15. Connolly, M., Science 221, 709 (1983).
16. Connolly, M., Quantum Chem. Program Exchange Bull. 1, 74 (1981).
17. Valcic, A.V., J. Crystal Growth 30, 129 (1975).
18. Aquilano, G.; Franchini-Angela, M.; Rubbo, M.; Mantovani, G.; Vaccari, G., J. Cryst. Growth, 61, 369 (1983).
19. Lifson, S. Hagler, A.T., Dauber, P. J. Am. Chem. Soc., 101, 5111 (1979).
20. Morrison, J.D. and Robertson, J.M., J. Chem. Soc. p.1001 (1949).

RECEIVED May 17, 1990

Chapter 6

Sucrose Crystal Growth

Theory, Experiment, and Industrial Applications

D. Aquilano[1], M. Rubbo[1], G. Mantovani[2], G. Vaccari[2], and G. Sgualdino[2]

[1]Dipartimento di Scienze della Terra, Università di Torino, Via Valperga Caluso, 37–10128 Torino, Italy
[2]Dipartimento di Chimica, Università di Ferrara, Via Borsari, 46–44100 Ferrara, Italy

The face-by-face (R, σ) isotherms on sucrose crystals growing from pure solutions allow us to determine the activation energies and, to some degree, the growth mechanisms for each of the F faces.
Furthermore, when crystals grow from doped solutions, two other phenomena can be qualitatively interpreted:
1. the specificity of impurity adsorption as a function of the concentration, and 2D-epitaxy, when it does occur, both affecting the growth rate of each face,
2. the coloring matter effects, due both to adsorption and liquid inclusion capture under critical kinetic conditions. This effects are strongly dependent on the growth rate of the faces.
By means of our experimental method (twin+single crystal kinetics) steady state growth morphology can be predicted as a function of supersaturation and temperature.

Sucrose crystal is monoclinic and polar crystal due to the crystalline arrangement of polar molecules: it generally grows from polar solvent, water in particular, and this means that both kinetic and structural problems are rather complicated. It is for these reasons that we have been trying for many years to give a contribution to the solution of a part of these problems starting from two considerations: 1. It is necessary to know all possible configurations of the surface structure of all faces of the crystal because we cannot obtain useful information from a crystal considered as a whole. We studied the theoretical growth morphology by means of the theory of Hartman and Perdok (1-2) and we stated

0097–6156/90/0438–0072$06.00/0

the character of the different faces (F and S) (3). 2. We need
measurements both at equilibrium and during kinetic experiments in
pure medium and in the presence of specific impurities. Unfortunately measurements about equilibrium forms are fairly
difficult to make whereas kinetic measurements are easier to
carry out.

Therefore a particular method was chosen (4). We worked on a
statistical population of crystals in order to minimize the
dispersion and on simultaneous measurement of all faces in order to
compare their growth rate under the same conditions of supersaturation and temperature. Therefore classical (R, σ) isotherms
were obtained. Experimentally we grew at the same time and in the
same solution a single crystal and twin. Whereas growth rate
measurements of the forms $\{ hOL \}$ are relatively simple (thanks to the
fact that the b axis is a binary axis) (Figure 1b), the kinetic
measurements of the p' $\{ 110 \}$ and p $\{ 1\overline{1}0 \}$ forms are more difficult.
We solved the problem by measuring the advancement rate along the b
axis of the twin and the single crystal, respectively. The R(p')
value is calculated by means of the equation shown in Figure 1,
whereas the R(p) value is given by the following equation:

$$Rp = (\Delta b_{single} - \Delta b_{twin} /2) \ \sin \gamma /2$$

This technique allowed us to give a very precise measurement on two
complementary forms (p' and p) and distinguish their growth rates
unambiguously.
In this way two objectives ensue: 1. We obtain the growth mechanism
of the most important flat and stepped faces of the crystal. 2. We
are able to foresee the global crystal morphology in a steady state
for all temperature and supersaturation values.

In Figure 2 we see the typical curve (R, σ) found for F faces.
The continuous curve represents the normal trend of our experimental
(R, σ) isotherms. Part (a), which represents the parabolic law at
very low σ is followed by a sudden increase in the growth rate due
to the exponential contribution of the two-dimensional nucleation in
between the steps, part (b), and at the end by the linear law, part
(c). From our isotherms we can calculate the activation energy for
crystallization and we are able to choose the growth mechanism. In
the past, many interesting data in the literature have given only
the average value for this energy because the crystal was considered
as a whole. From other calculations, Schliephake (5) determined the
activation energy for surface diffusion and Smythe (6) and
Kucharenko (7) for volume diffusion but it was impossible to
distinguish if a certain face advances under surface diffusion,
volume diffusion or both of them. In the left portion of Figure 3
the values of two important complementary forms are represented,

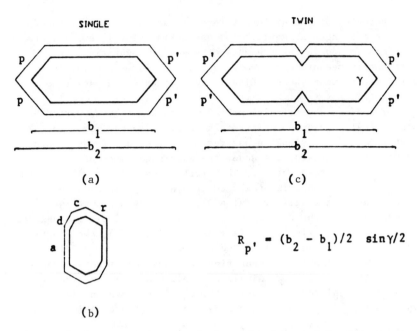

Figure 1. Cross section of single and twin crystal of sucrose:
a) single crystal: view along |001| direction; b) single crys-
tal: view along |010| direction; c) twin crystal: view along
|001| direction.

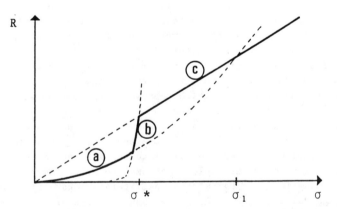

Figure 2. Typical (R,σ) growth isotherm for F faces.

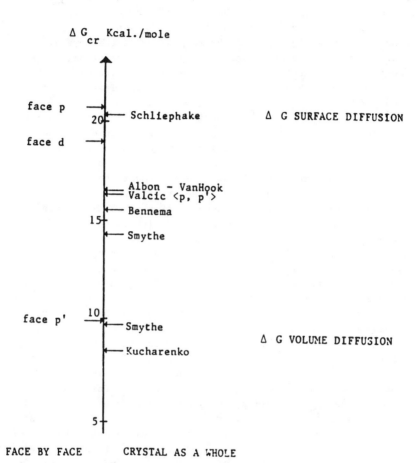

FACE BY FACE   CRYSTAL AS A WHOLE

Figure 3. Activation energy for the crystal as a whole (on the right: bibliographical data) and for different faces (on the left: our data).

that is p and p', and for one face which is not polar, the (101)-d
face. It must be observed that the average of our values is in a
very good agreement with the mean value obtained by Albon (8),
Valcic (9), Bennema (10), Smythe (11) and VanHook (12).

As a further comment to the advantage of the face by face
method we can see in Figure 4 the morphology of the crystal seen
along the c-axis. We can observe the dramatic variation of the shape
when the supersaturation is moving from very low values to the
highest ones and when the temperature of crystallization increases
by only 10°C.

Above we have taken into consideration the crystallization
from pure water solution only and we have not considered any
interpretation founded on structural consideration. Now the
crystallization in impure solution will be examined and only after
this an interpretation which takes into account the interaction
between the crystal structure and the growth medium may be
attempted. In Figure 5 we can see as a first example a comparison
between a sucrose crystal grown in pure solution (b), and one grown
in the presence of small amount (2-8 grams per one hundred grams of
water) of raffinose (a). It is clear the dramatic variation in the
growth morphology (13). As far as the effect of glucose and fructose
on sucrose crystal morphology is concerned, we observed that in the
presence of such low amounts as the ones quoted above for raffinose,
we were not able to observe evident effects. On the contrary, by
increasing the amounts of these impurities present in sucrose
solution to 150 grams per 100 grams of water the situation changes
dramatically. In particular, as far as glucose is concerned, Figure
6 shows that crystal stops growing along the -b axis under the
highest concentration of impurity. In particular the blocking
involves faces p so favoring the disappearance from the morphology
of the faces o, q, w on the left pole. Moreover, such effect becomes
more and more important as we increase the amount of glucose, that
is from zero to 150 grams per one hundred grams of water. As far as
fructose is concerned, phenomena of total blocking of one or the
other pole of the crystal were not observed but we noted a
macroscopic slowing down of (111)-o' and (011)-q' faces on the right
pole which normally do not appear (Figure 7). With the purpose of
confirming this effect, already surprising in itself, we grew twins
of the first type in the presence of various amounts of fructose. In
Figure 8 the sequence of these twins grown in the presence of
different amount of fructose, from 10 to 150 grams per one hundred
grams of water, is shown. Figure 9 shows a crystal grown in the
presence of glucose and fructose, each of them at a concentration of
100 grams per one hundred grams of water (b) in comparison with a
crystal grown in pure solution (a). It is quite clear that the
crystal grown in the presence of impurities does not show the small

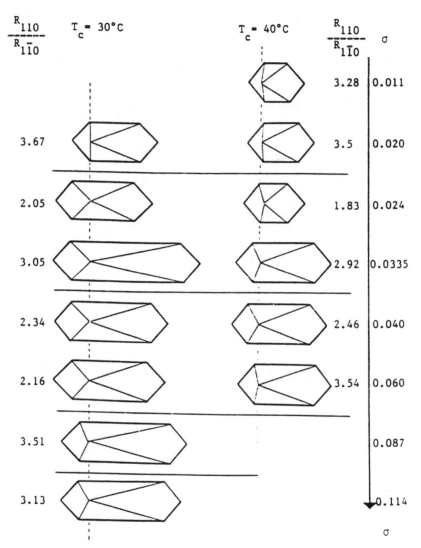

Figure 4. Effect of temperature and supersaturation on crystal morphology: | 001 | projection.

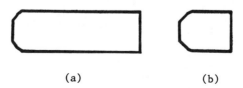

(a)   (b)

Figure 5. Morphology of sucrose crystal grown in pure solution (b) and in the presence of raffinose (a): | 100 | projection.

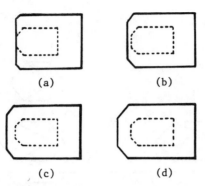

(a)   (b)

(c)   (d)

Figure 6. Morphology of sucrose crystals grown in the presence of different amount of glucose (a = 150; b = 100; c = 50; d = 0 grams per 100 grams of water). Dashed lines represent the initial stage.

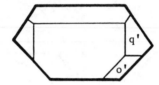

Figure 7. Appearance of the o' and q' faces on the right end of a sucrose crystal grown in the presence of fructose (100 grams per 100 grams of water).

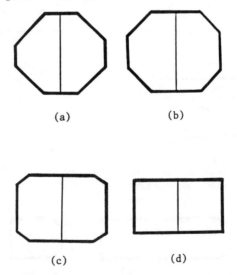

(a)   (b)

(c)   (d)

Figure 8. Sucrose twins (first type) seen perpendicularly to the a face grown in the presence of different amount of fructose (a = 150; b = 100; c = 50; d = 10 grams per 100 grams of water).

faces of the left pole but the o' and q' faces of the right pole,
that is the opposite to the situation for pure solution.

Above we have outlined the importance of the structure of the
crystal when we want to make a correlation between the kinetic data
and the surface of all the faces. So we considered the PBC's
analysis as a necessary tool to obtain the maximum of information on
all sites of each crystal surface (3). The PBC's analysis
specifically allows us to determine the polarity of the
complementary forms. As an example we consider the complementary
interface q and q' (Figure 10). The two opposite interfaces show
complementary behavior with respect to the hydrogen bond (HB)
pointing toward the mother solution. The q interface exposes 3 HB
donors and 4 acceptors whereas the opposite situation is set up on
the q' face. We fixed the ratio K between the number of donors and
the number of acceptors over one unit cell. Hence for q face K =
0.75 and for the q' face K = 1.33.

We will now consider the sequence of the growth rates in pure
solution of all faces as determined from literature and our
experimental data (Figure 11). If we neglect the behavior of non-
polar faces (a,d,c,r) we must observe that all faces belonging to
the left pole grow more slowly than those belonging to the right
pole. At the same time we observe that the K ratio increases from
the left to the right pole according to the experimental sequence.
K-ratio of a given face, roughly depends on whether the surface
dipole moments point outward or inward the face. So, for a donor OH-
bond, the dipole moment vector points outward and, for an acceptor,
inward; hence, when the number of donor exceeds the number of
acceptors (per unit cell) we can easily assume that over the unit
cell there is a residual component of the dipole moment pointing
outward the crystal surface. In other words a left pole surface (K
< 1) behaves as a dipole layer showing negative charges as the
outermost ones; the opposite occurs for the right pole (K > 1).
Therefore K ratio works as an indicator of the hierarchy of
polarization of the outermost layers of each face. This means only
that, if sucrose crystals might grow from vapor phase, we will
continue to observe the polarity effects (both for equilibrium and
growth forms). When considering now the water solution growth we
must take into account that a sucrose molecule, with its surrounding
water solvating molecules, behaves as a dipole surrounded by a given
number of dipoles associated to it; hence this complex dipolar
aggregate, when "landing" onto a dipole layer (as mentioned above)
adsorbs more or less strongly according to whether the outermost
charges are negative or positive. This is a first step which affects
the growth anisotropy between the two poles. A second step affecting
the growth anisotropy is that concerning the desorption of water
molecules (those adsorbed as "free" molecules onto free surface

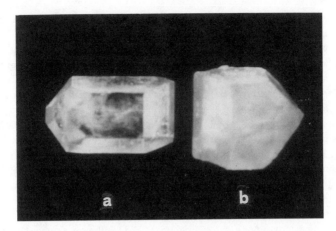

Figure 9. Sucrose crystal grown in pure solution (a) and in the presence of glucose and fructose (100 grams per 100 grams of water each).

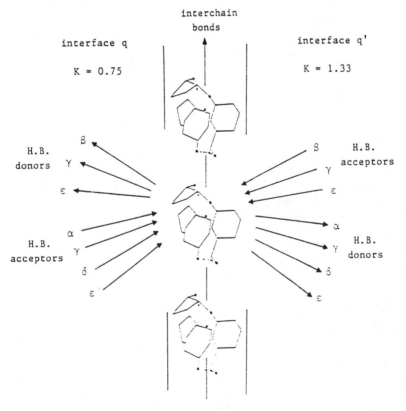

Figure 10. Complementary interface q and q'.

GROWTH RATE IN PURE MEDIUM

$$a \ll c < p < q < o < r < d \ll p' < q' \cong o'$$

============   ==============

left pole   righ pole

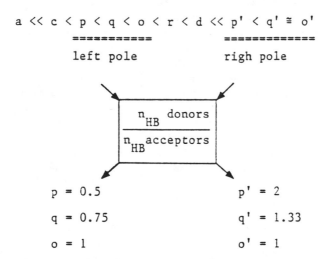

p = 0.5	p' = 2
q = 0.75	q' = 1.33
o = 1	o' = 1

Surface diffusion is the rate determining step:	Volume diffusion is the rate determining step:
e.g.	e.g.
$\Delta G_{cr}(p) \cong 21$ Kcal/mole	$\Delta G_{cr}(p') \cong 10$ Kcal/mole

Surface desolvation is
more difficult on the
left pole.

Figure 11. Sequence of growth rate of different faces and K
values for the different faces of the right and left pole.

sites and those "sandwiched" between sucrose surface and landing
sucrose molecules). For this step K-ratio is no longer working: in
fact the rate of desorption depends on desorption energy of water
molecules and, in turn, on the interaction between water, running
onto the surface, and all accessible atomic sites. A good way to
solve this problem seems to be (Saska, M., Audobon Sugar Institute,
personal communication, 1989.) the determination of the
accessibility of the Van der Waals-sized surface atoms for a
spherical probe representing the solvent molecule (14). By this way
we can consider in the adsorption mechanisms, not only those atoms
involved in sucrose-sucrose H-bonds (across the crystal solution
interface), but also all the remaining ones that may provide good
sites for surface-water contacts.
This model, applied till now to a, c, d, and p, p' forms, does
distinguish, at least qualitatively, the fast growing p' from the
slow growing p face. All that reasoning is in good agreement with
another consideration which ensues from our isotherms. When the
crystallization temperature is lower than 45°C all faces belonging
to the left pole are dominated by the surface diffusion whilst all
faces belonging to the right pole are dominated by volume diffusion.
This fact clearly emerges when considering the crystallization
energy for p and p' as an example. Finally we are induced to argue
that surface desolvation is more difficult on the left pole and that
water, together with the anisotropy of surface polarizability,is an
important agent responsible for the difference in the growth rate
between the two complementary forms.
       Now we can consider the effect of the impurities in a
qualitative way. As far as the raffinose effect is concerned we
have shown that the raffinose molecule is able to poison both the
kinks and the surface of the p' faces (13): this is due to the
geometry of this molecule in which the sequence fructose-glucose-
galactose is able to enter the kinks as a sucrose molecule and to be
adsorbed onto the surface. The galactose portion hinders the
entering of other sucrose molecules. For this reason, small
concentrations of raffinose are sufficient to block the right end
growth. The situation is completely different for fructose and
glucose molecules. The previously described mechanism does not occur
because no effect is shown when the concentration is small. The fact
that the morphological effect increases continuously when the
concentration of these impurities rises leads us to consider
alternative hypotheses: A. Both molecules can form a 2-D epitaxy
with residual water molecules as for KCl, NaBr, as shown in previous
work (4); B. Glucose and fructose molecules, as their concentration
increases, can give origin to a competition in the surface
adsorption with the sucrose molecules.
       Another effect must be taken into consideration: fructose

increases the solution viscosity more than glucose (at the same
concentration). Then it would be expected that all forms growing
through volume diffusion mechanism are more affected by fructose
than by glucose. In particular p' will reduce its advancement rate
with respect to p. Hence the consideration that fructose acts on the
right pole better than glucose and that the latter acts on the left
pole better than fructose is in favor of the B hypothesis and is in
good agreement with the considerations we made when geometrical
affinity of raffinose was taken into consideration.

The phenomena described and discussed above have a remarkable
importance from the industrial processing point of view both
concerning sugar beet and cane. The fact that the various faces of
sucrose crystal show different growth rates affects not only its
habit but also the inclusion of either inorganic or organic
impurities present in solution. In the case of inorganic impurities,
their inclusion inside growing sucrose crystals causes an increase
of the ash content. The inclusion of organic compounds, and in
particular of the coloring matters which abound in industrial
solutions, causes an increase of the crystal color showing even
preferential localization (15-19) (Vaccari, G.; Mantovani, G.; Morel
du Boil, P.G.; Lionnet, G.R.E. Proc. 20th I.S.S.C.T. Congress, 1989,
in press.), so decreasing the characteristics of the final product
from the trading point of view. The inclusion of coloring matters
inside sucrose crystals obviously causes also an increase of the
processing costs since needing onerous operations of washing and/or
recrystallization. The formation of needle shaped crystals, which,
as pointed out above, characterizes the presence of raffinose in
beet processing, can also occur as far as cane is concerned due to
the presence of organic non-sugars and in particular dextrans. It is
interesting to emphasize that, in the case of cane sugar, the
elongation of crystals does not occur along the b-axis, as in the
presence of raffinose, but along the c-axis. From the industrial
point of view, this particular habit modification causes the
formation of fragments due to the breakage of the fragile needles.
These fragments not only cause losses of sugar during the
centrifuging step but also difficulties inside the centrifuges owing
to clogging phenomena which hinder the separation between solid and
liquid phases. As far as the influence of glucose and fructose on
sucrose habit is concerned the modifications we have pointed out and
discussed above influence, though not remarkably, the trading
characteristics of the final product which must preferentially show
a stout compact shape. Again concerning the effect of glucose and
fructose on sucrose crystal habit, the amounts of these compounds we
have taken into consideration in the experiments we have discussed
above are not to be considered illogical. In fact, in the cane sugar
processing, high amounts of reducing sugars, that is glucose and

fructose are normal. During crystallization step, which concerns products which become more and more impure, sugar solutions have a very high dry substance content which can reach 95% and consequently very low water amounts. Obviously, in these conditions, the ratio between the various non-sugar constituents and water becomes very high and then reaches the high amounts we have taken into consideration. On the contrary, in the case of beet processing, the effect of raffinose can be remarkable also in the presence of very low amount, since raffinose, through an even temporary adsorption mechanism, can very remarkably slow down the growth rate of some sucrose crystal faces causing the formation of needles we have shown and discussed above.

Literature Cited

1. Hartman, P.;Perdok, W.G. Acta Cryst. 1955, 8, 49-52
2. Hartman, P. in "Crystal Growth: An introduction" - Hartman, P., Ed.; North Holland: Amsterdam, 1973; pp 367-402
3. Aquilano, D.; Franchini-Angela, M.; Rubbo, M.; Mantovani, G.; Vaccari, G. J. Crystal Growth 1983, 61, 369-376
4. Aquilano, D.; Rubbo, M.; Vaccari, G.; Mantovani, G.; Sgualdino, G. in "Industrial Crystallization 84" - Jancic, S.J.; de Jong, E.J., Eds.; Elsevier Science Pub. B.V.: Amsterdam, 1984, pp 91-96
5. Schliephake, D.; Austmeyer, K. Zucker 1976, 29, 293-301
6. Smythe, B.M. Sugar Technol. Rev. 1971, 1, 191-231
7. Kucharenko, J.A. Planter Sugar Mfg. 1928, 80, 463-464; 484-485
8. Albon, N.; Dunning, W.J. Acta Cryst. 1960, 13, 495-498
9. Valcic, A.V. J. Crystal Growth 1975, 30, 129-136
10. Bennema, P. J. Crystal Growth 1968, 3-4, 331-334
11. Smythe, B.M. Australian J. Chem. 1967, 20, 1087-1095
12. VanHook, A. Zuckerind. 1973, 17, 499-502
13. Vaccari, G.; Mantovani, G.; Sgualdino, G.; Aquilano, D.; Rubbo, M. Sugar Tech. Rev. 1986, 13, 133-178
14. Connolly, M.L. Science 1983, 221, 709-713
15. Mantovani, G.; Vaccari, G.; Sgualdino, G.; Aquilano, D.; Rubbo, M. Ind. Sacc. Ital. 1985, 78, 7-14
16. Mantovani, G.; Vaccari, G.; Sgualdino, G.; Aquilano, D.; Rubbo, M. Ind. Sacc. Ital. 1985, 78, 79-86
17. Mantovani, G.; Vaccari, G.; Sgualdino, G.; Aquilano, D.; Rubbo, M. Zuckerind. 1986, 111, 643-648
18. Mantovani, G.; Vaccari, G.; Sgualdino, G.; Aquilano, D.; Rubbo, M. Proc. 19th I.S.S.C.T Congress, 1986, pp 633-669 ; Ind. Sacc. Ital., 1986, 79, 99-107
19. Vaccari, G.; Mantovani, G.; Sgualdino, G.; Aquilano, D.; Rubbo, M. Gazeta Cukrownicza 1988, 95, 1-10
RECEIVED June 19, 1990

Chapter 7

Factors Affecting the Purity of L-Isoleucine Recovered by Batch Crystallization

Ronald C. Zumstein[1], Timothy Gambrel, and Ronald W. Rousseau

School of Chemical Engineering, Georgia Institute of Technology, Atlanta, GA 30332-0100

The purity of amino acids recovered by batch crystallization has been examined using L-isoleucine as a model system. The concentration of impurities in the feed solution were shown to affect crystal purity, as were variables that affect crystallization kinetics (e.g., agitation, precipitant addition rate, and cooling rate).

Crystallization processes can be used for separation, purification, or concentration of a solute, or because a particular product needs to be used in solid form, or as a component of an analytical procedure. Common requirements for accomplishing these functions are that the crystals must be produced with a particular size distribution and having a specified shape and purity. Almost all crystallizer operating problems are defined in terms of the product not meeting one of these criteria.

Phenomena, methods of operation, etc. have been studied extensively for the use of crystallization in separation processes. Although much remains to be learned about such processes, relatively little attention has been given to the other functions and the purpose of this work was to examine the role of various process variables in determining the purity of crystals recovered from a batch crystallizer. The system studied experimentally was a model system of amino acids, and the key variables were the composition of the liquor from which a key amino acid was crystallized, the rate at which supersaturation was generated by addition of an acid solution to reduce solubility, and the degree of mixing within the batch unit.

Key factors in solving problems associated with crystalline materials not meeting purity requirements are (a) the location of the impurities—i.e., on the surface or incorporated in the crystal—and (b) the nature of the impurity. Impurities are on the surfaces or, more generally, the exterior of host crystals due to adsorption, wetting by a solvent that contains the impurities, or through entrapment of impure solvent in cracks, crevices, agglomerates and aggregates. Incorporation

---

[1]Current address: Ethyl Corporation, P.O. Box 1028, Orangeburg, SC 29115

of impurities within crystals comes about through formation of occlusions (also referred to as inclusions) of solvent, lattice substitution or lattice entrapment. Obviously, the characteristics of an impurity determine whether it is positioned on the surface or the interior of host crystals. Three key impurity types are those similar to the product, those dissimilar from the product, and the solvent. In the present work the impurities have molecular structures similar to the product.

A common procedure in the use of crystallization for purification is shown schematically in Figure 1. As shown, product is recovered in an initial crystallization step, redissolved, and then recrystallized. Given enough of these steps, the purity of the final recovered crystals can be expected to meet demanding standards. Unfortunately, redissolution and recrystallization usually means loss of product, and caution must be exercised in using mother liquor recycle because of the potential for accumulation of impurities.

So that the purification of a given crystallization can be quantified, purification factors are defined with the following equation:

$$P_i = \frac{R_{i,c}}{R_{i,s}} \tag{1}$$

where

$$R_{i,c} = \left(\frac{\text{mols impurity}}{\text{mol reference}}\right)_{\text{crystal}} \tag{2}$$

and

$$R_{i,s} = \left(\frac{\text{mols impurity}}{\text{mol reference}}\right)_{\text{solution}} \tag{3}$$

It might be noted that P_i is similar to relative volatility used in the analysis of distillation processes. Clearly, when P_i is less than 1 the crystallization has resulted in purification, when it equals 1 no purification has occurred, and when it is greater than 1 the crystalline product has been further contaminated.

The notion of an ideal behavior also is defined here for those cases in which P_i is constant over a range of solution compositions, while variations with solution composition are said to characterize nonideal behavior. In the present studies values of purification factors are affected by the kinetics of the process. Accordingly, these quantities may not be true thermodynamic properties.

Experimental

The compounds investigated were the amino acids L-isoleucine, L-leucine, L-valine, and L-α-amino butyric acid. These compounds have similar molecular structures, as shown in Figure 2, and will be referred to throughout the present work as L-Ile, L-Leu, L-Val, and L-α-ABA. Where there is little likelihood of confusion, the designation L- will be omitted. Operations examined included crystallization of Ile through the addition of hydrochloric acid and through cooling. Under acidic conditions Ile crystallizes as a hydrochloride salt while in the vicinity of the isoelectric point (pH 5.2) it crystallizes as the neutral zwitterionic

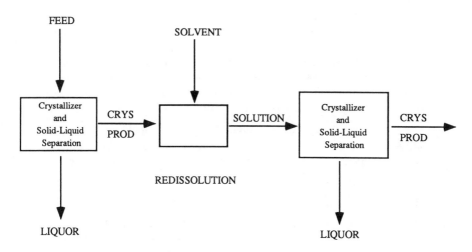

Figure 1. Crystallization-Redissolution-Recrystallization Process

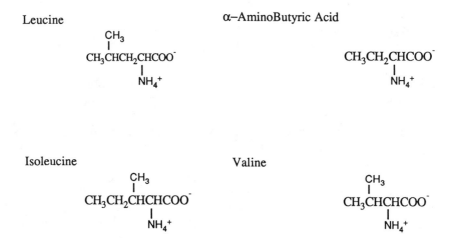

Figure 2. Molecular Formulas of L-isoleucine, L-leucine, L-valine, and L-α-amino butyric acid

species. Solubilities of L-isoleucine in acidic and neutral solutions were given by Zumstein and Rousseau (1).

The experimental crystallizer was a 350-mL jacketed glass unit that was provided with sufficient agitation to keep the contents well mixed. The compositions of the charges to the crystallizer were adjusted by adding the amino acids in predetermined ratios; concentrations of the amino acid impurities were maintained in ranges comparable to those found in the recovery and purification of L-isoleucine from industrial fermentation reaction masses. The experiments were divided according to the mode of crystallization:

1. For crystallization by acid addition, the effects of acid addition rate, agitation, and initial solution composition were examined. The experiments had constant initial and final temperatures, although there were some variations in temperature during a run because of the heat of crystallization of L-isoleucine hydrochloride (L-Ile·HCl·H$_2$O) and heat of solution of acid. The acid used was in the form of 37% HCl. The initial conditions of the batch corresponded to an HCl concentration giving the maximum solubility of L-Ile (1). A schematic of the batch crystallizer is shown in Figure 3.

2. For crystallization by cooling, the effects of agitation, impurity concentration, and time after nucleation were examined. The experimental apparatus used was the same as shown schematically in Figure 3 except that acid was not added to the solution.

Experimental Results and Discussion

All impurity determinations reported here are for crystals that have been washed with acetone prior to analysis. It is expected, therefore, that the impurities found were internal to the crystals.

Crystals Obtained by Acid Addition. Figure 4 shows the effect of initial solution composition on the impurity content of crystals obtained by acid addition. Clearly, this corresponds to the definition of an ideal system as presented above. These data show the order followed in impurity incorporation in the L-Ile crystals is L-Val > L-Leu > L-α-ABA, although there is only one data point on α-amino butyric acid. Also, the value of purification factors for all impurities is less than one. This means that purification by crystallization was indeed occurring.

A series of runs was performed in which the acid addition rate was varied while holding the solution compositions and agitation constant: $R_{Val,s} = 0.023$, $R_{Leu,s} = 0.021$, and 1000 RPM. The temperature was 25°C in all runs. Figure 5 shows that the purification factors were impacted by acid addition rate, and increased with the rate at which HCl was added to the system. The greatest effects are noticed below acid addition rates of about 5 g/min; as the initial charge to the batch crystallizer was 150 g of solution, this corresponds to an addition rate of about 3.3% by mass per minute.

Agitation was also varied while holding other variables constant. The results are shown in Figure 6. Clearly, different results were observed for L-Val

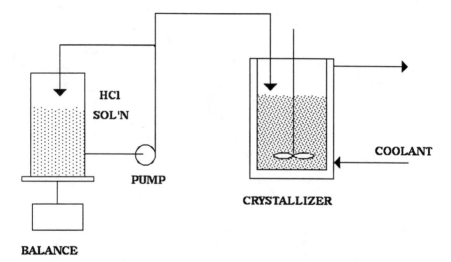

Figure 3. Batch Crystallizer Used for Acid-Addition Experiments

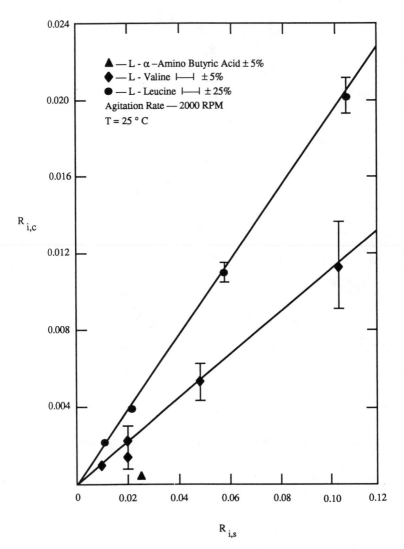

Figure 4.   Effect of Solution Composition on the Purity of Recovered L-Ile·HCl·H$_2$O Crystals

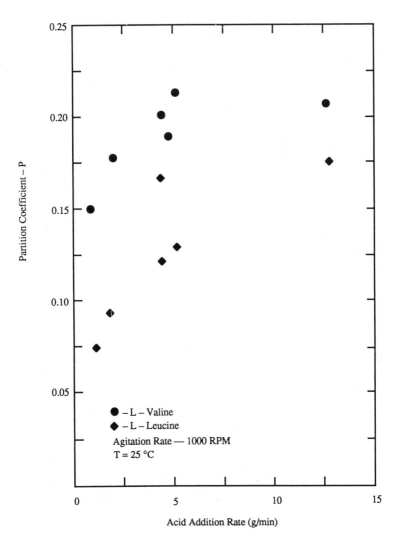

Figure 5. Effect of Acid Addition Rate on L-Leu and L-Val Purification Factors in Recovered L-Ile·HCl·H$_2$O Crystals

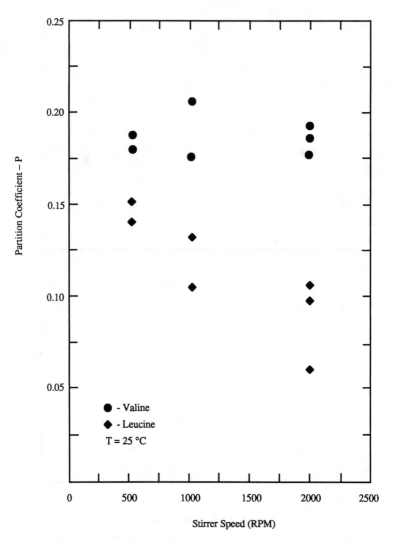

Figure 6. Effect of Agitation on L-Leu and L-Val Purification Factors in Recovered L-Ile · HCl · H$_2$O Crystals

and L-Leu; agitation had no noticeable impact on the purification factor of valine, while it decreased with increased agitation for L-Leu. This would seem to indicate differing mechanisms of incorporation for the two impurities.

<u>Crystals Obtained by Cooling</u>. Crystallizations in which cooling was the mode of supersaturation were obtained from solutions near the isoelectric point of L-Ile. Such crystals are of zwitterionic form and are referred to as neutral. They are structurally different from the crystals obtained by acid addition because they are free of HCl and do not have a water of hydration. It is not surprising then that different behavior may be observed.

In examining the effect of solution composition on the purity of recovered L-Ile crystals some unusual observations have been noted. Figure 7 shows data similar to that presented in Figure 4 for L-Ile·HCl·H$_2$O with one major difference: L-Valine is relatively unimportant as an impurity in comparison to L-Leu. Even though there is greater scatter in the data than was observed in the acid-addition experiments, purification factors for the two impurities are less than one and nearly constant.

Since the recovery of neutral L-Ile may be performed after crystallization, redissolution and recrystallization, the concentrations of impurities in the solution were reduced by an order of magnitude in an additional series of experiments. Figure 8 shows the results. Once again L-Val is relatively unimportant and P_{Leu} appears constant, but note that the data do not go through the origin. Moreover, close examination shows that $P_{Leu} > 1$ which means that purification by crystallization has not occurred. Figure 9 shows that the purification factor for L-Leu is not constant and, therefore, the system is nonideal.

Data on the effect of agitation on the purity of L-Ile crystals are shown in Figure 10. Although less convincing than those shown in Figure 6 for the effect of agitation on the crystallization of L-Ile·HCl·H$_2$O, the importance of agitation is demonstrated.

It was suspected that two mechanisms could lead to an increase in the impurity content of crystals as supersaturation increased. The first is through nucleation; i.e., an increase in supersaturation leads to greater nucleation rate and, concomitantly, larger numbers of crystals and higher crystal surface areas. As the crystals recovered in the present experiments were washed with acetone, surface impurities are not thought to have been important in the results presented here.

The second mechanism by which supersaturation was thought to influence crystal purity is through crystal growth; an increase in supersaturation leads to higher growth rates which means greater likelihood of impurity entrapment. This is considered to be the most probable cause of the relationship between supersaturation and purity for crystallizations of the type examined here.

An experiment was performed to examine the supposition that increases in supersaturation lead to greater impurity content in crystals. In a batch unseeded cooling crystallizer supersaturation is expected to be high at the point of nucleation, diminish rapidly after nucleation and then approach zero as the batch is

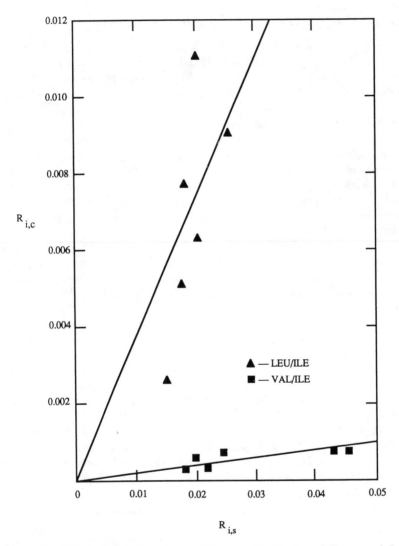

Figure 7. Effect of Solution Composition on the Purity of Recovered L-Ile Crystals (High-Impurity Solutions)

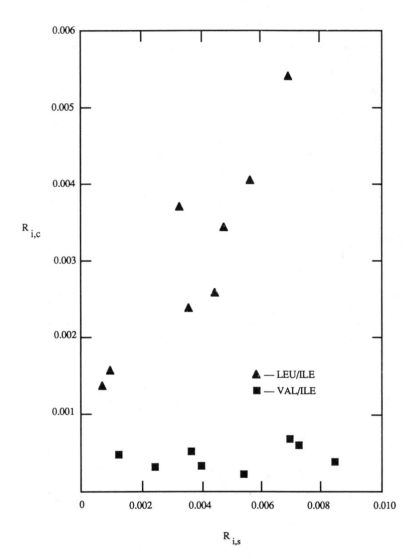

Figure 8. Effect of Solution Composition on the Purity of Recovered L-Ile Crystals (Low-Impurity Solutions)

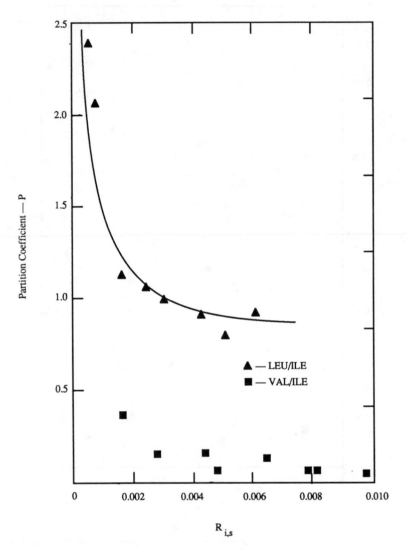

Figure 9. Effect of Solution Composition on the Purification Factors for L-Leu and L-Val for Low-Impurity Solutions

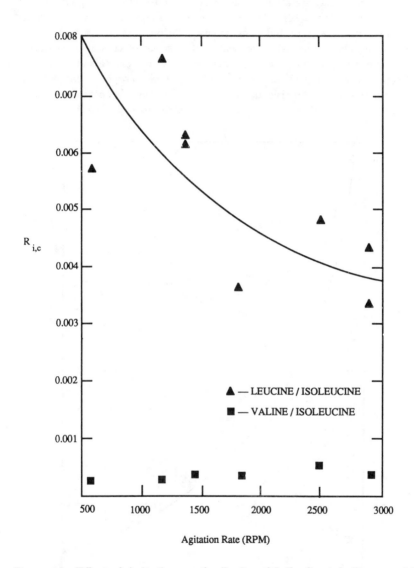

Figure 10. Effect of Agitation on the Purity of L-Ile Crystals Recovered by Cooling Crystallization

maintained at constant temperature for a long period (2). Accordingly, crystals recovered at varying times after will have been exposed to varying supersaturation: high at time zero, lower with increasing time. Figure 11 shows the impurity ratio for L-Ile crystals recovered under such conditions. As suggested by the hypothesis stated above, the L-Leu content of the recovered crystals diminishes as the crystals are allowed to grow. In other words, young crystals are exposed to higher supersaturation, grow at more rapid rates, and incorporate larger quantities of impurities than older crystals. The net effect of this phenomenon is that the purity of a given crystal increases as it is allowed to grow in a batch crystallizer.

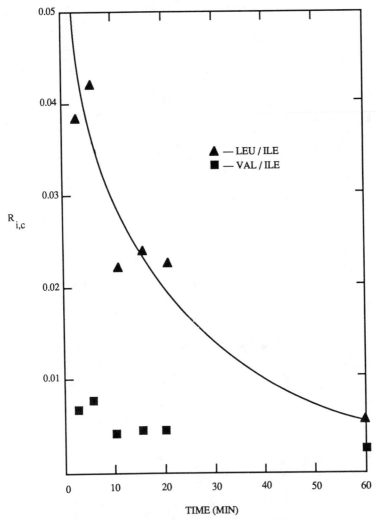

Figure 11. Effect of Growth Period on the Purity of L-Ile Crystals Recovered by Cooling Crystallization

Conclusions

The key factors controlling the purity of L-Ile recovered from batch crystallizers are shown in the study to be the composition of the solution from which the crystals are recovered, agitation, and the rate at which supersaturation is generated. Also, the molecular form of the recovered amino acid determines which of the impurities investigated, L-Leu and L-Val, is the more plentiful impurity in the recovered crystals.

Acknowledgment

Support of the National Science Foundation through grant number CBT-8722281 is acknowledged with gratitude. Ajinomoto U.S.A. provided the amino acids used in the study.

Literature Cited

1. Zumstein, R. C.; Rousseau, R. W. *Ind. Eng. Chem. Res.*, 1989, **28**, 1226–1231.

2. Zumstein, R. C.; Rousseau, R. W. *Chemical Engineering Science*, 1989, **44**, 2149–2155.

RECEIVED May 30, 1990

CRYSTALLIZER OPERATION AND CONTROL

Chapter 8

Light-Scattering Measurements To Estimate Kinetic Parameters of Crystallization

W. R. Witkowski, S. M. Miller, and J. B. Rawlings[1]

Department of Chemical Engineering, The University of Texas at Austin, Austin, TX 78712

This paper presents further development of the ideas presented by Witkowski and Rawlings (1) for improving kinetic parameter estimation of batch crystallization processes using nonlinear optimization techniques and particle size distribution measurements. As shown by Witkowski and Rawlings (1), desupersaturation information from a seeded, batch crystallization contains only enough information to determine the growth parameters reliably. It is shown that the addition of the obscuration, a light scattering measurement, along with the solute concentration data, allows accurate identification of all of the kinetic parameters. The obscuration is related to the second moment of the crystal size distribution (CSD). A Malvern 3600Ec Particle Sizer is used to obtain the obscuration measurement. Through numerical and experimental studies it is verified that both growth and nucleation parameters are identifiable using these measurements. It is also shown that accurate knowledge of the shape of the crystals is essential when analyzing CSD information from light scattering measurements.

Batch crystallizers are often used in situations in which production quantities are small or special handling of the chemicals is required. In the manufacture of speciality chemicals, for example, it is economically beneficial to perform the crystallization stage in some optimal manner. In order to design an optimal control strategy to maximize crystallizer performance, a dynamic model that can accurately simulate crystallizer behavior is required. Unfortunately, the precise details of crystallization growth and nucleation rates are unknown. This lack of fundamental knowledge suggests that a reliable method of model identification is needed.

Identification of a process involves formulating a mathematical model which properly describes the characteristics of the real system. Initial model forms are developed from first principles and a priori knowledge of the system. Model parameters are typically estimated in accordance with experimental observations. The method in which these parameters are evaluated is critical in judging the reliability and accuracy of the model.

The crystallization literature is replete with theoretical developments of batch crystallizer models and techniques to estimate their parameters. However, most of the schemes are constrained to specific crystallizer configurations and model formulations.

---

[1]Address correspondence to this author.

Therefore, a flexible method to evaluate physical and chemical system parameters is still needed (2, 3).. The model identification technique presented in this study allows flexibility in model formulation and inclusion of the available experimental measurements to identify the model. The parameter estimation scheme finds the optimal set of parameters by minimizing the sum of the differences between model predictions and experimental observations. Since some experimental data are more reliable than others, it is advantageous to assign higher weights to the dependable data.

A lack of measurable process information is the primary problem to be overcome in developing a reliable model identification and verification scheme. A measure of the solute concentration is commonly available. To provide another process measurement, a Malvern 3600Ec Particle Sizer, which is based on Fraunhofer light scattering theory, is used to determine the CSD. A primary drawback of this CSD measurement technique is that the inversion of the scattered light data is an ill-posed problem. Also, Fraunhofer diffraction theory is accurate only for a dilute collection of large spherical particles. Through the typical operating region of a seeded, batch crystallization experiment performed in this work, the particle concentration stays well within the manufacturer's specifications. However, this is not common for all crystallizers where the particle concentration can change from essentially no particles to a very dense slurry. These high particulate concentrations increase the probability of multiple scattering which would corrupt the CSD approximation (4).

Dynamic Batch Crystallization Model

The dynamic model used in predicting the transient behavior of isothermal batch crystallizers is well developed. Randolph and Larson (5) and Hulburt and Katz (6) offer a complete discussion of the theoretical development of the population balance approach. A summary of the set of equations used in this analysis is given below.

$$\frac{\partial n(L,t)}{\partial t} + G\frac{\partial n(L,t)}{\partial L} = 0 \tag{1}$$

$$\frac{dc}{dt} = -3\rho k_v G \int_0^\infty n(L,t)L^2 dL \tag{2}$$

$$G = k_g(c - c_s)^g \tag{3}$$

$$B = k_b(c - c_s)^b \tag{4}$$

in which $n(L,t)$ is the CSD at time t. The population balance, Equation 1, describes the CSD dynamics. The solute concentration, c, on a per mass of solvent basis, is described by Equation 2. Constitutive relationships modelling particle growth rate, G, and new formation (nucleation) rate, B, are given by Equations 3 and 4, respectively. These processes are typically formulated as power law functions of supersaturation, $c - c_s$.

The description is completed with the statement of the necessary initial and boundary conditions,

$$n(L, t = 0) \;\; = \;\; n_0(L) \tag{5}$$
$$n(L = 0, t) \;\; = \;\; B/G \tag{6}$$
$$c(t = 0) \;\; = \;\; c_0 \tag{7}$$

One of the most popular numerical methods for this class of problems is the method of weighted residuals (MWR) (7, 8). For a complete discussion of these schemes several good numerical analysis texts are available (9, 10, 11). Orthogonal collocation on finite elements was used in this work to solve the model as detailed by Witkowski (12).

Parameter Estimation

The parameter estimation approach is important in judging the reliability and accuracy of the model. If the confidence intervals for a set of estimated parameters are given and their magnitude is equal to that of the parameters, the reliability one would place in the model's prediction would be low. However, if the parameters are identified with high precision (i.e., small confidence intervals) one would tend to trust the model's predictions. The nonlinear optimization approach to parameter estimation allows the confidence interval for the estimated parameter to be approximated. It is thereby possible to evaluate if a parameter is identifiable from a particular set of measurements and with how much reliability.

Several investigators have offered various techniques for estimating crystallization growth and nucleation parameters. Parameters such as g, k_g, b, and k_b are the ones usually estimated. Often different results are presented for identical systems. These discrepancies are discussed by several authors (13, 14). One weakness of most of these schemes is that the validity of the parameter estimates, i.e., the confidence in the estimates, is not assessed. This section discusses two of the more popular routines to evaluate kinetic parameters and introduces a method that attempts to improve the parameter inference and provide a measure of the reliability of the estimates.

A survey of crystallization literature shows that the most popular technique to estimate kinetic parameter information involves the steady-state operation of a mixed suspension, mixed product removal (MSMPR) crystallizer (15, 5). The steady-state CSD is assumed to produce a straight line when plotted as $\log n$ vs. L. The slope and intercept of this line determine the steady-state G and B values. The major disadvantages of this method are that the experiments can be time consuming and expensive in achieving steady state, especially when studying specialty chemicals, and that valuable dynamic information is ignored. Also, maintaining a true steady state is often difficult in practice. The simplicity of the method is the primary advantage.

An alternative scheme, proposed by Garside et al. (16, 17), uses the dynamic desupersaturation data from a batch crystallization experiment. After formulating a solute mass balance, where mass deposition due to nucleation was negligible, expressions are derived to calculate g and k_g in Equation 3 explicitly. Estimates of the first and second derivatives of the transient desupersaturation curve at time zero are required. The disadvantages of this scheme are that numerical differentiation of experimental data is quite inaccurate due to measurement noise, the nucleation parameters are not estimated, and the analysis is invalid if nucleation rates are significant. Other drawbacks of both methods are that they are limited to specific model formulations, i.e., growth and nucleation rate forms and crystallizer configurations.

In the proposed technique, the user is free to pick the model formulation to suit his needs. The optimal set of parameters are found by solving the nonlinear optimization problem

$$\min_{\theta} \quad \Phi(y, \hat{y}; \theta)$$

$$\text{subject to} \quad \text{Equations 1--4} \tag{8}$$

in which $\theta^{\mathbf{T}} = [g, k_g, b, k_g]$.

The objective function, Φ, is typically formulated as the summed squared error between experimental measurements, y, and model predictions, $\hat{y}$. The relationship should properly describe the experimental error present and best utilize available experimental data. In the common least squares estimation, the measurement error is assumed to be normally distributed and Φ takes the form

$$\Phi = \sum_{i=1}^{n} \sum_{j=1}^{m} \omega_j (\hat{y}_{ij} - y_{ij})^2 \tag{9}$$

where n represents the number of experimental time records and m is the number of independent measurements at each sample time. The weighting factors, ω_j, are used to scale variables of different magnitudes and to incorporate known information about measurement uncertainty. The details of the calculations involved to estimate the confidence intervals are given in Bard (18) and Bates and Watts (19). Briefly, the confidence interval is estimated by looking at the curvature of the objective function at the optimal set of parameters.

The crystallizer model was solved using orthogonal collocation as previously discussed to provide the model predictions during the nonlinear optimization problem. Caracotsios and Stewart's (20, 21) nonlinear parameter estimation package, which solves the nonlinear programming (NLP) problem using sequential quadratic programming, was used in this work to estimate the parameters and compute the approximate uncertainties. There are several sources for discussions of NLP (22, 23), and there are several established codes available for the solution of this problem. In 1980 Schittkowski (24) published an extensive evaluation of 26 different NLP codes available at that time.

Experimental Studies

In this section, a brief description of the necessary experiments to identify the kinetic parameters of a seeded naphthalene-toluene batch crystallization system is presented. Details about the experimental apparatus and procedure are given by Witkowski (12). Operating conditions are selected so that the supersaturation level is kept within the metastable region to prevent homogeneous nucleation. To enhance the probability of secondary nucleation, sieved naphthalene seed particles are introduced into the system at time zero.

Available process measurements include the dissolved solute concentration and CSD information. To estimate the concentration, a PAAR vibrating U-tube densitometer is used to measure frequency as a function of the fluid's density and temperature. This information is used to interpolate a concentration value from an experimentally constructed database.

The Malvern Particle Sizer 3600Ec uses light scattering measurements to infer the CSD. The device has a He-Ne laser that illuminates the particles suspended in a stream that is moving continuously through a flow cell. The diffracted light is then focused through a lens onto a solid-state photodiode detector. The CSD inversion problem is performed using Fraunhofer diffraction theory. The primary assumption of the Fraunhofer theory is that the scattering pattern is produced by spherical particles that are large with respect to the wavelength of the light. Manufacturer's specifications show that the instrument is capable of measuring particles with a diameter of approximately 0.5 to 564 μm.

The Malvern also measures the fraction of light that is obscured by the crystals in the flow cell. The obscuration is defined as

$$\text{obscuration} = 1 - \frac{I}{I_0}$$

where I is the intensity of undiffracted light after passing through the suspension of crystals, I_0 is the intensity of unobstructed light. Note that the obscuration is measured directly and does not suffer from the ill-posed nature of the CSD measurement. The obscuration can be predicted by the model for comparison with experimental data using the following relations:

$$\frac{I}{I_0} = \exp\left(-\tau l\right) \tag{10}$$

$$\tau = \int_0^\infty n(L) A(L) Q(L) \, dL \tag{11}$$

where τ is the turbidity, l is the flow cell width, $A(L)$ and $Q(L)$ are the projected cross-sectional area and extinction efficiency factor, respectively, of crystals of characteristic length L. Fraunhofer diffraction theory gives $Q = 2$ for all L (25). From Equations 10 and 11 it can be seen that the obscuration provides a measure of the second moment of the CSD.

Model Identification Results

Numerical Analysis. It is difficult to determine which measurements contain sufficient information to allow the independent determination of all model parameters. This issue can be studied by assessing the impact of the use of various measurements on the parameter estimation problem using pseudo-experimental data.

Pseudo-experimental data can be generated by solving the model, Equations 1 – 4, for a chosen set of parameters and initial conditions, and then adding random noise to the model solution. For a given choice of measurement variables, the simulated data is then used in the parameter estimation problem. This procedure provides a means by which to evaluate the measurements that are required and the amount of measurement noise that is tolerable for parameter identification.

First, for the case in which only solute concentration measurements are available, pseudo-experimental data are simulated and used in the parameter estimation scheme. Even with noise-free data, the recovered parameters differ greatly from the true parameters and the uncertainties of the nucleation parameters are large. This indicates that there exists a large set of quite different b and k_b pairs that would lead to very similar solute concentration profiles. The insensitivity of the objective function to the nucleation parameters can be attributed to the fact that the mass of a nucleated particle is almost negligible and the change in the solute concentration is primarily due to seed growth.

The Malvern uses light scattering measurements to determine the weight percent of "spherical equivalent" crystals in each of 16 size classes and the mean size diameter.

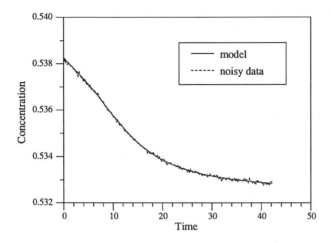

Figure 1: Simulated concentration data.

Using the analysis technique described above, it was determined that while the addition of the weight percent information narrowed the parameter confidence intervals, this additional measurement does not allow reliable estimation of all kinetic parameters.

Using the concentration and obscuration measurements allow all of the kinetic parameters of interest to be identified. The simulated data is shown in Figures 1 - 2. The parameter estimation results corresponding to these measurements are given in Table 1. These results indicate that these measurements may provide enough process information to allow identification, even in the presence of measurement noise. This hypothesis is investigated experimentally in the next section.

Table 1: Parameter estimation results utilizing simulated process measurements

	Kinetic Parameters			
	g	$ln\ k_g$	b	$ln\ k_b$
Mean Parameter Estimate	1.62	9.78	0.83	13.13
2-Sigma Confidence Interval	±0.015	±0.091	±0.50	±1.67
True Parameter	1.61	9.74	1.01	14.06

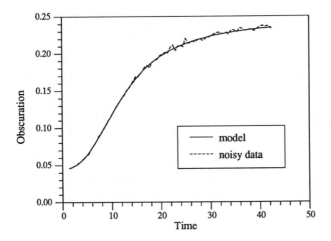

Figure 2: Simulated obscuration data.

Experimental Analysis. The most reliable process measurement is the oscillator frequency from the PAAR densitometer. Along with the frequency, the temperature is also measured ($\pm$ 0.05 ^oC). These two states are used to interpolate the solute concentration. CSD weight percent information and obscuration measurements were obtained from the Malvern Particle Sizer. Approximately 500 concentration data points and 200 CSD and obscuration measurements were recorded during a run of about 80 - 100 minutes. Therefore, the dynamics of the system were well monitored, i.e., the time constant of the crystallizer is much larger than the sampling time. We have performed 25 experimental runs. This section summarizes the analysis of a single, typical experiment.

The experimental operating conditions for the case study on which the parameter analysis was performed are given in Table 2. In most previous studies, the kinetic parameters of interest (g, k_g, b, and k_b) have been estimated using only desupersaturation data or steady-state CSD information. To judge the parameter reliability using only the desupersaturation data, the nonlinear optimization routine was run using only the solute concentration information. Table 3 shows the resulting parameter estimates and confidence intervals. The predicted concentration profile is shown by the solid line in Figure 3, which fit the experimental data quite well. The growth rate parameters are evaluated easily and with reasonable confidence intervals. However, as predicted from the pseudo-experimental data, calculating the nucleation parameters with certainty using only the concentration information is not possible. Even though mean parameter values were calculated for b and k_b, no confidence is placed in their values; the large confidence intervals indicate that the objective function surface was almost completely insensitive to these particular parameters. The mean parameter values are actually very dependent on their initial guesses in the optimization, again indicating the objective function insensitivity to these parameters.

To investigate whether the jagged character of the data would cause an unnecessary

Table 2: Case study operating conditions

Variable	Value
Temperature, T	24.0°C
c_0	0.5386 g Naph./g Tol.
c_s	0.5325 g Naph./g Tol.
Seed size (sieve tray range)	180 - 212 μm
Seed amount	1.3 g
Stirrer speed	500 rpm

Table 3: Parameter estimation results utilizing only concentration data

	Kinetic Parameters			
	g	$ln\ k_g$	b	$ln\ k_b$
Mean Parameter Estimate	1.23	8.43	2.89	12.88
2-Sigma Confidence Interval	±0.028	±0.16	±5.02	±35.88

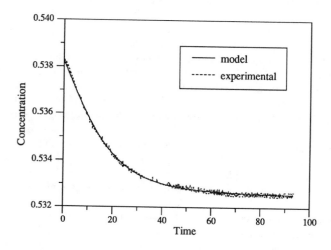

Figure 3: Predicted concentration profile using concentration data only.

Table 4: Parameter estimation results utilizing both concentration and obscuration data

| | Kinetic Parameters | | | |
	g	$ln\ k_g$	b	$ln\ k_b$
Mean Parameter Estimate	1.66	10.04	1.23	14.74
2-Sigma Confidence Interval	±0.011	±0.063	±0.040	±0.21

burden on the optimization routine or be a source of uncertainty, the concentration measurements were smoothed with a first order filter and the curve's jaggedness was removed. The re-evaluated parameters were the same as those using the raw data, indicating that the relative noise level in the concentration data was low and data smoothing offers no advantages.

As shown in the previous section, the use of concentration and obscuration measurements should contain enough information to identify all of the kinetic parameters of interest. Table 4 displays the parameters with corresponding confidence intervals determined using experimental concentration and obscuration data. Since the noise-to-signal ratio was much larger for the light scattering data than the concentration data, the latter was weighted more heavily (100:1) in the objective function. Details concerning the weight factor assignments are given by Witkowski (12). The model's fit to the experimental solute concentration data is shown in Figure 4. The fit to the obscuration is also very good as displayed in Figure 5.

While the CSD weight percent information was not used in the parameter estimation, Figures 6 and 7 demonstrate that the particle shape is important when comparing model predictions of CSD to those determined from light scattering methods. The circular points represent the weight percent information obtained from the light scattering device. The two curves represent the theoretical weight percents for crystals with two different shape factors. It appears that the results for spherical particles fit the experimental data better than for cubic particles. This is probably because the light scattering instrument determines the CSD by assuming *spherical* particles. It is clear that the "spherical equivalent" weight percent information from the light scattering instrument must be properly treated if it is used in the parameter estimation.

Fortunately, the particle shape has little effect in the resolved parameter values when using solute concentration and obscuration measurements to identify the model. The parameters shown in Table 5 are estimated assuming spherical crystals. Comparing these values to those for cubic particles (Table 4) shows little difference between the two sets of parameters.

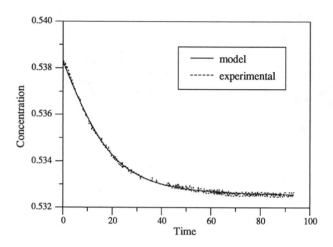

Figure 4: Predicted concentration profile using concentration and obscuration data.

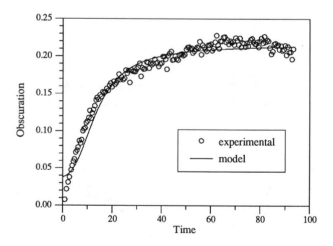

Figure 5: Predicted obscuration profile using concentration and obscuration data.

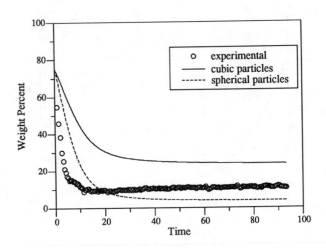

Figure 6: Crystal weight percent for the 160–262 μm size class.

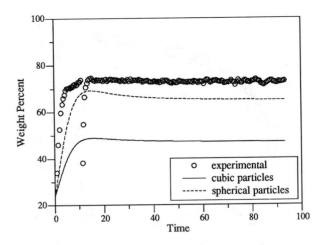

Figure 7: Crystal weight percent for the 262–564 μm size class.

Table 5: Parameter estimation results utilizing both concentration and obscuration data assuming spherical particles

	Kinetic Parameters			
	g	$ln\ k_g$	b	$ln\ k_b$
Mean Parameter Estimate	1.63	10.46	1.37	14.08
2-Sigma Confidence Interval	±0.010	±0.058	±0.028	±0.15

Conclusions

This study demonstrates that the nonlinear optimization approach to parameter estimation is a flexible and effective method. Although computationally intensive, this method lends itself to a wide variety of process model formulations and can provide an assessment of the uncertainty of the parameter estimates. Other factors, such as measurement error distributions and instrumentation reliability can also be integrated into the estimation procedure if they are known. The methods presented in the crystallization literature do not have this flexibility in model formulation and typically do not address the parameter reliability issue.

It is shown that while solute concentration data can be used to estimate the kinetic growth parameters, information about the CSD is necessary to evaluate the nucleation parameters. The fraction of light obscured by an illuminated sample of crystals provides a measure of the second moment of the CSD. Numerical and experimental studies demonstrate that all of the kinetic parameters can be identified by using the obscuration measurement along with the concentration measurement. It is also shown that characterization of the crystal shape is very important when evaluating CSD information from light scattering instruments.

Acknowledgments

The authors would like to thank the Eastman Kodak Company for financial support of this work. The first author also gratefully acknowledges Shell Oil Company for a Shell Foundation Fellowship.

Literature Cited

1. Witkowski, W. R. and Rawlings, J. B.   Kinetic Parameter Estimation of Naphthalene-Toluene Crystallization.   National AIChE Meeting, Houston, TX, 1989.

2. Halfon, A. and Kaliaguine, S. *Can. J. Chem. Eng.*, 1976, 54:160–167.

3. Tavare, N. S. and Garside, J. *Chem. Eng. Res. Des.*, 1986, 64:109–118.

4. Weiner, B. B.   Particle and Droplet Sizing Using Fraunhofer Diffraction.   In

Barth, H. G., editor, *Modern Methods of Particle Size Analysis*, pages 135–172. John Wiley & Sons, 1984.

5. Randolph, A. D. and Larson, M. A. *Theory of Particulate Processes*. Academic Press, New York, 1988.

6. Hulburt, H. and Katz, S. *Chem. Eng. Sci.*, 1964, 19:555–574.

7. Singh, P. N. and Ramkrishna, D. *Comput. Chem. Eng.*, 1977, 1:23–31.

8. Subramanian, G. and Ramkrishna, D. *Math. Bio.*, 1971, 10:1–23.

9. Becker, E. B.; Carey, G. F., and Oden, J. T. *Finite Elements: An Introduction, Volume I*. Prentice-Hall, Englewood Cliffs, NJ, 1981.

10. Reddy, J. N. *Applied Functional Analysis and Variational Methods in Engineering*. McGraw-Hill, New York, 1986.

11. Villadsen, J. and Michelson, M. L. *Solution of Differential Equation Models by Polynomial Approximation*. Prentice-Hall, 1978.

12. Witkowski, W. R. PhD thesis, Univeristy of Texas at Austin, 1990.

13. Garside, J. and Shah, M. B. *Ind. Eng. Chem. Proc. Des. Dev.*, 1980, 19:509–514.

14. Tavare, N. *Chem. Eng. Commun.*, 1987, 61:259–318.

15. Chen, M. R. and Larson, M. A. *Chem. Eng. Sci.*, 1985, 40(7):1287–1294.

16. Garside, J.; Gibilaro, L., and Tavare, N. *Chem. Eng. Sci.*, 1982, 37(11):1625–1628.

17. Tavare, N. *AIChE J.*, 1985, 31(10):1733–1735.

18. Bard, Y. *Nonlinear Parameter Estimation*. Academic Press, New York, 1974.

19. Bates, D. M. and Watts, D. G. *Nonlinear Regression Analysis and Its Applications*. John Wiley & Sons, 1988.

20. Caracotsios, M. and Stewart, W. E. *Comput. Chem. Eng.*, 1985, 9(4):359–365.

21. Caracotsios, M. PhD thesis, University of Wisconsin-Madison, 1986.

22. Gill, P. E.; Murray, W., and Wright, M. H. *Practical Optimization*. Academic Press, 1981.

23. Edgar, T. F. and Himmelblau, D. M. *Optimization of Chemical Processes*. McGraw-Hill, 1988.

24. Schittkowski, K. Nonlinear Programming Codes: Information, Tests, Performance. In *Lecture Notes in Economic and Mathematical Systems*. vol. 183, Springer-Verlag, 1980.

25. van de Hulst, H. C. *Light Scattering by Small Particles*. Dover, 1981.

RECEIVED May 12, 1990

Chapter 9

Double Draw-Off Crystallizer

Major Player in the Acid Rain Game?

Alan D. Randolph, S. Mukhopadhyay[1], B. C. Sutradhar, and Ross Kendall[2]

Department of Chemical Engineering, University of Arizona, Tucson, AZ 85721

A novel Double Draw-Off (DDO) crystallizer has been designed in order to improve the particle size distribution in the precipitation of $CaSO_3 \cdot \frac{1}{2}H_2O$ from simulated Flue Gas Desulfurization (FGD) liquor. The effects of DDO ratio and residence time on the mean particle size were studied. Industrial conditions were maintained in all experiments as far as practical. Significant improvement in mean particle size was achieved. The performance of an actual industrial DDO crystallizer (DuPont) for gypsum crystallization was reported.

Acid rain control is an important issue in the U.S. today. One of the major pollutants that has been linked to acid rain is sulfur dioxide. SO_2 control measures (e.g. Clean Air Act Amendment of 1977) for single source power plants have been passed and are likely to be strengthened in the future. Considerable technology has been developed for efficient and cost-effective SO_2 removal. The rapid growth of coal-fired power plants has provided incentives for work in this area. Thus, a number of SO_2 removal methods have been developed, including new space-saving dry processes as well as traditional wet scrubbing processes. However, the most widely-used method is still the wet-scrubbing lime/limestone process.

The lime/limestone wet scrubbing process involves the reaction of acidic SO_2 in the flue gas with alkaline lime and/or limestone in the scrubbing liquor to form the solid products $CaSO_3 \cdot 1/2H_2O$ or $CaSO_4 \cdot 2H_2O$ (gypsum). Sulfate can be the major product if forced oxidation is used. There is always some excess air present in the flue gas and, depending on its amount, different proportions of sulfite or sulfate are formed. The solid precipitates of

---

[1]Current address: Chemistry Department, State University of New York at Buffalo, Buffalo, NY 14214
[2]Current address: E. I. du Pont de Nemours and Company, Wilmington, DE 19714–6090

$CaSO_3 \bullet 1/2H_2O$ and/or $CaSO_4 \bullet 2H_2O$ are filtered off as a waste product. The cost-effectiveness of this process is greatly influenced by the cost of the dewatering step which in turn is strongly influenced by the mean particle size and particle size distribution. The crystallization kinetics of calcium sulfite and calcium sulfate are such that the former is crystallized as a finer product and filters poorly. The sulfate form (gypsum) usually gives larger crystals that filter easier, but exhibits severe fouling (e.g. wall scaling) problems. Good dewatering properties not only ease solid liquid separation, but also decrease water loss, waste disposal load and reduce free moisture content of the filter cake. Research on particle size improvement of the precipitate from flue gas desulfurization (FGD) liquor has been conducted at the University of Arizona for several years. Crystallization kinetics studies of both calcium sulfate and sulfite have been reported (1)(2).

Etherton studied the growth and nucleation kinetics of gypsum crystallization from simulated stack gas liquor using a one-liter seeded mininucleator with a Mixed Suspension Mixed Product Removal (MSMPR) configuration for the fines created by the retained parent seed. The effect of pH and chemical additives on crystallization kinetics of gypsum was measured. This early fundamental study has been the basis for later CSD studies.

Thus, the CSD of gypsum from simulated FGD liquors was studied in a bench-scale Double-Draw-Off (DDO)-type crystallizer (3). They studied the effect of variables such as overflow/underflow ratio, agitator rpm, overall liquid phase retention time, pH, rate of make, additive (citric acid) and crystallizer configuration (DDO or MSMPR) on the product CSD. The results were simulated using Etherton's reported kinetics. Experimental and simulated results were in reasonable agreement. Chang and Brna also studied gypsum crystallization in a DDO configuration from a forced oxidation FGD liquor in pilot plant scale (4). Predicted increased particle size and improved dewatering properties were confirmed.

Nucleation/growth rate kinetics and growth and nucleation modification by use of chemical additives on the crystallization of $CaSO_3 \bullet 1/2H_2O$ from simulated FGD liquors have been studied by Keough and Kelly, respectively (2)(5). An MSMPR crystallizer was used for all their studies. The effect of total dissolved solids composition and concentration on the crystallization kinetics of $CaSO_3 \bullet 1/2H_2O$ was studied by Alvarez-Dalama (6). The presence of high total dissolved solids from Cl^- solutions (> 100K ppm) was found to increase nucleation whereas high total dissolved solids from $SO_4^=$ solutions were found to inhibit nucleation.

Benson et al. (7) showed that high Mg^{++} ion concentration also inhibits nucleation during the precipitation of calcium sulfite hemihydrate from simulated FGD liquors.

However, the largest mass mean particle size reported thus far for the $CaSO_3 \bullet 1/2H_2O$ system is only about 32 microns. The objectives of the present work were to demonstrate a method of achieving improved calcium sulfite particle size (by use of the DDO configuration) while showing the industrial practicality of this crystallizer configuration. The particular industrial case illustrated was the production of large gypsum crystals with minimal process fouling from an in-process stream of weak sulfuric acid.

CSD Improvement: The DDO Crystallizer

The CSD studies of $CaSO_3 \cdot 1/2H_2O$ mentioned earlier all used the MSMPR configuration. In low solids systems, e.g. sulfite from FGD, the DDO crystallizer configuration is useful to increase particle size (3)(4)(8)(9)(10). These studies demonstrated that the mean size from a continuous crystallizer can be significantly increased using the DDO configuration.

Among the advantages of a DDO design are:

1. Increased mean particle size.
2. Less vessel fouling.
3. High per-pass yields of solute.

The present study demonstrates significant particle size increases for the sulfite system using the DDO crystallizer. In addition, the industrial success of this configuration to improve gypsum size while reducing vessel fouling is reported.

Experimental

A novel laboratory DDO crystallizer has been designed in order to obtain an overflow containing a small amount of fines (Figure 1). (In fact, fines overflow is necessary for the DDO configuration to produce larger-size crystals). Also shown in Figure 1 is the MSMPR configuration and the characteristic form of the population density-size plot for each configuration. The configuration embodies a conical section attached to a cylindrical bottom. The bottom contains a draft tube and four baffles. Agitation is provided by an upflow marine propeller. The total volume of the crystallizer is 10 liters. The objective of using the conical upper part is to aid the settling of fines that must be removed in this configuration. The rpm of the propeller is adjusted to confine agitation mainly to the cylindrical section.

The detailed experimental set-up is shown in Figure 2. Simulated flue gas (92% nitrogen and 8% SO_2) is passed through a sparger of sintered glass placed inside an absorption tank, where it comes into contact with the liquor from the crystallizer. Liquor continuously recirculates from the crystallizer through the adsorption tank. Lime is pumped in slurry form directly into the crystallizer.

Precipitation of $CaSO_3 \cdot 1/2H_2O$ under normal oxidation conditions usually results in 10-15% of the sulfite hemihydrate being oxidized to sulfate hemihydrate, incorporated in the crystal product as a crystalline solid solution. In order to obtain a representative SO_3/SO_4 crystal product, a sulfate concentration of 15,000 ppm was maintained in the feed liquor, thus simulating industrial conditions.

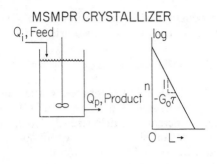

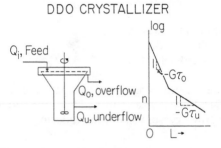

Figure 1. Comparison of MSMPR and DDO Configurations.

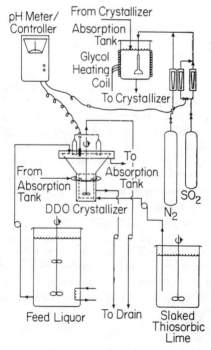

Figure 2. Schematic Diagram of Experimental Setup for FGD Process.

In process operations with high magnesium-containing lime (thiosorbic lime of Dravo Lime Company, U.S.A.), a high magnesium ion concentration develops. Therefore, in the present work a Mg^{2+} concentration of approximately 5,000 ppm was used in the feed liquor. The sulfate and magnesium ion concentration are maintained by using a solution of $MgSO_4 \bullet 7H_2O$ and $MgCl_2 \bullet 6H_2O$ in the make-up feed tank.

The present crystallizer set-up is equipped with a pH indicator-controller and a temperature indicator. The temperature in the crystallizer is controlled by heat supplied to the feed and liquor recirculated through the absorption tank.

In our laboratory runs, recirculation of the liquor from the absorption tank to the crystallizer using a 3,600 rpm centrifugal pump adversely affected particle size. This problem was overcome by use of a peristaltic pump with liquor withdrawal from and return to the crystallizer as shown in Figure 3. Liquor is withdrawn through four fines-traps placed at the surface of the liquor and returned through four manifold ports. This arrangement plus use of the peristaltic pump minimized the amount of crystals contacted by high shear and/or high velocity mechanical parts, thus reducing secondary nucleation.

The experimental DDO crystallizer was shown in Figures (2-3). A run was made for each set of conditions, e.g. DDO ratio, residence time and recycle ratio. The run was continued until a steady state CSD was obtained. (Size analysis was made using a PDI ELZONE 80 XY Particle Counter). Steady state was ascertained by analyzing particle size in sample intervals of one hour.

Results and Discussion

The main operating parameters which were varied were residence time (based on feed rate) and DDO ratio. Table I indicates the range of operating conditions used for all the experiments performed. Table II summarizes the particle size obtained using these conditions.

Table I. Operating Conditions

Residence Time	32-80 min
DDO ratio	9-21.4
Recycle Rate (through absorption tank)	2.9 ℓ/min
Temperature (crystallizer)	55 ± 1°C
Stirrer RPM	210 ± 5 RPM
Crystallizer pH	7.0 ± 0.2
Fines Cut Size	20-28μm
Rate of Make (ROM)	0.121 g $CaSO_3 \bullet \frac{1}{2}H2O$ per liter per min.

Table II. Mass Mean Particle Size Obtained for Different Residence
Time and DDO Ratio (Mixed Product)

Residence Time, min	DDO Ratio	Mass Mean Size, Micron
80	9	53.2
70	9	59.85
60	21.4	81.6
60	16	74.0
60	12.8	72.6
60	11.3	75.5
45	11	71.2
45	9	59.8
30	9	55.0

Both residence time and DDO ratio have significant effects on the particle size. Figure 4 shows the effect of DDO ratio on the mass mean size of the mixed product from the crystallizer for an overall residence time of 60 min. Similar behavior was observed for other residence times (not shown on plot). It was found that the particle size increased with an increase in DDO ratio, as expected. However, high DDO ratios in the laboratory size crystallizer lead to solids accumulation in the cylindrical section of the crystallizer and in plugging of the inlet and outlet ports. Figures 5-8 show strange behavior of the last (largest) five population density points. These points indicate that the population density not only doesn't decrease with size, but actually increases at a high DDO ratio. This behavior, highlighted by the dashed lines in the population density plot, might be qualitatively explained with conventional population balance mechanics (e.g. size-dependent growth rate) wherein the growth rate for crystals above about 60μm is severely retarded. A necessary condition for positive-sloped semi-log population density plots to occur would be:

$$\frac{d(\log G(L))}{dL} < \frac{1}{G(L)\tau} \tag{1}$$

No parameterization of G(L) was found which satisfactorily fit these data. A possible mechanism for this apparent declining growth rate at the larger sizes might be micro attrition. The observed effect is exacerbated at high DDO ratios and hence higher slurry densities. Regardless of the mechanism, larger and better filtering crystals were produced with higher DDO ratios in this laboratory apparatus. The same result would be expected in full-scale operation.

The effect of overall liquid phase residence time on particle size is presented in Figure (9). It is observed that there is an optimum residence time for the system for each DDO ratio. This has to do with poor particle settling at

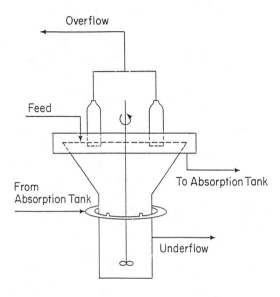

Figure 3. Improved DDO Crystallizer.

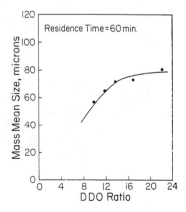

Figure 4. Effect of DDO Ratio on Mass Mean Size of Mixed Product.

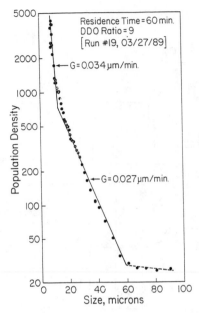

Figure 5.   Population Density Distribution of Mixed Product from the Crystallizer.

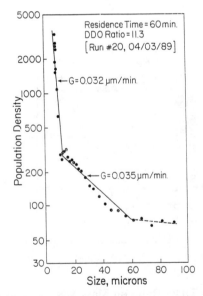

Figure 6.   Population Density Distribution of Mixed Product from the Crystallizer.

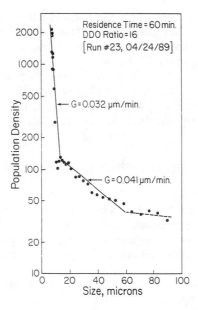

Figure 7. Population Density Distribution of Mixed Product from the Crystallizer.

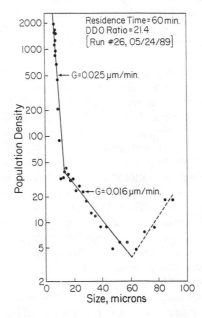

Figure 8. Population Density Distribution of Mixed Product from the Crystallizer.

the higher overall flow rates. This effect would not be expected as the process was increased in size. Keogh's kinetic's predict a smaller average particle size at longer retention times. Indeed, this inverse size response is observed industrially as well as in the laboratory.

The particle sizes ($L_{4,3}$) at different experimental conditions were calculated using the equations developed by White and Randolph (8). Experimental $L_{4,3}$ values are plotted against calculated values in Figure (10). It is observed that the experimental values were always larger than the calculated ones.

Performance of an Industrial Gypsum Crystallizer of the DDO-type

As stated earlier in this paper, FGD wet scrubbers can produce either calcium sulfite (the typical product) or calcium sulfate. The DDO crystallizer is advantageous for either product. The following industrial case history describes the production of calcium sulfate dihydrate (or gypsum) product from an industrial in-plant weak sulfuric acid liquor using a DDO crystallizer configuration.

The DDO mode is recognized as an effective method to reduce encrustation often experienced with dilute feeds in MSMPR crystallization processes. Three DDO gypsum crystallization processes have been developed, built and operated in DuPont in the last twenty years. The streams ranged in size from fifty to over one thousand gallons per minute. The midsize of the three converts a five hundred gallon per minute dilute sulfuric acid by-product stream to a useful twelve percent moisture gypsum centrifuge cake, which is used for natural gypsum substitutes. This DuPont gypsum plant (DGP) will be discussed below.

The DDO DGP is built around the standard DuPont draft tube crystallizer design which was developed in the early sixties and is illustrated in Figure 11. This design features an equal area inside and outside the draft tube and has a slow turning, low shear, shrouded, high flow agitator. This DGP crystallizer is sixteen feet in diameter by thirty-four feet deep and has a sixty horsepower, sixteen rpm drive. The circulation rate is over 100,000 gpm. This high circulation rate is desired to dilute and distribute the incoming feeds which this slow turning, large diameter agitator achieves while minimizing crystal attrition. Low DDO ratios (~ 3-4) accomplished with a primary settler followed by six-inch diameter hydrocyclones. The settler and hydrocyclone unders are returned to the crystallizer. The product is removed and dewatered through thirty-six inch by seventy-two inch horizontal, solid-bowl centrifuges. The purpose of DDO operation was to increase slurry density, thus reducing fouling rate. Modest size increases was observed using the low DDO ratios mentioned above.

The calcium source for the crystallization is the aragonite form of calcium carbonate which is wet ballmilled with closed-loop hydrocyclone classification to ninety-five percent minus two hundred mesh. The thirty weight percent aragonite slurry is fed to the crystallizer by pH control. A very pure carbon dioxide stream evolves from the crystallizer.

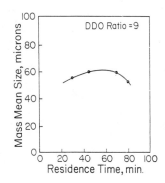

Figure 9. Effect of Overall Liquid Phase Residence Time on Particle Size.

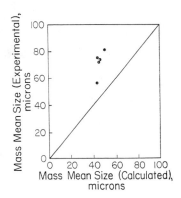

Figure 10. Experimental vs. Calculated Mass Mean Sizes.

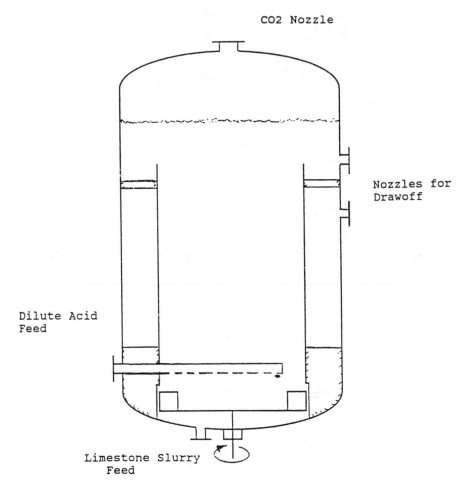

Figure 11. Standard Dupont Draft tube crystallizer design which was developed in the early sixties.

The DDO mode of operation yields a reduced crystal growth rate by increasing the crystal concentration independent of the feed concentration. As the DDO ratio (clarified water to product slurry ratio) increases, the system crystal concentration increases. The DGP operates with a thirty-five weight percent gypsum slurry. Two important benefits are gained by this mode of operation. First, a very low encrustation rate is obtained due to the lowered growth rate. Gypsum is well-known for its tendency to grow on process vessel or pipe walls and DDO was chosen to prevent encrustation from impacting plant utility. This goal has been achieved---the crystallizer and process pipes did not require cleaning in the first five years of continuous operation. Gypsum encrustation is not a factor in the overall process utility. Second, an improved crystal habit results from the DDO mode of operation (Figure 12). The scanning

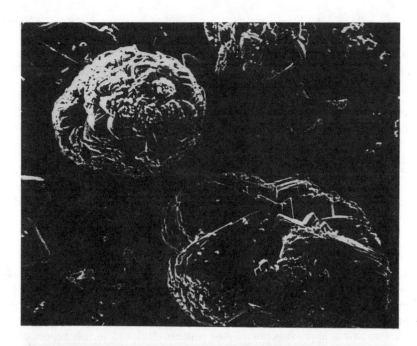

Figure 12. Gypsum crystals produced in the same crystallizer using DDO (top) or MSMPR (bottom) modes of operation.

electron micrographs show crystals which resulted from the DGP in its normal DDO mode versus the MSMPR mode run at equal feed rates. The DDO product dewaters to about twelve percent moisture and yields an easily handeable cake; whereas the MSMPR process yields a twenty percent moisture product which fractures easily, producing fines when centrifuged.

In summary, the DDO DGP is effective for low feed concentration, dilute systems which would experience high growth and high encrustation rates in finite size MSMPR crystallizers.

Conclusion

This paper has described advantages of the simple DDO crystallizer for production of calcium sulfite in a bench-scale study and calcium sulfate dihydrate (gypsum) in an industrial process.

Acknowledgements

This report was prepared by the University of Arizona as an account of work sponsored by the Electric Power Research Institute, Inc. (EPRI). Neither EPRI, members of EPRI, the University of Arizona, nor any person acting on their behalf: (a) makes any warranty, express or implied, with respect to the use of any information, apparatus, method or process disclosed in this report or that such use may not infringe privately owned rights; or (b) assumes any liabilities with respect to the use of, or for damages resulting from the use of, any information, apparatus, method or process disclosed in this report.

Literature Cited

1.   Etherton, D.L., "Experimental Study of Calcium Sulfate (Gypsum) Crystallization from Stack-Gas Liquors," M.S. Thesis, University of Arizona, Tucson (1980).

2.   Keough, B.K. "Influence of Additives and Particle-Size Classification on the Continuous Crystallization of Calcium Sulfite Hemihydrate," M.S. Thesis, University of Arizona, Tucson (1983).

3.   Randolph, A.D., D.E. Vaden and D. Stewart, "Improved Crystal Size Distribution of Gypsum from Flue Gas Desulfurization Liquors," Chemical Engineering Symposium Series, No. 240, 80:110 (1984).

4.   Chang, J.C.S. and T.G. Brna, "Gypsum Crystallization for Limestone FGD," Chemical Engineering Progress, November 1986, p.51.

5.   Kelly, B.J., "Study of $CaSO_3 \cdot \frac{1}{2}H_2O$ Nucleation and Growth Rates in Simulated Flue-Gas Desulfurization Liquors," M.S. Thesis, University of Arizona, Tucson (1983).

6.   Alvarez-Dalama, A., "Calcium Sulfite Hemihydrate Crystallization in Liquors with High Total Dissolved Solids," M.S. Thesis, University of Arizona, Tucson (1986).

7. Benson, L.B., Ray D'Alesandro, J. Wilhelm and A.D. Randolph, *"Improving Sludge Dewatering in Wet Lime FGD,"* Paper presented at First Combined FGD and Dry SO_2 Control Symposium, October 25-28, St. Louis, Missouri (1988).

8. White, E.T. and A.D. Randolph, *"Optimum Fines Size for Classification in Double Draw-Off Crystallizers,"* Industrial and Engineering Chemistry Research, No. 3, 28, 276 (1989).

9. Randolph, A.D. and M.A. Larson, *"Theory of Particulate Processes,"* Second Edition, Academic Press, New York, NY (1988).

10. Hulburt, H.M. and D.G. Stephango, *"Design Models for Continuous Crystallizers with Double Draw-Off,"* CEP Symposium Series No. 95, 65:50 (1969).

RECEIVED June 4, 1990

Chapter 10

Hydrocyclones for Size Classification in Continuous Crystallizers

Johan Jager[1], Sjoerd de Wolf[2], Herman J. M. Kramer[1], and Esso J. de Jong[1]

[1]Laboratory for Process Equipment and [2]Laboratory for Measurement and Control, Delft University of Technology, Mekelweg 2, Delft, Netherlands

The experimental results reported in this paper demonstrate the ability of a flat-bottom hydrocyclone to separate the coarse fraction of ammonium sulfate crystals from a slurry which contains crystals of a wide size range. It appears that the grade efficiency curve, which predicts the probability of a particle reporting to the underflow of the cyclone as a function of size, can be adjusted by a change in the underflow diameter of the hydrocyclone. These two observations lead to the suggestion to use hydrocyclone separation to reduce the crystal size distribution which is produced in crystallisers, whilst using a variable underflow diameter as an additional input for process control.

The number of inputs which are available for controlling crystallisation processes is limited. Possible inputs for a continuous evaporative crystallisation process are, crystalliser temperature, residence time and rate of evaporation. These inputs affect the crystal size distribution (CSD) through overall changes in the nucleation rate, the number of new crystals per unit time, and the growth rate, the increase in linear size per unit time, and therefore do not discriminate directly with respect to size. Moreover, it has been observed that, for a 970 litre continuous crystalliser, the effect of the residence time and the production rate is limited. Size classification, on the other hand, does allow direct manipulation of the CSD.

 (1) Fines removal, the selective removal of small crystals from a well-mixed crystalliser, is generally used to remove excessive fines, thus increasing the average size. An increase in the fines removal rate immediately reduces the number of small crystals contained in the reactor.

 (2) Product classification, the selective removal of the large crystals as the product, is generally used to reduce the coefficient of variation (CV) of the crystals produced.

0097–6156/90/0438–0130$06.00/0
© 1990 American Chemical Society

$$CV = \sqrt{[\frac{m_5 \, m_3}{m_4^2} - 1]} \; 100 \; \%$$

where

$m_j = j^{th}$ moment of the crystal size distribution

Usually, this also results in a smaller mass based average size (L50m) (1).

$$L50m = \frac{m_4}{m_3}$$

As a result of the selective removal of the largest crystals, the specific surface area tends to increase, which imposes a decrease in the crystal growth rate and eventually causes a decrease in the average crystal size. Therefore a product classification step should preferably be combined with fines removal.

Conventionally, an annular settling zone separated by a baffle from the well-mixed region of the crystalliser is used for fines removal (2). A large settling area is usually needed for the combination of a low cut size and a large fines removal rate, whereas a minimum height is required to minimise the influence of turbulence in the mixed region of the crystalliser (3). These requirements increase the crystalliser volume significantly. Furthermore the separation characteristics cannot easily be changed after the plant construction.

Product classification is generally realised by product discharge through an elutriation leg which also serves as a washing step to remove contaminated mother liquor. The stable operation of such an elutriation leg is difficult and the separation efficiency is unsatisfactory. Previous results (1) show that values of both the average size and the coefficient of variation in the product stream delivered from the elutriation leg differed only slightly from those in the circulation mains in a forced circulation pilot plant crystalliser, as illustrated in Table 1.

There is a clear need for other size classifiers which combine a high separation efficiency with flexibility and compactness. Hydrocyclones have a small volume, are simple in operation and are standard size classification equipment, for example in closed circuit grinding applications. The recent development of the flat-bottom hydrocyclone, which permits classification in the coarse size range, creates an additional motive to study the use of hydrocyclones for Crystal Size Distribution (CSD) control. Furthermore, throttling of a flat botom hydrocyclone does not necessarily provoke blockage but allows continuous control of its cut size when a controlled throttling valve is used. There is a clear incentive for its use in this application since it may provide an additional process input.

This paper presents the grade-efficiency curves of a 75 mm flat bottom cyclone (RWB 1613) provided by the Amberger Kaolin Werke (AKW). It is tested for the ammonium sulfate-water system for both fines removal and product classification. Its results will be compared with the results for fines removal obtained when using an

Table 1: Classification results by Grootscholten

RUN	L50m product μm.	CV product %	L50m mains μm.	CV mains %
80.4.1	380	22	370	24
80.4.2	431	21	412	34
80.9.1	367	24	350	26
80.9.2	397	23	388	24
80.17.1	415	23	400	26
80.17.2	435	28	405	32
80.18.1	420	26	395	35
80.19.1	405	24	382	29
80.19.2	410	30	365	35
80.20.1	435	31	415	30
80.20.2	380	22	362	29
80.21.1	400	24	365	30
80.21.2	395	23	385	27
80.21.3	350	25	340	27

SOURCE: Data are from ref. 1.

annular zone separator. In addition, the effect of varying the cyclone feed flow as well as the apex diameter of the cyclone are investigated in order to predict whether or not these could be active inputs in CSD-control.

Theory

Principle Of Operation. Figure 1 demonstrates schematically the working principles of a hydrocyclone. The feed flow is fed tangentially to the upper part of the hydrocyclone where it forms a primary vortex along the inside of the wall and leaves through the underflow. Since the underflow is throttled, only part of the stream is discharged as underflow, carrying the coarse solids or even all of the solids with it. Most of the liquid from which the coaser solids have been removed by the centrifugal action of the primary vortex, leaves through the overflow forming an upward spinning secondary vortex. In this secondary vortex a separation again takes place and the ejected fine particles move radially, re-join the primary vortex, and are mainly discharged through the underflow. The separating characteristics are influenced by both the cyclone geometry and the operating conditions. Throttling the underflow or increasing the overflow diameter usually results in a coarser cut size for a given cyclone whereas increasing the feed flow tends to decrease the cut size with a corresponding increase in the pressure drop. A number of theories have been developed to describe and predict this behaviour, but experimental tests are necessary to determine the operating characteristics for a specific application. The presentation of such experimental data is the purpose of this paper. For background information the reader is referred to reviews given in (4) (5) (6) and in three conference proceedings (7) (8) (9).

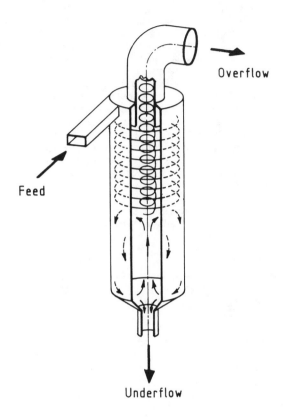

Figure 1: The working principles of a flat bottom hydrocyclone.

Circulating Bed Classifier. A flat bottom hydrocyclone or
circulating bed classifier (CBC) is chosen for this application
because of its reported capability to separate at a coarse cut size
and its improved controlability (10). The flat bottom (10) reduces
the velocity of the vortex by wall friction. In addition, the
increased solids concentration near to the bottom of the
hydrocyclone (11) causes an increase in the effective viscosity.
These effects create convection near to the cyclone bottom which is
downwards at the wall and upwards in the centre and which
facilitates the discharge of solids and prevents blockage when the
cyclone is choked. The solids concentrate in the circulating flow,
thus forming a circulating fluidised bed. By varying the height of
this bed a change in the cut size, by a factor as large as 6, can be
achieved for a given cyclone length (10). According to its inventor
(10) this fluid bed acts as a rotating fluid bed in which the coarse
material is concentrated at the wall whereas other investigators
(12) modify this opinion by including the enhanced solids mixing
effects which adversely effect the separation efficiency.

Definitions. The performance of a hydrocyclone is generally
characterised by means of a grade efficiency or Tromp-curve which is
the fractional mass recovery expressed as a function of particle
size.

Thus, solids recovery is defined as:

$$\theta = \frac{\text{solids underflow}}{\text{solids feedflow}} \tag{1}$$

The Tromp-curve for the removal of coarse particles with the
underflow is defined by:

$$\text{efficiency for interval } i = \frac{\theta \; \Delta V(i)_u}{\theta \; \Delta V(i)_u + (1-\theta) \; \Delta V(i)_o} \tag{2}$$

where

$\Delta V(i)_o$ = volume fraction in the ith interval reporting to the
overflow

$\Delta V(i)_u$ = volume fraction in the ith interval reporting to the
underflow

The efficiency for the removal of fines with the overflow is defined
by:

$$\text{efficiency at interval } i = \frac{(1-\theta) \; \Delta V(i)_o}{\theta \; \Delta V(i)_u + (1-\theta) \; \Delta V(i)_o} \tag{3}$$

From these grade efficiency curves the nominal cut size, that is the
size with an efficiency of 50%, and the classifier imperfection can
be determined. The imperfection is defined by:

$$\text{Imperfection} = \frac{d_{75} - d_{25}}{2\ d_{50}} \tag{4}$$

A small imperfection corresponds to a sharp separation. Within the field of crystallisation the so-called classification function is used to define solids classification in terms of number population densities.

n = number of crystals per unit size per unit crystalliser volume).

$$\lambda(i) = \frac{n_{product}(i)}{n_{feed}(i)} \tag{5}$$

or

$$\lambda(i) = \frac{(\Delta V(i)\ Msl)/(\rho_c\ k_v\ L^3(i)\ \Delta L(i))_{product}}{(\Delta V(i)\ Msl)/(\rho_c\ k_v\ L^3(i)\ \Delta L(i))_{feed}}$$

or

$$\lambda(i) = \frac{(\Delta V(i)\ Msl)_{product}}{(\Delta V(i)\ Msl)_{feed}}$$

where

Msl	=	kilogrammes solids per cubic meter slurry
$L(i)$	=	average size in interval i
$\Delta L(i)$	=	interval width
k_v	=	volume of crystal/L^3
ρ_c	=	crystal density

The shape of this classification function is identical to the shape of the grade efficiency curve for product classification but the ordinate values are changed.

Experimental

The experiments were carried out in a 75 mm flat bottom cyclone provided by the Amberger Kaolin Werke (RWB 1613). Apexes and vortex finders of several diameters were available. The cyclone length could be chosen as 0.25 m, 0.45 m or 0.65 m (Figure 2). The cyclone was fed from a thermostatically controlled storage tank using a variable speed mohno pump. The feed flow was measured by a magnetic flow meter, whereas the underflow and overflow were measured manually. Samples from the feed, under- and over-flow were taken manually, filtered, dried and subsequently sieved using a set of 20 Veco micro-precision sieves. The experimental set-up is shown in Figure 3. Prior to a set of experiments, the storage tank was filled with the product from a 20 litre continuous crystalliser in order to ensure similarity between the separation experiments and the actual working conditions of the cyclone. The batch was reproduced for every set of experiments because in the course of time, the fines were preferentially dissolved. The average solids concentration was of the order of 1 - 10 % (m^3/m^3). Sampling was only started after about 20 minutes of operation in order to ensure steady-state conditions of the cyclone.

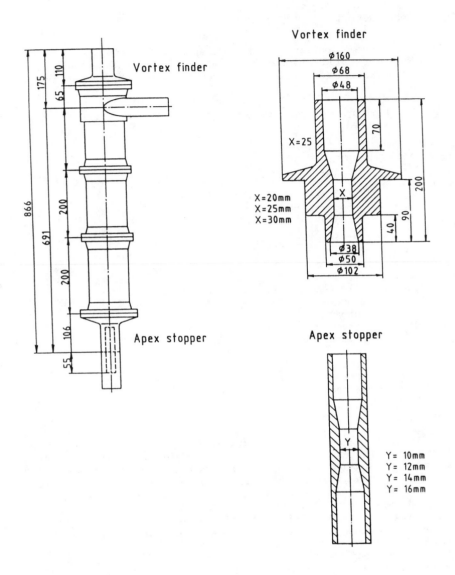

Figure 2: The dimensions (in mm) of a RWB 1613 flat bottom hydrocyclone.

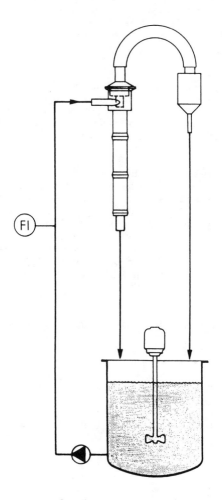

Figure 3: The experimental set-up.

The results which represent the use of an annular zone were obtained using a 970 litre continuous evaporative crystalliser. This crystalliser was equipped with a fines dissolution and a recycle system, which allows the fines dissolution rate to be changed without changing the cut size, by varying the recycle rate (13).

Results

A typical grade efficiency curve for the product classification step is given in Figure 4. A value of nearly 100 percent is attained at large sizes, whereas normally a value equal to or larger than the so-called dead flux is attained at small sizes. This is caused by the diluted discharge of the coarse fraction. It represents the minimum amount of residual fines in the product after one separation stage.

In fines removal, both the cut size and the grade efficiency are difficult to assess because of the limited accuracy of the sieve analysis technique and the problems involved in the determination of the solids concentration in the overflow. For a .65 m cyclone, whilst using a 20 mm vortexfinder diameter, an apex diameter of 16 mm and a feed flow of 1.6 l/s, solids recovery is over 99 %. This recovery corresponds to a cut size between 50 - 100 μm. Typical distributions of size by weight, for the feed flow as well as the overflow are shown in Figure 5. Results are summarised in Table 2.

Table 2: Experimental results and conditions

Fines removal

feed				underflow			
flow [l/s]	conc. [%]	L50m [um]	CV [%]	flow [l/s]	conc. [%]	L50m [um]	CV [%]
1.6	6.6	347	36	0.4	28	351	35

Product classification

feed				underflow				apex
flow [l/s]	conc. [%]	L50m [um]	CV [%]	flow [l/s]	conc. [%]	L50m [um]	CV [%]	
0.7	2.2	454	46	0.10	15	468	43	10
0.7	3.8	485	47	0.23	11	496	46	16
1.4	2.4	361	47	0.07	24	446	38	10
1.4	3.4	373	48	0.15	21	425	41	16
2.0	2.7	364	47	0.10	29	438	39	10
2.0	3.0	371	46	0.16	31	424	39	16

In the 970 litre pilot plant crystalliser, whilst using an annular zone, fines are removed at an approximate size of 50 μm at a concentration of 0.03 % (m^3/m^3) at a flowrate of 1.7 l/s. Again,

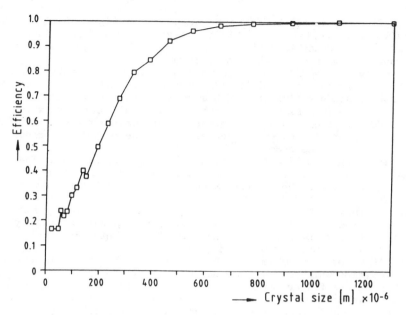

Figure 4: A typical grade efficiency curve for the product classification step.

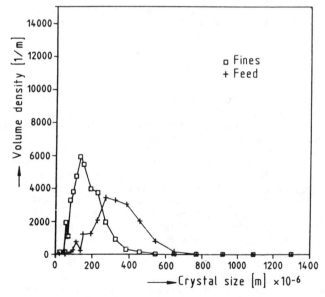

Figure 5: Typical size distributions by volume, for the feed flow and the overflow, if the hydrocyclone is operated for fines removal. The ordinate value is defined by volume percentage divided by interval width.

the efficiency cannot be assessed accurately. Figure 6 shows the typical distributions of size by weight for the fines removed as well as the product flow of the crystalliser. The separation is much sharper than when using the hydrocyclone.

However, the differences between an experiment whereby 1.4 l/s of the total fines flow of 1.7 l/s is dissolved and an experiment at the same residence time of 1.25 hrs without fines dissolution, are negligible. This is probably due to the small cut size which is attained with the present configuration. Increasing the cut size by using an insert in the annular zone, may be one method of increasing the effect of fines removal.

In product classification, a 0.25 m cyclone was used. The cut size is considerably increased to a maximum cut size of about 400 μm. Figure 7 shows the efficiency curves determined at a flowrate of 0.7 l/s, 1.4 l/s and 2.1 l/s using an apex diameter of 10 mm. These results indicate that, at flowrates which are sufficiently high to transport the particles in the overflow, the grade efficiency is relatively insensitive for changes in the flowrate. Figure 8 shows the efficiency curves determined at a flowrate of 1.4 l/s using the apex diameters of 10 mm and 16 mm respectively, which illustrates the variation which can be obtained by varying the apex-diameter. A moving average filter, with a window of three points, was applied to reduce the measurement noise in the smaller size range in both Figures 7 and 8. Figure 9 shows the weight distributions for the feedflow, the underflow and the overflow, at an apex diameter of 10 mm and a flowrate of 1.4 l/s. At these conditions the difference in size between the feed flow and the underflow is most pronounced at a value of 100 μm. The other conditions are listed in Table 2.

The effect of throttling, is significant, as indicated in Figure 8, which indeed suggests that an additional process-input can be created. A shift of the grade efficiency curve of about plus or minus 50 μm is found.

Discussion

Fines removal. The annular zone enables a sharp separation to be made because of the low solids concentration which prevails in the upper part of this zone. Furthermore, it is a simple separation technique and, as such, is very successful. As a result of the low cut size, its effect on the average size produced in the crystalliser is negligible. At a higher solids concentration a reasonable separation is also attained using the hydrocyclone.

The disadvantage of the annular zone is its large volume, which is .6 m³ in this application as opposed to 0.003 m³ for the hydrocyclone. If, size classification is to be applied in a batch crystallisation operation (14) this might be a serious disadvantage. Furthermore, a hydrocyclone is more easily exchanged if a change in cut size is required.

Product classification. In product classification promising results have been obtained. The differentiation between underflow and feedflow is considerable, especially if the performance of an elutriation leg is considered. Furthermore, an additional degree of

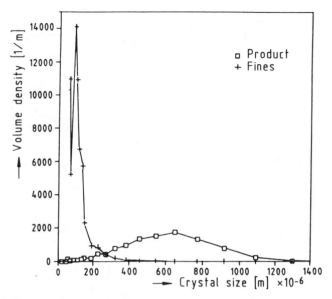

Figure 6. Typical size distributions by volume for the fines flow and the product flow obtained from a 970-L pilot plant crystallizer operated with fines removal. The ordinate value is defined by volume percentage divided by interval width.

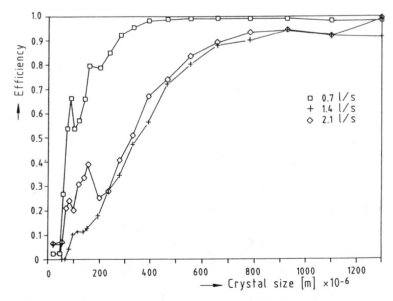

Figure 7. Three grade efficiency curves for the product classification step at the flow rates indicated, while using an apex diameter of 10 mm.

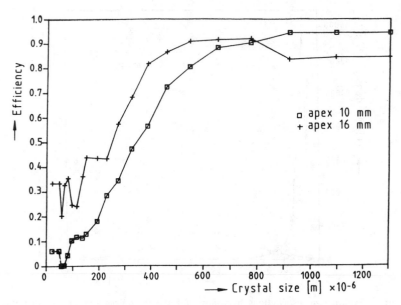

Figure 8. Two grade efficiency curves for product classification at the apex diameters shown at a flow rate of 1.4 L/s.

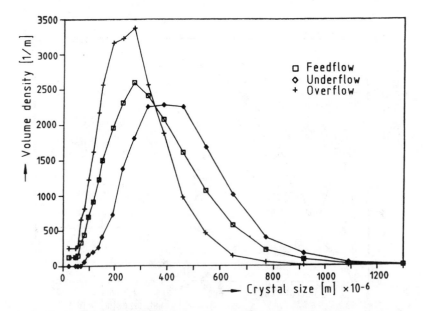

Figure 9. The volume distributions at a flow rate of 1.4 L/s and an apex diameter of 10 mm. The ordinate value is defined by volume percentage divided by interval width.

freedom is introduced by a controllable apex diameter, which possibly enables a better control of crystallisers to be achieved. Product degradation, after 25 average residence times in the combined system of cyclone, pump and storage tank is negligible. However, a disadvantage of the hydrocyclone can be that it produces a large amount of small nuclei by attrition. In the present investigations these fines dissolve preferentially in the storage tank and can, therefore, not be detected. Under crystallising conditions the effect of these fines might be very disadvantageous.

Acknowledgments

The outhors wish to thank the Netherlands Technology Foundation (STW), AKZO, DOW, DSM, DUPONT de NEMOURS, RHONE POULENC, SUIKER UNIE and UNIELEVER for their financial support of the UNIAK-research programme. The support of Prof.Dr. H. Trawinsky in selecting and supplying the AKW hydrocyclone, and the support of his stimulating discussions is gratefully acknowledged. Ir. M. Wouters contributed to this paper during his final year study as an MSc-student, while J. Koch and J. Weergang from the Hogeschool Utrecht carried out a large number of the experiments reported.

Literature Cited

1. Grootscholten, P.A.M.; Jancic S.J., In Industrial Crystallizaton 84, Eds. Jancic S.J.; DeJong E.J, Elsevier, Amsterdam, 1984; pp 203 - 210.
2. DeLeer, B.G.M., Ph.D. Thesis, Delft University of Technology, Delft, 1981.
3. Grootscholten, P.A.M., Ph.D. Thesis, Delft University of Technology, Delft, 1982.
4. Rietema, K.; Verver C.G. (Eds.), Cyclones In Industry, Elsevier, Amsterdam, 1961.
5. Bradley, D, The Hydrocyclone, Pergamon, Oxford, 1965.
6. Svarovsky, L., Hydrocyclones, Technomic, London, 1985.
7. Proc. International Conference on hydrocyclones, Cambridge, 1980.
8. Proc. 2nd International Conference on hydrocyclones, Bath, 1984.
9. Proc. 3rd International Conference on hydrocyclones, Oxford, 1987.
10. Trawinsky, H.F., Filtration & Separation, Jan/Feb 1985.
11. VanDuijn, G.; Rietema, K., Chem. Eng. Science, 1983, 38, pp 1651 - 1673.
12. VanDuijn, G., Ph.D. Thesis, Delft University of Technology, Delft, 1982.
13. Jager, J.; DeWolf, S.; Kramer, H.J.M.; DeJong, E.J., AIChE Annual Meeting, San Francisco, 1989.
14. Zipp, G.L.; Randolph, A.D., Ind. Eng. Chem. Res., 1989, 28, pp 1446 - 1448.

RECEIVED May 12, 1990

Chapter 11

Derivation of State Space Models of Crystallizers

Sjoerd de Wolf[1], Johan Jager[2], B. Visser[1], Herman J. M. Kramer[2], and O. H. Bosgra[1]

[1]Laboratory for Measurement and Control and [2]Laboratory for Process Equipment, Delft University of Technology, Mekelweg 2, Delft, Netherlands

For effective control of crystallizers, multivariable controllers are required. In order to design such controllers, a model in state space representation is required. Therefore the population balance has to be transformed into a set of ordinary differential equations. Two transformation methods were reported in the literature. However, the first method is limited to MSMPR crystallizers with simple size dependent growth rate kinetics whereas the other method results in very high orders of the state space model which causes problems in the control system design. Therefore system identification, which can also be applied directly on experimental data without the intermediate step of calculating the kinetic parameters, is proposed.

Crystallization from solution is a widely utilized separation and purification technique in chemical industry. It is characterized by the formation of a spectrum of differently sized crystals. This spectrum, called the Crystal Size Distribution or CSD, is highly important for the performance of the crystallizer, the crystal handling equipment like centrifuges and dryers, and the marketability of the produced crystals. However, in many industrial crystallizers, the observed CSD's show large transients due to disturbances or are unstable because of the internal feedback mechanisms of the crystallization process (1). The main limitation for effective CSD control was the lack of a good on-line CSD measurement device, but recent developments show that this hurdle is taken (2).

With the on-line CSD-measurement equipment available, it becomes relevant to investigate and design control schemes for CSD's. A number of schemes were investigated by simulation in the past. Most of these control schemes use one proportional single-input-single-output (SISO) controller with the nucleation rate (or a

0097–6156/90/0438–0144$06.00/0

variable proportional to the nucleation rate) as the variable to be controlled, see for example (3,4). The main drawback of these simple proportional controllers is that transients and offsets of the CSD are still possible. In order to improve the control performance, more process inputs and outputs can be used in the controller and, to cope with unwanted interactions due to simultaneous variations of different process inputs by the controller, 'the more powerfull multivariable controllers are to be used.

Most design methods for multivariable controllers require a dynamic model of the process in the linear state space representation. Such a model is given by

$$
\begin{aligned}
\dot{\underline{x}}(t) &= A\,\underline{x}(t) + B\,\underline{u}(t) \\
\underline{y}(t) &= C\,\underline{x}(t) + D\,\underline{u}(t)
\end{aligned}
\tag{1}
$$

where

$\underline{x}(t)$ = state vector   dimension $(n \times 1)$
$\underline{u}(t)$ = vector of process inputs   dimension $(m \times 1)$
$\underline{y}(t)$ = vector of process outputs   dimension $(\ell \times 1)$
A = state matrix   dimension $(n \times n)$
B = input matrix   dimension $(n \times m)$
C = output matrix   dimension $(\ell \times n)$
D = direct transfer matrix   dimension $(\ell \times m)$
n = number of states (order of the state space model)
m = number of process inputs
ℓ = number of process outputs

which is a set of linear first-order ordinary differential equations in matrix-notation. As discussed in (5 and De Wolf, S.; Ph.D. Thesis, Delft University of Technology, in preparation) the physical model for the CSD in evaporative and non-evaporative crystallizers is given by the set of differential equations

- population balance
- concentration balance   (2a)
- energy balance   (for non-evaporative systems)

and the set of algebraic equations

- algebraic relation for the vapour flow rate
                     (for evaporative systems)
- relation between the input and output flow rates   (2b)
- growth rate kinetics
- nucleation kinetics.

The problem in obtaining a state space model for the dynamics of the CSD from this physical model is that the population balance is a (nonlinear) first-order partial differential equation. Consequently, to obtain a state space model the population balance must be transformed into a set of ordinary differential equations. After this transformation, the state space model is easily obtained by substitution of the algebraic relations and linearization of the ordinary differential equations.

In this paper, three methods to transform the population balance into a set of ordinary differential equations will be discussed. Two of these methods were reported earlier in the crystallizer literature. However, these methods have limitations in their applicabilty to crystallizers with fines removal, product classification and size-dependent crystal growth, limitations in the choice of the elements of the process output vector $\underline{y}(t)$ that is used by the controller or result in high orders of the state space model which causes severe problems in the control system design. Therefore another approach is suggested. This approach is demonstrated and compared with the other methods in an example.

Transformation methods

The discussion of the transformation methods will be based on the population balance for a crystallizer with fines removal and product classification:

$$V \frac{\partial n(L,t)}{\partial t} + V \frac{\partial [G(L,t)n(L,t)]}{\partial L}$$

$$+ Q_p(t)h_p(L,t)n(L,t) + Q_f(t)h_f(L,t)n(L,t) = 0 \qquad (3)$$

with boundary condition

$$n(L_0,t) = B(t)/G(L_0,t) = K_B \ G(L_0,t)^{i-1} \ m_x(t)^j \qquad (4)$$

where
$\quad n(L,t)$ = population density $\qquad\qquad [\#/(m.m^3)]$
$\quad V$ $\quad$ = crystallizer volume $\qquad\qquad\quad [m^3]$
$\quad G(L,t)$ = crystal growth rate $\qquad\qquad\quad [m/s]$
$\quad Q(t)$ $\quad$ = flow rate $\qquad\qquad\qquad\qquad [m^3/s]$
$\quad h(L,t)$ = classification function $\qquad\qquad [-]$
$\quad B(t)$ $\quad$ = nucleation rate $\qquad\qquad\qquad [\#/(m^3 s)]$
$\quad m_x(t)$ = xth moment of $n(L,t)$ $\qquad\quad [m^x/m^3]$
indices:
$\quad$ p = product flow
$\quad$ f = fines removal flow.

Size-dependent crystal growth is included in the model because it can be important to describe diffusion limited growth rates or crystal attrition. As discussed in ($\underline{6},\underline{7}$), the size reduction by attrition can be modelled by an effective growth rate $G_e(L,t)$ which is the difference between the kinetic growth rate $G_k(L,t)$ and an attrition rate $G_a(L,t)$:

$$G_e(L,t) = G_k(L,t) - G_a(L,t) \qquad (5)$$

Inclusion of attrition can be important because the effect of process inputs, as for example the fines flow rate and the residence time on for example the average crystal size, is strongly affected by attrition.

Method of Moments. The moments of the population density can be used
to characterize the CSD. The first four moments have physical
interpretations and the mean crystal size (L_{50}) and the Coefficient
of Variation (CV) based on the mass distribution are functions of
the moments:

$$L_{50} = m_4/m_3 \qquad\qquad\qquad [m]$$
$$CV = (m_5 m_3/m_4^2 - 1)^{1/2}. \qquad [-]$$

(6)

When control is required for moments, L_{50} or CV only, a model for
the dynamics of the moments is sufficient. Such a model can be obtained by multiplying the population balance with L^j and integration
over L:

$$\frac{dm_j(t)}{dt} = B(t)L_0^{\,j} + j \int_{L_0}^{\infty} G(L,t)n(L,t)L^{j-1}dL$$

$$- \frac{Q_p(t)}{V} \int_0^{\infty} h_p(L,t)n(L,t)L^j dL - \frac{Q_f(t)}{V} \int_0^{\infty} h_f(L,t)n(L,t)L^j dL. \quad (7)$$

This results in a set of first-order ordinary differential equations
for the dynamics of the moments. However, the population balance is
still required in the model to determine the three integrals and no
state space representation can be formed. Only for simple MSMPR
(Mixed Suspension Mixed Product Removal) crystallizers with simple
crystal growth behaviour, the population balance is redundant in the
model. For MSMPR crystallizers, $Q_f=0$ and $h_p(L)=1$, thus:

$$\frac{dm_j(t)}{dt} = B(t)L_0^{\,j} + j \int_{L_0}^{\infty} G(L,t)n(L,t)L^{j-1}dL - \frac{Q_p(t)}{V} m_j(t). \quad (8)$$

The final integral disappears when the size-dependent growth rate is
written as a polynomial of the crystal size:

$$G(L,t) = G_\sigma(t)G_L(L) = G_\sigma(t)[a_0 + a_1 L + a_2 L^2 + \ldots] \quad (9)$$

where $G_\sigma(t)$ and $G_L(L)$ are the supersaturation dependent part and the
size-dependent part of the crystal growth rate respectively. This
results for Equation 8:

$$\frac{dm_j(t)}{dt} = B(t)L_0^{\,j} + jG_\sigma(t) \sum_{i=0}^{q} a_i m_{i+j-1}(t) - \frac{Q_p(t)}{V} m_j(t) \quad (10)$$

and the population balance is no longer required in the model.
However, there is another problem. As already discussed in (8), for
$a_i \neq 0$ for i=2,3,... the equation for $dm_j(t)/dt$ contains higher moments than $m_j(t)$ and the set of equations is not a closed model set.
In order to solve this problem, an approximation of the population
density by orthogonal Laguerre polynomials[1], i.e.

$$n(L,t) \approx \sum_{k=0}^{p} b_k(t) \, Lag_k(L) \tag{11}$$

was investigated in (10). However, this results in unstable approximations because it is not based on the minimization of the error involved in Equation 11 (11).

In conclusion, the method of moments can be used to obtain a state space model for the dynamics of the moments of the CSD. The method is limited to MSMPR crystallizers with size-independent growth or size-dependent growth described by

$$G(L,t) = G_\sigma(t)G_L(L) = G_\sigma(t)[a_0 + a_1 L] \tag{12}$$

and the resulting moment equations are given by:

$$\frac{dm_j(t)}{dt} = B(t)L_0^{\,j} + jG_\sigma(t)[a_0 m_{j-1}(t)+a_1 m_j(t)] - \frac{Q_p(t)}{V} m_j(t). \tag{13}$$

In applying the resulting state space model for control system design, the order of the state space model is important. This order is directly affected by the number of ordinary differential equations (moment equations) required to describe the population balance. From the structure of the moment equations, it follows that the dynamics of $m_j(t)$ is described by the moment equations for $m_0(t)$ to $m_j(t)$. Because the concentration balance contains $\varepsilon(t)=1-k_v m_3(t)$, at least the first four moments equations are required to close off the overall model. The final number of equations is determined by the moment $m_x(t)$ in the equation for the nucleation rate (usually $m_3(t)$) and the highest moment to be controlled.

Method of Lines. The method of lines is used to solve partial differential equations (12) and was already used by Cooper (13) and Tsuruoka (14) in the derivation of state space models for the dynamics of particulate processes. In the method, the size-axis is discretized and the partial differential $\partial[G(L,t)n(L,t)]/\partial L$ is approximated by a finite difference. Several choices are possible for the accuracy of the finite difference. The method will be demonstrated for a fourth-order central difference and an equidistant grid. For non-equidistant grids, the Lagrange interpolation formulaes as described in (15) are to be used.

Using an equidistant discretization in K+1 points (L_0, L_1, L_2,...L_{K-1}, L_K) and the fourth-order central difference

---

[1] A different approach in the use of orthogonal polynomials as a transformation method for the population balance is discussed in (8,9). Here the error in Equation 11 is minimized by the Method of Weighted Residuals. This approach releases the restrictions on the growth rate and MSMPR operation, however, at the cost of the introduction of numerical integration of the integrals involved, which makes the method computationally unattractive. The applicability in determining state space models is presently investigated and results will be published elsewere.

$$(\frac{df(x)}{dx})_{x=x_i} = \frac{1}{12\Delta x} \left(f_{i-2}-8f_{i-1}+8f_{i+1}-f_{i+2}\right) + O(\Delta x^4) \quad (14)$$

the population balance changes into a set of ordinary differential equations for the population densities $n(L_i,t)$ in the grid points L_i:

$$V \frac{dn(L_i,t)}{dt} = \frac{-V}{12\Delta L} \left[G(L_{i-2},t)n(L_{i-2},t)-8G(L_{i-1},t)n(L_{i-1},t)\right.$$
$$\left. +8G(L_{i+1},t)n(L_{i+1},t)-G(L_{i+2},t)n(L_{i+2},t)\right]$$
$$- Q_p(t)h_p(L_i,t)n(L_i,t) - Q_f(t)h_f(L_i,t)n(L_i,t) \quad (15)$$

The central difference Equation 14 can not be used in the grid points L_0, L_1, L_{K-1} and L_K at the boundaries of the grid, as the points L_{-2}, L_{-1}, L_{K+1} and L_{K+2} do not exist. As a rule of thumb, the order of accuracy of the finite differences used at the boundaries must be the same as in the middle of the grid. In the boundary points we therefore use (15):

$$(\frac{df(x)}{dx})_{x_0} = \frac{1}{12\Delta x} \left(-25f_0+48f_1-36f_2+16f_3-3f_4\right) +O(\Delta x^4)$$

$$(\frac{df(x)}{dx})_{x_1} = \frac{1}{12\Delta x} \left(-3f_0-10f_1+18f_2-6f_3+f_4\right) + O(\Delta x^4)$$

$$(\frac{df(x)}{dx})_{x_{K-1}} = \frac{1}{12\Delta x} \left(-f_{K-4}+6f_{K-3}-18f_{K-2}+10f_{K-1}+3f_K\right) + O(\Delta x^4) \quad (16)$$

$$(\frac{df(x)}{dx})_{x_K} = \frac{1}{12\Delta x} \left(3f_{K-4}-16f_{K-3}+36f_{K-2}-48f_{K-1}+25f_K\right) + O(\Delta x^4)$$

Discretizing the population balance in K+1 grid points results in K ordinary differential equations as the population density $n(L_0,t)$ is determined by the algebraic relation Equation 4. The equation for $dn(L_0,t)/dt$ is therefore not required. In the overal model of the crystallizer, the population density must be integrated in the calculation of $\varepsilon(t)$ and in the calculation of $m_x(t)$ in the nucleation rate. As the population balance is discretized, these integrals have to be replaced by numerical integration schemes.

The method of lines can handle size-dependent growth rates, fines removal and product classification and is not restricted in the choice of the elements of the output vector $\underline{y}(t)$. The population densities at the grid points are system states, thus moments, L_{50}, CV, population densities at the grid points and the number or mass of crystals in a size range can be elements of $\underline{y}(t)$.

A limitation in the application of the method of lines is that (as will be demonstrated in the example given below) the number of grid points required to describe the dynamics of the population balance and its moments with sufficient accuracy, is high. Especially when the resulting state space model is used for control system design, high model orders are a serious problem. The accuracy problem is connected with the width of the grid and the number of grid points. Because the steady state population density is of the form a.exp(-bL), a wide grid results in large errors in the lower moments, whereas the position of the last grid point determines the error in the higher moments. In order to solve this problem, a non-equidistant grid with a grid width that increases with the crystal size can be used. The reduction of the required number of grid points for a certain level of accuracy is however limited, because the width at the larger sizes determines the accuracy of the higher moments and thereby the accuracy of the mass-based mean crystal size and the coefficient of variation, and still unacceptable high orders of the state space model are obtained.

System Identification Techniques. In system identification, the (nonlinear) responses of the outputs of a system to the input signals are approximated by a linear model. The parameters in this linear model are determined by minimizing a criterion function that is based on some difference between the input-output data and the responses predicted by the model. Several model structures can be chosen and depending on this structure, different criteria can be used (16,17). System identification is mainly used as a technique to determine models from measured input-output data of processes, but can also be used to determine compact models for complex physical models. The input-output data is then obtained from simulations of the physical model.

Several identification methods result in a state space model, either by direct identification in the state space structure or by identification in a structure that can be transformed into a state space model. In system identification, discrete-time models are used. The discrete-time state-space model is given by

$$\begin{aligned} \underline{x}_{k+1} &= A_d \, \underline{x}_k + B_d \, \underline{u}_k \\ \underline{y}_k &= C_d \, \underline{x}_k + D_d \, \underline{u}_k \end{aligned} \qquad (17)$$

where $\underline{x}_k = \underline{x}(k\Delta T)$ and ΔT is the sample interval. The transformation between the discrete-time description and an equivalent continuous-time model and visa versa is straight forward and described in for example (18).

In the current investigation, a two step identification procedure is used. In the first step, an ARX model, i.e. a model of the form (16)

$$H(z^{-1})\underline{y}_k = F(z^{-1})\underline{u}_k \qquad (18)$$

where

$$H(z^{-1}) = I_\ell + H_1 z^{-1} + H_2 z^{-2} + \ldots + H_p z^{-p} \quad \text{dimension } (\ell \times \ell)$$

$$F(z^{-1}) = F_0 + F_1 z^{-1} + F_2 z^{-2} + \ldots + F_q z^{-q} \quad \text{dimension } (\ell \times m)$$

I_ℓ is the identity matrix and z^{-v} is defined by $z^{-v}\underline{y}_k = \underline{y}((k-v)\Delta T)$, is determined. In the second step, this model is transformed into a discrete-time state space model. This is achieved by making an approximate realization of the markov parameters (the impulse responses) of the ARX model (19). The order of the state space model is determined by an evaluation of the singular values of the Hankel matrix (19).

In using system identification to determine state space models for crystallizers, moments, L_{50}, CV, population densities at sizes L_i and the number or mass of crystals in a size range $[L_i, L_{i+1}]$ can be chosen as outputs. An advantage of using system identification is that identification of a simulation of the nonlinear model results directly in a linear state space model. Transformation of the population balance to ordinary differential equations, substitution of algebraic relations and separate linearization (including the concentration balance and the energy balance) is not required.

Discussion

Three methods to derive a state space model for a crystallizer were discussed. The choice of a method in a specific situation depends on the crystallizer configuration, the growth rate kinetics and the variables to be controlled.

A special point of interest is the order of the resulting state space model. This order is quite different for the three methods. For MSMPR crystallizers with simple size-dependent growth rate kinetics the method of moments is preferred if moments, L_{50} or CV are to be controlled, because it is expected to result in the lowest model order. In the presence of complicated size-dependent growth rate kinetics, fines removal or product classification, the method of lines or system identification is to be used. The lowest model order will be obtained in using system identification. A disadvantage of system identification is that the relation between the physical parameters and the parameters in the state space model is not transparent. This can be a disadvantage, especially in case of controllability and observability studies for changing operating conditions or different kinetic parameters. Furthermore, when the operating conditions are drastically changed, the identification procedure has to be repeated whereas in the other methods the linearization step around the operating point has to be repeated only.

An advantage of system identification is that it can directly be applied on experimental data. In the two other methods, knowledge about the kinetic parameters is required. To substract these parameters from dynamic data requires complicated analysis, see for example (De Wolf, S.; Ph.D. Thesis, Delft University of Technology, in preparation). Furthermore, the theoretical model is based on assumptions with respect to growth rate dispersion, the equation for the nucleation kinetics, etc., and modelling errors, as for example neglecting attrition and agglomeration, will influence the accuracy of the theoretical model. In applying system identification on experimental data, these assumptions and the kinetic parameters are not required because the method considers the measured process inputs and outputs only.

The use of system identification and the method of lines will now be illustrated in an example.

Example. Consider an evaporative, isothermal, Class II crystallizer with fines removal. It is assumed that the growth rate is size-independent and that there is no growth rate dispersion. Because we assume fines removal, the method of moments can not be applied.

The nucleation rate is described by

$$B(t) = K_B \, G(t)^i m_3(t)^j \tag{19}$$

with $K_B = 7.30 \, 10^{25}$, $i=2.53$ and $j=1.80$. All densities, specific heats, etc. are chosen for the ammonium sulfate water system operating at $50°C$ with a saturated feed of $60°C$. The crystallizer volume is $.97$ m^3 and the fines classification function is defined as

$$h_f(L) = \begin{cases} 1 & L \leq 100.10^{-6} \, m \\ 0 & L > 100.10^{-6} \, m \end{cases} \tag{20}$$

The process inputs are defined as the heat input, the product flow rate and the fines flow rate. The steady state operating point is $P_h = 120$ kW, $Q_p = .215$ l/s and $Q_f = .8$ l/s. The process outputs are defined as the third moment $m_3(t)$, the (mass based) mean crystal size $L_{50}(t)$ and the relative volume of crystals $vr_f(t)$ in the size range $0.-10^{-4}$ m. In determining the responses of the nonlinear model the method of lines is chosen to transform the partial differential equation in a set of (nonlinear) ordinary differential equations. The time responses are then obtained by using a standard numerical integration technique for sets of coupled ordinary differential equations. It was found that discretization of the population balance with 1001 grid points in the size range 0. to 5.10^{-3}m results in very accurate solutions of the crystallizer model.

To determine the state space model by the method of lines, first the number of grid points required for an accurate state space model is determined. This is achieved by simulating the nonlinear set of ordinary differential equations (Equation 15) with a varying number of grid points (the nonlinear equations are used as the linearization will introduce extra errors). The responses of L_{50} obtained with 51, 101, 201, 501 and 1001 equidistant grid points in the size range 0. to 5.10^{-3}m to a step of $.4$ l/s on Q_f at $t=1$ hour are given in Figure 1. As can be seen, the absolute values of the responses determined with different numbers of grid points vary. However, as a linear state space model only describes the dynamics of the outputs around a steady state, it is allowed to correct for these differences by substracting the initial steady states; see Figure 2 for L_{50} and Figure 3 for the response of vr_f to a step of +20 kW on P_h at $t=1$ hour. From these figures, it is found that at least 501 grid points are required for an accurate solution as otherwise the number of grid points in the small size range (the range where the fines are removed) is too low. However, due to computer memory limitations, the calculation of the state space model was limited to 201 grid points. With these 201 grid points, the order of the resulting state space model is 200.

To determine the state space model with system identification, responses of the nonlinear model to positive and negative steps on the inputs as depicted in Figure 4 were used. Amplitudes were 20 kW for P_h, $.4$ l/s for Q_f and $.035$ l/s for Q_p. The sample interval for the discrete-time model was chosen to be 18 minutes. The software described in (20,21) was used for the estimation of the ARX model, the singular value analysis and the estimation of the approximate

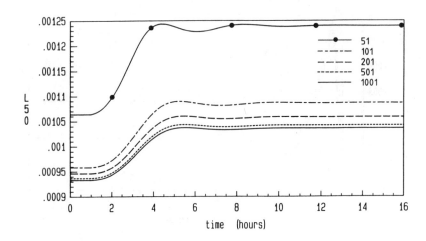

Figure 1: Response of L_{50} to a step of $+.4$ 1/s on Q_f obtained with the nonlinear model with 51, 101, 201, 501 and 1001 grid points.

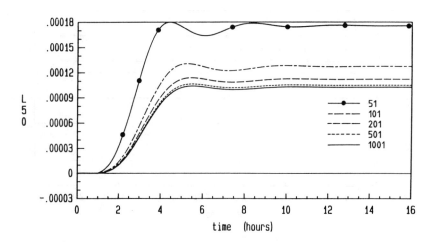

Figure 2: Responses as in Figure 1 with initial steady states subtracted.

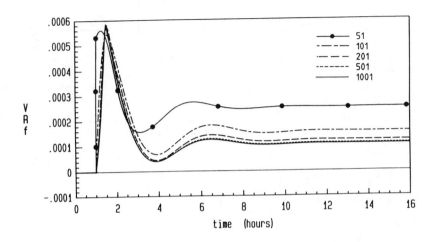

Figure 3: Response of vr_f to a step of +20 kW on P_h obtained with the nonlinear model with 51, 101, 201, 501 and 1001 grid points (initial steady states subtracted).

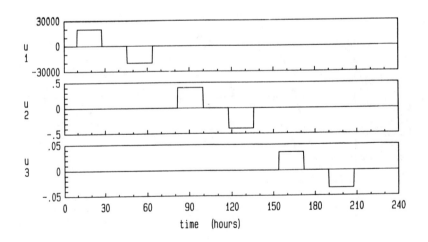

Figure 4: Inputs as used for system identification.

realization. Several orders (values of p and q in Equation 18) of H and F in the ARX model were investigated. The results with p and q equal to 30 were superior. From the singular value analysis in the second step of the identification procedure, the order of the state space model was chosen to be 11.

Step responses of the state space model obtained by the method of lines with 201 grid points and by system identification are compared with the nonlinear responses obtained with 1001 grid points in Figure 5 to 7. From the nonlinear model, both the responses to a positive step and a negative step multiplied by -1, are given to illustrate the nonlinearity. It is expected that the linear models will have an intermediate response. From Figure 5 and 6, it appears that system identification results in more accurate solutions. It is expected, however, that the results of the method of lines will improve for an increasing number of grid points (i.e. order of the state space model).
□

Conclusions

The problem in deriving a state space model for crystallizers is that the equation describing the dynamics of the CSD (the population balance) is a partial differential equation. To obtain a state space model, this equation has to be transformed into a set of ordinary differential equations. Three methods to transform a population

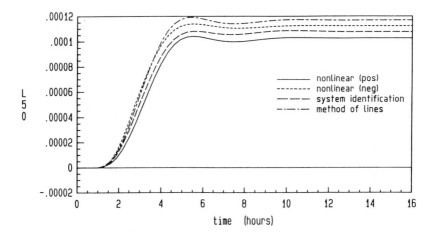

Figure 5: Response of L_{50} to a step of $+.4$ l/s on Q_f obtained with the nonlinear model, system identification and the method of lines (201 grid points).

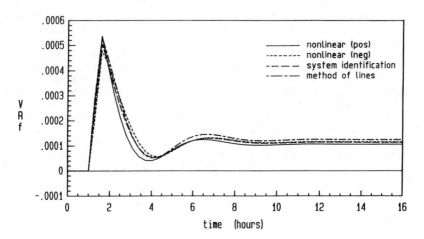

Figure 6: Response of vr_f to a step of +20 kW on P_h obtained with the nonlinear model, system identification and method of lines (201 grid points).

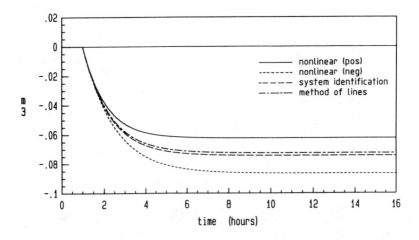

Figure 7: Response of m_3 to a step of +.035 l/s on Q_p obtained with the nonlinear model, system identification and the method of lines (201 grid points).

balance into a set of ordinary differential equations were discussed.

The first method, the method of moments, is restricted to MSMPR crystallizers with size-independent growth or very simple size-dependent growth rate kinetics. Depending on the control demand, several process outputs can be chosen for the control algorithm. When population densities, or the number or mass of crystals in a size range are to be controlled, the method of moments can not be used because it reveals information on the dynamics of the moments of the crystal size distribution only.

The method of lines and system identification are not restricted in their applicability. System identification is preferred because the order of the resulting state space model is significantly lower. Another advantage of system identification is that it can directly be applied on experimental data without complicated analysis to determine the kinetic parameters. Furthermore, no model assumptions are required with respect to the form of the kinetic expressions, attrition, agglomeration, the occurence of growth rate dispersion, etc.

Acknowledgments

The authors would like to express their gratitude to the Netherlands Technology Foundation (STW), AKZO, DOW, DSM, Dupont de Nemours, Rhone Poulenc, Suiker Unie and Unilever for their financial support of the research program.

Literature Cited

1. Randolph, A.D.; Beckman, J.R.; Kraljevich, Z.I. AIChE J., 1977, 23, 500-510.
2. Jager, J.; De Wolf, S.; De Jong, E.J.; Klapwijk, W. In Industrial Crystallization 87; Nyvlt, J.; Zacek, S., Ed.; Elsevier: Amsterdam, 1989; p 415-418.
3. Randolph, A.D.; Chen, L.; Tavana, A. AIChE J., 1987, 33, 583-591.
4. Rohani, S.; Lee, K.K. Proc. Int. Symp. on Crystallization and Precipitation, Saskatchewan, Canada, 5-7 October 1987; p 11-20.
5. De Wolf, S.; Jager, J.; Kramer, H.J.M.; Eek, R.; Bosgra, O.H. Proc. IFAC Symp. on Dynamics and Control of Chemical Reactors, Distillation Columns and Batch Processes, Maastricht, The Netherlands, 1989.
6. Juzaszek, P.; Kawecki, W. 5th Symp. on Industrial Crystallization, Czechoslovakia, 1972.
7. Pohlisch, J.; Mersmann, A. Chem. Engng. Technol., 1988, 11, 40-49.
8. Ramkrishna, D. Chem. Engng. Sci., 1971, 26, 1134-1136.
9. Wiskowski, W.R.; Rawlings, J.B. Proc. 1987 American Control Conference, Minneapolis, 1987; p 1400-1405.
10. Hulburt, H.M.; Katz, S. Chem. Engng. Sci., 1964, 19, 555-574.
11. Randolph, A.D.; Larson, M.A. Theory of particulate processes; analysis and techniques of continuous crystallization; Academic Press: New York, 1971.

12. Vemuri, V.; Karplus, W.J. Digital computer treatment of partial differential equations; Prentice-Hall: Englewood Cliffs, New Jersey, 1981.
13. Cooper, D.J.; Clough, D.E. AIChE J., 1985, 31, 1202-1212.
14. Tsuruoka, S. Ph.D. Thesis, University of Arizona, Dep. of Chem. Engng., Arizona, 1986.
15. Perry, R.H.; Green, D.W.; Maloney, J.O. Perry's chemical engineerings' handbook; McGraw-Hill Book Co: New York, 1984, sixth edition.
16. Ljung, L. System identification, theory for the user; Prentice-Hall: Englewood Cliffs, New Jersey, 1987.
17. Söderström, T.; Stoica, P. System identification; Prentice Hall: New York, 1989.
18. Ogata, O. Discrete-time control systems; Prentice-Hall: Englewood Cliffs, New Jersey, 1987.
19. Damen, A.A.H.; Hajdasinski, A.K. Proc. 6th IFAC Symp. Identification and system parameter estimation, Washington DC, 1982; p 903-908.
20. Aling, H. RPEPAC User's manual; Report N-304, Laboratory for Measurement and Control, Delft University of Technology, Delft, 1989.
21. Van Den Hof, P.J.M. DUMSI-package for off line multivariable system identification; Laboratory for Measurement and Control, Delft University of Technology, Delft, 1989.

RECEIVED May 12, 1990

Chapter 12

Simulation of the Dynamic Behavior of Continuous Crystallizers

Herman J. M. Kramer[1], Sjoerd de Wolf[2], and Johan Jager[1]

[1]Laboratory for Process Equipment and [2]Laboratory for Measurement and Control, Delft University of Technology, Mekelweg 2, Delft, Netherlands

The simulation of a continuous, evaporative, crystallizer is described. Four methods to solve the nonlinear partial differential equation which describes the population dynamics, are compared with respect to their applicability, accuracy, efficiency and robustness. The method of lines transforms the partial differential equation into a set of ordinary differential equations. The Lax-Wendroff technique uses a finite difference approximation, to estimate both the derivative with respect to time and size. The remaining two are based on the method of characteristics. It can be concluded that the method of characteristics with a fixed time grid, the Lax-Wendroff technique and the transformation method, give satisfactory results in most of the applications. However, each of the methods has its own particular draw-back. The relevance of the major problems encountered are dicussed and it is concluded that the best method to be used depends very much on the application.

The observed transients of the crystal size distribution (CSD) of industrial crystallizers are either caused by process disturbances or by instabilities in the crystallization process itself (1). Due to the introduction of an on-line CSD measurement technique (2), the control of CSD's in crystallization processes comes into sight. Another requirement to reach this goal is a dynamic model for the CSD in industrial crystallizers. The dynamic model for a continuous crystallization process consists of a nonlinear partial difference equation coupled to one or two ordinary differential equations(3,4) and is completed by a set of algebraic relations for the growth and nucleation kinetics. The kinetic relations are empirical and contain a number of parameters which have to be estimated from the experimental data. Simulation of the experimental data in combination with a nonlinear parameter estimation is a powerfull technique to determine the kinetic parameters from the experimental

0097–6156/90/0438–0159$06.00/0

data under instationary conditions (5). Experimental data from our
crystallizer, indicate that for a correct description of the CSD of
the crystallizer, a simple model with size independent crystal
growth is not satisfactory and that other phenomena, like growth
dipersion and size dependent growth or crystal attrition have to be
incorporated into the model. Implementation of these phenomena in
the simulation however increases the required CPU time for
simulation, and more parameters are to be estimated. Nonlinear
multiparameter optimization techniques combined with simulation
requires a large amount of computing power. It is therefore
important to choose the optimal simulation algorithm for the
parameter estimation program. In this paper a comparison between
four algorithms, which have been implemented and tested in our
simulation programs is presented.
    Industrial crystallizers are often equiped with an annular
zone for fines removal. This fines removal system gives rise to a
mass accumulation in the large annular zone (4), which affects the
process dynamics at least under unsteady conditions. It is shown,
that implementation of this fines destruction system into the
simulation program has some implications for the algorithm used. The
influence of the mass accumulation in the annular zone on the
process dynamics is also discussed.

The Model

In this section the model for a continuous evaporative crystallizer
is discussed. The crystallizer is of the draft tube baffled (DTB)
type and is equiped with a fines removal system consisting of a
large annular zone on the outside of the crystallizer (see Figure
1). In order to vary the dissolved fines flow without changing the
cut-size of the fines removal system, the flow in the annular zone
is kept constant and the flow in the dissolving system is varied by
changing the recycle flow rate. The model assumptions are:
- The crystallizer is well mixed.
- The crystallizer has a constant temperature and volume.
- Only secondary nucleation takes place at size L=0.
- The feed flow is crystal free.
- All fines are dissolved in the dissolved fines flow.
- The fines flow in the annular zone is a plug flow and the
  classification of the fines flow is immediate at the entrance of
  the annular zone.
- The removal of the fines and the product is size dependent and can
  be described by the product of a size dependent classifcation
  function h(L) and the population density n(L,t) in the
  crystallizer.
The model for this crystallizer configuration has been shown to
consist of the well known population balance (4), coupled with an
ordinary differential equation, the concentration balance, and a set
of algebraic equations for the vapour flow rate, the growth and
nucleation kinetics (4). The population balance is a first-order
hyperbolic partial differential equation:

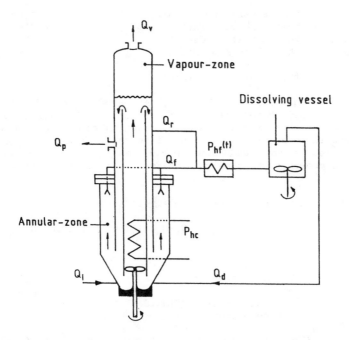

Figure 1. Schematic picture of the continuous, evaporative crystallizer with fines removal system.

$$\frac{\partial n(L,t)}{\partial t} + \frac{\partial (G(L,t)n(L,t))}{\partial L} + (\frac{Q_p(t)h_p(L)+Q_fh_f(L)}{V}) \, n(L,t) -$$

$$\frac{Q_r(t).h_f(L)}{V} \, n(L,t-\tau) \; = 0 \tag{1}$$

with the boundary condition

$$n(0,t) = B(t)/G(0,t) \tag{2}$$

with G   the crystal growth rate,   [m/s]
      B   the nucleation rate,   [#/(m³ s)]
      n   the population density,   [#/m⁴]
      Q   the volume flow rate,   [m³/s]
      h   the classification function,
      V   the crystallizer volume.   [m³]

The indices refer to the fines (f), the product (p) and the fines recycle (r) flows. Note that part of the removed fines flow is fed back in the crystallizer with a delay τ, which depends on the volume of the annular zone and on the fines flow rate. The nucleation rate is detemined by the well known power law (3,4). For a class II system the growth rate is calculated from the concentration balance (4).

$$_0\int^{\infty} G(L,t)n(L,t)L^2\,dL = [Q_i(t)C_i(t)+Q_d(t)C_d(t)+Q_r(t)C_r(t) -$$

$$C\{Q_p(t)+Q_f-Q_r(t)(1-\varepsilon_r(t))\}]/(\rho_c-C)3k_vV \tag{3}$$

with C   the concentration,   [kg/m³]
      ρ   the density,   [kg/m³]
      ε   the fraction clear liquid of the slurry,
      k_v a shape factor.

The indices i,d,c refer to the feedflow, the dissolved fines flow and crystal substance respectively. The energy balance, then gives a relation for the vapour flow rate. The solution of the population balance is strongly affected by growth rate function used in the model. It has been shown that size dependent growth rate is important to describe the diffusion limited crystal growth. Moreover it has been shown that the size reduction by attrition can also be described as a negative size dependent growth rate (6). For the simulations described here two types of size dependent growth rates are used:

$$G_k(L,t) = G(0,t)(1 + aL) \tag{4}$$

for simple size dependent crystal growth and

$$G(L,t) = G_k(L,t) + G_a(L,m_3,t) \tag{5}$$

to describe the crystal attrition.

Simulation Algorithms.

Simulation of the dynamic behaviour of industrial crystallizers is mainly concerned with the solution of the partial differential equation, which is solved either directly or by a transformation method. Four algorithms have been incorporated in our simulation program. Here a comparison will be given of the different algorithms in terms of:
- applicabillity
- efficiency with respect to calculation time
- accuracy
- robustness
First the different methods will be discussed briefly.

Method of lines. This method, which was introduced in the crystallization research by Tsuruoka and Randolph (7), transforms the population balance into a set of ordinary differential equations by discretization of the size axis in a fixed number of grid points. The differential $\partial\{G(L,t)n(L,t)\}/\partial L$ is then approximated by a finite difference scheme. As has been shown (7) a fourth-order accurate scheme using an equidistant grid (i= 1....z) results in:

$$\left.\frac{\partial F}{\partial L}\right|_{L_i} = (F_{i-2} -8F_{i-1} +8F_{i+1} -F_{i+2})/12\Delta L + O(\Delta L)^4 \qquad (6)$$

where $F_i= G(L_i,t)n(L_i,t)$. At the boundaries this central difference Equation 6 cannot be used because the grid points L_{-1},L_0, L_{z+1} and L_{z+2} do not exist. A five point Lagrange interpolation is used to estimate these grid points, leading to the following equations:
i=2,

$$\left.\frac{\partial F}{\partial L}\right|_{L_i} = (-3F^0 -10F_i +18F_{i+1} -6F_{i+2} +F_{i+3}) / 12\Delta L \qquad (7)$$

i=z-1,

$$\left.\frac{\partial F}{\partial L}\right|_{L_i} = (-F_{i-3} +6F_{i-2} -18F_{i-1} +10F_i +3F_{i+1}) / 12\Delta L \qquad (8)$$

i=z,

$$\left.\frac{\partial F}{\partial L}\right|_{L_i} = (+3F_{i-4} -16F_{i-3} +36F_{i-2} -48F_{i-1} +25F_i) / 12\Delta L \qquad (9)$$

with $F^0 = G(0,t)n(0,t)$. As the population density at L=0 is determined by the boundary condition, the scheme leads to a set of z-1 ordinary differential equations, which are solved using a Runge Kutta algorithm. Due to the fixed simulation grid and the numerical approach, this scheme is general applicable in the sence, that there are no limitations in the crystallizer configuration or the crystal growth rate function. However, the method is sensitive to discontinuities in Gn as is shown in the next section. To examine these oscillations in more detail, also lower order versions of this scheme have been implemented. The second-order version is given by:

$$\left.\frac{\partial F}{\partial L}\right|_{L_i} = (- F^0 + F_{i+1}) / 2\Delta L \qquad i = 2 \qquad (10)$$

$$\left.\frac{\partial F}{\partial L}\right|_{L_i} = (- F_{i-1} + F_{i+1}) / 2\Delta L \qquad i = 3,..z-1 \qquad (11)$$

$$\frac{\partial F}{\partial L}\bigg|_{L_i} = (- F_{i-2} - 4F_{i-1} + 3F_i) / 2\Delta L \qquad i = z \qquad (12)$$

and the first-order

$$\frac{\partial F}{\partial L}\bigg|_{L_i} = (- F^0 + F_i) / \Delta L \qquad i = 2 \qquad (13)$$

$$\frac{\partial F}{\partial L}\bigg|_{L_i} = (- F_{i-1} + F_i) / \Delta L \qquad i = 3,..z \qquad (14)$$

Lax-Wendroff. This is a well known method to solve first-order hyperbolic partial differential equations in boundary value problems. The two step Richtmeyer implementation of the explicit Lax-Wendroff differential scheme is used (8).

$$n_{i+.5,t+.5} = .5(n_{i,t} + n_{i+1,t}) - \frac{\Delta t}{2\Delta L}(F_{i+1,t} - F_{i,t}) \qquad (15)$$

$$n_{i,t+1} = n_{i,t} - \frac{\Delta t}{\Delta L}(F_{i+.5,t+.5} - F_{i-.5,t+.5}) \qquad (16)$$

This implementation is second-order accurate with respect to the time and the size step. The scheme is general applicable and as shown in the next section, this scheme is also sensitive for discontinuities in Gn as caused by the R-Z model for fines removal. The oscillations are however less severe than for the method of lines. Also for this method a first-order scheme was implemented. Here the so-called Lax scheme was chosen (8):

$$n_{j,t+1} = \frac{1}{2}(n_{j-1,t} + n_{j+1,t}) + \frac{\Delta t}{2\Delta L}(F_{j+1,t} - F_{j-1,t}) \qquad (17)$$

Method of Characteristics. A different approach is used in the method of characteristics, which was introduced by De Leer (9) as the fraction trajectory concept. In this method, the convective terms in the population balance are removed by a separate treatment of the crystal growth and the number affecting events. This is realised by following the number of crystals in a size interval during their growth. The mean size of the crystals in this interval depends only on their age and on the growth rate during their lifespan. The number affecting events are described by an ordinary differential equation for each of the size intervals. The timestep in the simulation and the width of the first size class are coupled to the mean growth rate of the size class by the following equation:

$$\Delta t(t) = 2(L_2(t) - L_1(t)) / (G(L_1,t)+G(L_2,t)) \qquad (18)$$

where L_1 and L_2 are the boundaries of the smallest size class. The coupling implicates that there is only one degree of freedom in the choice of the simulation grid. Fixation of the timestep forces a variable size grid, while fixation of the size grid results in a variable time step. Both versions of the method of characteristics were implemented.

Method of Characteristics with Fixed Size grid (MCFS). In this implementation, the size grid is fixed in the beginning of the simulation. The timestep is variable and is determined by Equation 18. The method is a simplified version of the fraction trajectory concept, in which provisions were taken for grid refinement when the timestep becomes too large, due to a decrease in the growth rate, which can be caused by a process disturbance (9). The scheme is very simple and due to the fixed size grid during the simulation, the technique can be used to simulate the affects of the mass accumulation in the fines removal system on the crystallization dynamics. However, more complicated size dependent growth rate functions like crystal attrition cannot be handled by this technique because the crystal attrition introduces a term in the growth rate, which is not only dependent on the size but also on the $m_3(t)$. Thus the size dependency of the grid must be variable in the time.

Method of Characteristics with Fixed Time grid (MCFT). This approach is similar to that of the MCFS, except that in this method the timestep of the simulation is fixed. Therefore the position of the size intervals along the size axis varies during the simulation as a function of the growth rate and are recalculated after each timestep. The scheme is still relative simple. Yet it is capable to handle al sorts of size dependencies in the growth rate function including attrition. On the other hand, the coupling of the size grid with the growth rate, makes this method less appropriate for systems with fines recycle, because the position of the size intervals from the crystallizer volume and from the delayed recycle flow do not match.

Results and Discussion.

In this section, first the simulation results are discussed of a simple evaporative, class II crysallizer without fines removal using a length independent growth rate. The cystallizing substance was ammonium sulphate and the number of gridpoints along the size axis was 1000. The used parameters for the nucleation kinetics ars given in the following equation.

$$B(t) = .73 \ 10^{26} \ G_k(t)^{2.533} \ m_3(t)^{1.8} \tag{19}$$

In Table I the accuracy under stationary conditions of the four simulation algorithms in a simulation of the MSMPR crystallizer are compared. The method of lines algorithm appears to be the most accurate of the four schemes. However all the schemes used show an acceptable low deviation from the analytical solution. Table II shows, that the calculation time needed for a typical simulation is for three of the four algorithms comparable. The calculation time needed for the method of lines algorithm however, is much higher. In Figure 2 the respons of G is given on a negative step in the heat input of the crystallizer for the Lax-Wendroff (line) and the MCFS (dots) calculated with 200 gridpoints. The respons of the Lax-Wendroff simulation is correct and almost identical to the respons found with the method of moments (3). As shown in Figure 2, large dynamic errors can occur using the MCFS scheme due to the direct

Table I. Stationary values of Growth rate (G) and second moment (m_2) and the deviation (s) from the analytical solution in percentages for the MSMPR crystallizer. The theoretical values were 0.400742×10^{-7} and 712.222 for G and m_2, respectively

	G 1.E-7	s (%)	m_2	s(%)
Method of Lines	.400741	2.4E-4	712.232	1.4E-3
Lax-Wendroff	.400751	2.2e-3	712.169	7.4E-3
MCFS	.400738	1.1E-3	712.235	1.8E-3
MCFT	.400738	1.1E-3	712.235	1.8E-3

Table II. Relative calculation times for a typical simulation and applicability for the different algorithms

Simulation Algorithm	Time (arbitrary units)	Applicabillity
Method of Lines	7.	General
Lax- Wendroff	1.	General
MCFS	.6	no attrition
MCFT	.9	no fines delay

coupling of the timestep of the simulation to the growth rate. These errors can be reduced by an increase of the number of grid points. The calculation time is however a quadratic function of the number of grid points used.

In the rest of this section simulations will be discussed of a similar crystallizer, now equipped with a fines removal system. The well known R-Z fines removal function (3) has been used:

$$h_f = 1 \qquad L \leq L_f$$
$$\quad = 0 \qquad L > L_f \qquad\qquad (20)$$

This removal function gives rise to a discontinuity in the population density at the cutsize of the fines. The nucleation parameters are given in equation 19. In Figure 3 the responses are shown of the population density at 120 μm and of the growth rate after a step in the heat input to the crystallizer from 120 to 170 kW for three simulation algorithms. The cut-size of the fines L_f was 100 μm, a size dependent growth rate was used as described by Equation 4 with a= -250 and the number of grid points was 400. When the simulation was performed with the method of lines, severe oscillations are present in the response of the population density at 120 μm, which dampen out rather slowly. Also the response of the Lax-Wendroff method shows these oscillations to a lesser extend.

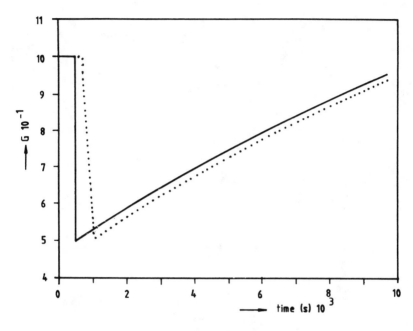

Figure 2. Response of the growth rate G on a negative step in the heat input of the crystallizer simulated with the Lax–Wendroff (line) and the MCFS (dots) algorithm.

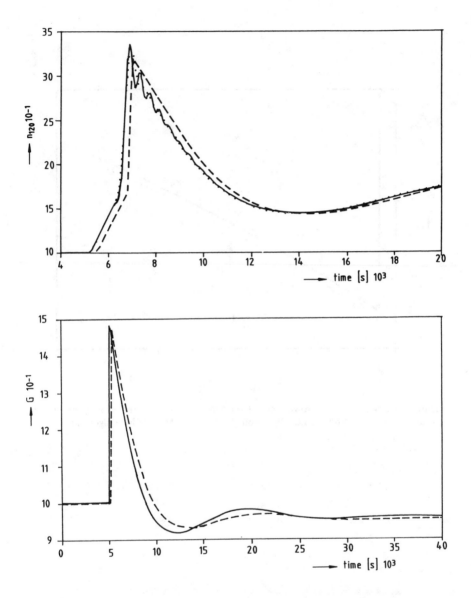

Figure 3. Response of the population density at 120 μm (top) and of the crystal growth rate (bottom) on a step in the heat input of the crystallizer simulated with method of lines (line), Lax–Wendroff (dots), and the MCFS (dashes) algorithm.

This unstable behaviour is not seen in the responses of the integrated process variables like the moments and the growth rate (see Figure 3B). In the responses of the two implementations of the methods of characteristics, no unstable behaviour was found. The response of the MCFS, however again shows the dynamic error, which results from the variable timestep. The delay in the responses could almost completely be removed by a doubling of the number of grid points (not shown). To investigate whether the oscillations in the response of the population density after a step in the heat input are caused by the order of accuracy of the algorithm, simulations were performed using lower order implementations of the method of lines and of the Lax-Wendroff algorithm. The conditions were similar to those in Figure 3. The results are given in Figure 4 and 5, while in Table III the stationary values of m_0 and the growth rate before and 35000 seconds after the step in the heat input are given. The results indicate that there is a clear relationship between the intensity of the oscillations and the order of the algorithm. In the response of the first-order method of lines simulations no oscillations are present. However the response (Fig. 4) and the stationary values (Table III) deviate somewhat from the higher order solution, which cannot be compensated completely by an increase in the number of grid points used, as is shown in Figure 6. The first-order Lax scheme gives too high deviations and is not reliable.   The stationary values as given in Table III give a much higher spread than for the MSMPR crystallizer, which is due to a combination of the discontinuous fines removal function and the difference schemes used in the different methods. The relevance of the oscillations seen in the method of lines and the Lax-Wendroff techniques are not easy to describe. In most cases one is not interested in the population density at a single crystal size, but in integrated proces variables like the number of crystals in a certain size range, the moments or the crystal growth rate, in which these oscillations are not present. On the other hand, the oscillations will affect the accuracy of these responses. The lower order schemes of the method of lines form a reasonble alternative to decrease the effect of the unstable behaviour. The two methods of characteristics do not show this unstable behaviour. However, the MCFS technique suffers from large dynamic errors and is limitted in its application. The MCFT seems to form a good alternative.

As discussed in the previous section and summarized in Table II, a drawback of the MCFT method is that the mass accumulation in the fines removal system cannot be simulated. therefore we examined whether this mass accumulation has a noticable effect on the process dynamics. In the simulation the fines removal is simulated with a cut sizes of 150 μm. The fines flow rate Q_f and the recycle flow rate Q_r were 1.25 and .75 liter per second. The results are shown in Figure 7. It is clear that the mass accumulation has indeed an effect on the process dynamics. Even on the mean size of the crystals a clear shift in the response is seen. It appears that the effect is strongly dependent on the value of the recycle flow rate (not shown). The conclusion from these results is that the effects of mass accumulation in the fines system are present, and can only be neglected at low cutsizes and low fines recycle rates.

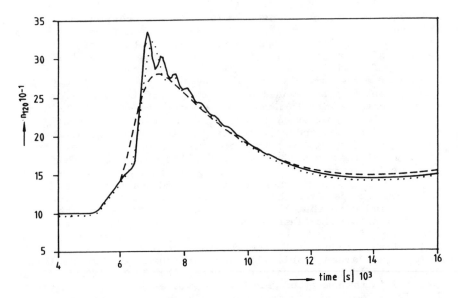

Figure 4. Response of the population density at 120 μm on a step in the heat input of the crystallizer simulated with method of lines, fourth-order (line), second-order (dots), and first-order (dashes).

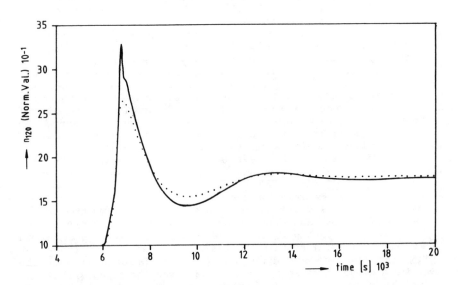

Figure 5. Response of the population density at 120 μm on a step in the heat input of the crystallizer simulated with Lax–Wendroff, second-order (line), and the first-order Lax scheme.

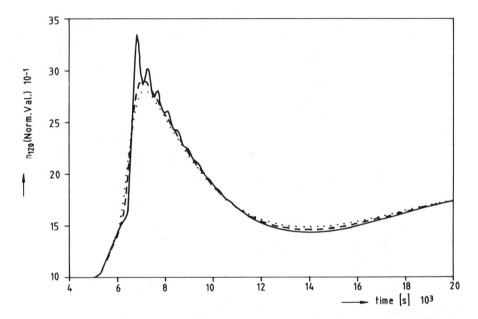

Figure 6. Effect of the number of grid points on the response of the population density at 120 μm on a step in heat input; fourth-order (line), first-order 400 grid points (dots), and first-order 800 grid points (dashes).

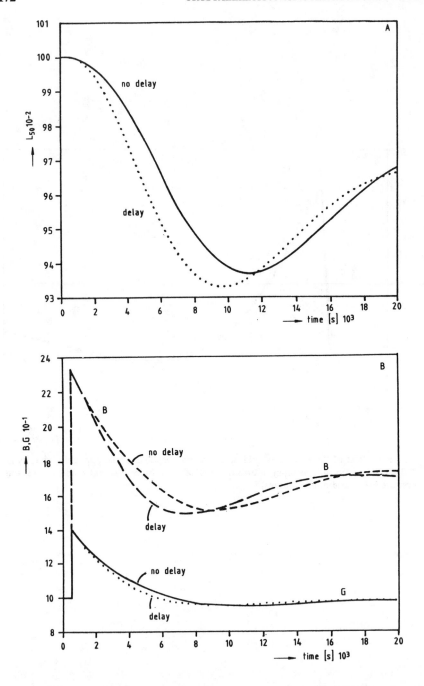

Figure 7. Effect of the mass accumulation on the response of the L_{50} (A) and on that of the growth rate G and the nucleation rate B (B) on a step in the heat input of the crystallizer. The cutsize of the fines was 150 μm.

Table III. Stationary values of G and m_2 for a DTB crystallizer operated with a fines removal system and a size dependent growth regime. The values are given before and 35000 second after a step in the heat input of the crystallizer from 120 to 170 kW. The simulation has been performed with 400 gridpoints

	Order	before step G *1.E-7	before step m0 *1.E10	35000 s after step G *1.E-7	35000 s after step m0 *1.E10
Method of Lines	4	.5620	.8416	.5363	1.491
" "	2	.5624	.8381	.5364	1.485
" "	1	.5688	.8531	.5226	1.511
Lax-Wendroff	2	.5628	.8370	.5369	1.484
" "	1	.5332	.8740	.5055	1.552
MCFS	1	.5587	.8368	.5328	1.482
MCFT	1	.5582	.8361	.5294	1.478

Conclusions.

Four simulation algorithms have been discussed in terms of applicability, accuracy, calculation time and robustness. Two of the methods described are based on numerical methods, while the other two are based on the method of characteristics. To simulate a simple MSMPR crystallizer, all four algorithms give reasonable results. The methods were also applied to more complex crystallizer configurations and on different growth rate kinetics. Results show that the best method to be used depends on the application, while all the methods studied are unsatisfactory in one or more aspects studied. The most important drawbacks of the method of lines algorithm are its long calculation time and the unstable behaviour in the response of population density in the presence of discontinuities in Gn. Lower order implementations can be an alternative. In this case more grid points will be necessary to maintain the required accuracy.

The Lax-Wendroff method is less accurate, but is faster than the method of lines. The scheme is also sensitive for discontinuities in Gn, although the oscillations are less severe. The first-order Lax scheme is not reliable.

The MCFS method is simple to use and fast, The technique is however not general applicable and shows severe dynamic errors.

The MCFT algorithm is the most favorable of the studied simulation methods. The only draw-back of the method is that it cannot be used to simulate the effects of the mass accumulation in the fines removal system. In this paper it has been shown however, that the effects of the mass accumulation in the fines removal system on the process dynamics can not be neglected, unless low a cutsize for the fines removal and low fines recycle rates are used.

Acknowledgments

The authors would like to express their gratitude to the Netherlands Technology Foundation (STW), AKZO, DOW, DSM, Dupont de Nemours, Rhone Poulenc , Suiker Unie and Unilever for their financial support of the reseach program.

References

1. Randolph, A.D., Beckman, J.R., Kraljevich, Z.I., (1977) AIChE J.,23, 500-510
2. Jager, J., De Wolf, S., De Jong, E.J. and Klapwijk, W. in Industrial Crystallization 87 , Nyvilt, J., Zacek, S. Eds, Elsevier Amsterdam, 1989 p 415 418.
3. Randolph, A.D. and Larson, M.A. Theory of particulate processes; analysis and techniques of continuous crystallization; Academic Press New York, 1971. enz
4. De Wolf, S., Jager, J., Kramer, H.J.M., Eek, R. and Bosgra, O.H. Proc. IFAC Symp. on Dynamics and Control of Chemical Reactors, Destillation Columns and Batch Processes, Maastricht, The Netherlands, 1989.
5. Frank, R., David, R, Villermaux, J. and Klein, J.P. Chem. Engng. Sci., 43, 69-77
6. Polisch, J. and Mersman, A. Chem. Engng. Sci. Technol.,1988, 11, 40-49
7. Tsuruoka, C. and Randolph, A.D., AIChE Symposium series, no. 253,1987, vol.83, pp. 104-109
8. Vemuri, V. and Karplus, W.J. (1981) Digital computer treatment of partial differential equations. Prentice Hall, Engelood Cliffs, New Jersey.
9. De Leer, B. G. M., Ph.D. Thesis Delft University of Technology, Delft 1981.

RECEIVED May 12, 1990

Chapter 13

Crystal Size Distribution and Characteristics Associated with a Continuous Crystallizer

Yoshio Harano, Shogo Sudo, and Yoshio Aoyama[1]

Department of Applied Chemistry, Osaka City University, Sugimoto 3, Sumiyoshi-ku, Osaka 558, Japan

Both the number and weight basis probability density functions of final product crystals were found to be expressed by a χ^2-function, under the assumption that the CSD obtained by continuous crystallizer is controlled predominantly by RTD of crystals in crystallizer, and that the CSD thus expressed exhibits the linear relationships on Rosin-Rammler chart in the range of about 10-90 % of the cumulative residue distribution.

The method to inspect the crystallizer characteristics and to estimate the crystallization rate parameters from the characteristics of CSD was presented and shown to be useful, by applying this method to many CSD data of CEC type crystallizer. It was found that there exists a tendency, where with increase in η, f_v, ρ and L and with decrease in production rate, the tank number j increases and the crystallizer is expected to act as more classified one.

The crystal size distribution (CSD) is one of important characteristics concerning qualities of final product crystals and is controlled by the operational conditions and many dynamic crystallization processes taking place simultaneously in crystallizer, such as secondary nucleation, growth, agglomeration, breakage, ripening, etc. In spite of these complexities, most of the CSDs obtained with the industrial continuous crystallizers have been known to be presented by the weight basis Rosin-Rammler's expression, regardless of crystal substances and of crystallizers. This fact may means that there exists a certain common key to control CSD.

From this point of view, it was tried previously to correlate the CSD to the characteristics of continuous crystallizers, under the assumption that the CSD is controlled predominantly by the flow pattern in crystallizer, namely the residence time distribution (RTD) of the crystals, and to examine the correlation thus obtained, by applying to the real data on two kinds of compounds in CEC type crystallizer. As the results, the procedures was found to be useful to inspect the characteristics of crystallizer (1). Whereupon, in order to get the correlation, the following additional assumptions were made: 1) the crystal flow in the real crystallizer is expressed by the

---

[1]Deceased

tank-in-series (j-equal size tank) model, 2) the supersaturation is constant throughout the crystallizer, 3) the growth rate of crystal is given by the Bransom's relation

$$G \equiv dL/dt = k_g L^B \Delta C^l = k_G L^B \tag{1}$$

In this paper, the results obtained by applying this method to another 23 kinds of crystals have been shown, after a brief survey on the analytical procedures.

CSD Functions and Its Characteristics.

CSD Functions. From the RTD function based on the tank-in-series model, both the number basis and weight basis probability density functions of final product crystals, $f_n(L)$ and $f_w(L)$, are given by a following chi-square (χ^2) distribution function, only with different parameter of v [see Appendix 1].

$$f_n(L) \equiv \frac{n(L)}{N_T} \frac{dL}{dy} = \frac{1}{\Gamma(\alpha_n)} y^{\alpha_n - 1} \exp[-y]$$

$$= 2F_n(\chi^2/2, v_n) \tag{2n}$$

$$f_w(L) \equiv \frac{w(L)}{W_T} \frac{dL}{dy} = \frac{1}{\Gamma(\alpha_w)} y^{\alpha_w - 1} \exp[-y]$$

$$= 2F_w(\chi^2/2, v_w) \tag{2w}$$

where j is number of tank. n(L) and w(L) are the number and weight density of size L, and, N_T and W_T are the total number and weight of crystals, per unit volume of exit stream from crystallizer, respectively, and

$$\chi^2/2 = y = j\rho^\beta, \quad \rho = L/\overline{L}, \quad \beta = 1 - B$$

$$\overline{L} = (\theta k_G \beta)^{1/\beta}, \quad N_T = n_0 G \theta / j = n^0 G \theta \tag{3}$$

$$v_n/2 = \alpha_n = j, \quad v_w/2 = \alpha_w = j + 3/\beta$$

n_0 and n^0 are the number density of nuclei in the 1st tank and in the whole crystallizer, respectively. $\overline{L}$ is the size of crystal grown for the mean residence time $\overline{\theta}$.

From Equation 2, cumulative residue distribution function, R is given by,

$$R = 1 - \int_0^{\chi^2} \frac{1}{2\Gamma(v/2)} (\chi^2/2)^{v/2 - 1} \exp[-\chi^2/2] d\chi^2 \tag{4}$$

where number basis : R_n, $v/2 = \alpha_n = j$, weight basis : R_w, $v/2 = \alpha_w = j + 3/\beta$.

$$j = 1 \quad ; \text{ MSMPR} \quad : R_n = \exp[-\rho^\beta] \tag{4_1}$$

$$j = 1 \text{ and } \beta = 1; \text{ MSMPR and } \Delta L \text{ law } : R_w = \exp[-\rho](1 + \rho + \rho^2/2 + \rho^3/6) \tag{4_2}$$

Equation 4 indicates that R does not obey the Rosin-Rammler's expression, except for R_n with j=1 (MSMPR) and R_n, R_w at j=∞ (=ideal classifying crystallizer) [see Appendix 2].

Characteristics of CSD. Table I shows some characteristics of CSD expressed by Equations 2 and 4 [see Appendix 3].
The value of R and characteristics of CSD are easily obtained by using the numerical table of probability integral of χ^2-distribution (3).

Table I Characteristics of CSD

	Average size, y_{av}	Variance, σ^2	Dominant size, y_D
$GF^{*1)}$	$y_{av}=j(\rho_{av})^\beta=\alpha$	$\sigma^2=\alpha$	$y_D=j(\rho_D)^\beta=\alpha-1$
$NB^{*1)}$	$\alpha=j, L_{av,n}/\overline{L}=1$	$\sigma_n^2=j$	$y_{D,n}=j-1$
$WB^{*1)}$	$\alpha=j+3/\beta,$	$\sigma_w^2=j+3/\beta$	$y_{D,w}=j+(3/\beta)-1$
	$L_{av,w}/\overline{L}=(j+3/\beta)^{1/\beta}$		

	$\sigma^2/(L_{av})^2$	Median size, y_m (at R=50 %)
GF	$\Gamma(\alpha)\Gamma(\alpha+2/\beta)/(\Gamma^2(\alpha+1/\beta))-1$	$y_m\fallingdotseq\alpha-0.33^{*2)}$
NB	$(\alpha=j)$	$y_{m,n}\fallingdotseq j-0.33$
WB	$(\alpha=j+3/\beta)$	$y_{m,w}\fallingdotseq j+(3/\beta)+0.33$

	Slope: (dR/dy) at $y=y_m$	W_T/N_T
GF	Slope$=-2F(\chi_m^2/2,\nu),\chi_m^2/2=y_m$..(5)	$W_T/N_T=\rho_c f_v \Gamma(\alpha)L_D^3/((\alpha-1)^{3/\beta}\Gamma(j))$
NB	$(y_{m,n}\fallingdotseq j-0.33^{*2)}, \nu_n=2j)$	$(\alpha=j, L_D\rightarrow L_{D,n})$
WB	$(y_{m,w}\fallingdotseq j+(3/\beta)-0.33, \nu_w=2(j+3/\beta))$	$(\alpha=j+3/\beta, L_D\rightarrow L_{D,w})$

*1 GF=General Form, NB=Number Basis, WB=Weight Basis
*2 error (%): -7 ($\alpha=1$), -2.6 (2), ngl. (≥3)

Table II Example of R-value (%) at L=$\overline{L}$

j	1	2	3	4	5	7	10	12	15
R_n	36.8	40.6	42.3	43.3	44.0	45.0	45.8	46.2	46.6
$R_w(\beta=1.0)$	98.1	94.7	91.6	88.9	86.7	83.1	79.2	77.2	77.2

β	0.5	0.75	1.0	1.5	3.0
$R_w(j=1)$	99.9	99.6	98.1	92.0	73.6
$R_w(j=5)$	98.6	93.2	86.7	76.2	61.6
$R_w(j=8)$	96.6	88.8	81.6	71.7	59.3

Table II illustrates R-value (%) at $\rho=1$ (L=$\overline{L}$), indicating that it is impossible to estimate the value of L graphically, from the data on the relation between R and ρ or L, since j and β are unknown.

CSD on Rosin-Rammler Chart, log(log(1/R))-log ρ. Figure 1 shows some examples of CSD on R-R chart. As mentioned above, R based on Equation 4 does not obey the R-R's expression except for R_n at j=1 and R_n, R_w at j=∞. Nevertheless, as can be seen from Figure 1, it was found that there exists an approximate linear relationship between R and ρ (or L) on R-R graph paper in the range of about 10 to 90 % of R, regardless of j and β values. This fact agrees with the empirical information, that is, most of CSDs data display the linear relationships on R-R chart.

From Equation 5 (in Table I), the slope of CSD on R-R chart at the dimensionless median size, S is given by,

$$S=5.771\,(\chi_m^2/2)\,F\,(\chi_m^2,\nu)\,\beta \tag{6}$$

$$\chi_m^2/2=y_m \doteqdot \alpha-0.33, \quad \nu=2\alpha$$

and approximately by,

$$S \doteqdot \alpha^{0.55}\beta \tag{7}$$

$$S_n \doteqdot j^{0.55}\beta, \quad S_w \doteqdot (j+3/\beta)^{0.55}\beta \tag{7'}$$

with 4.0-0 % error for j=1-20. The slopes of straight lines in Figure 1 approximately satisfy Equation 7 [see Appendix 4].

From these results, it was found that it should be carefully to estimate j to be unity and/or $\overline{L}$ to be L at R=36.8 %, even if the linear relationship is obtained on R-R chart, since both the slopes and R values at ρ=1 much depend on j and β values, as illustrated in Table II and Figure 1.

Estimation Method of Characteristic Parameters from CSD data.

The desired parameters to estimated from CSD data are the tank number j (the characteristics of crystallizer) and the rate parameter β. From these, the others; $\overline{L}$, k_G, $\dot{N}^0$ (nucleation rate, $=N_T/\overline{\theta}$) are estimated easily.

In the distribution function derived here, the dimensionless size ρ (=$L/\overline{L}$) is introduced as $y=j\rho^\beta$ and $\overline{L}$ is unmeasurable and is difficult to estimate as mentioned above. Therefore, it is necessary to estimate j and β simultaneously, by using two or three independent characteristics of distribution function, such as those shown in Table I. Among many approaches, the convenient ones were found to use the following relations: the slope of the weight basis CSD on R-R chart at the dimensionless median size y_m, S_w vs. that of the number basis CSD at y_m, S_n (Figure 2), the slope ratio, $\sigma_{w/n}(\equiv S_w/S_n)$ vs. the dominant size ratio of weight basis to number basis, $(\rho_{w/n})_D(\equiv L_{D,w}/L_{D,n})$ (Figure 3) and the slope ratio, $\sigma_{w/n}$ vs. the median size ratio, $(\rho_{w/n})_m$ ($\equiv L_{m,w}/L_{m,n}$) (Figure 4).

Results and Consideration.

All the CSD data examined in this work are those obtained by CEC (Crystal Engineering Corp.) style Krystal-Oslo type crystallizer (Figure 5), except sample no.22 by DTB type crystallizer. Figure 6 and Table III illustrate some examples of CSD on R-R chart and j, β values estimated by different ways (number indicates sample no. in Table IV), respectively.

Table IV shows whole data relating to CSD and j, β values, the latters of which were mainly estimated by the relation between S_w-S_n (Figure 2), because that almost the same values were obtained with different methods, as shown in Table III. The data in Table IV are arranged in order of volumetric shape factor of crystals, f_v. f_v is unity for cubic and $\pi/6$ (=0.52) for spherical. Thus, 0.08 (no.24, 25) mean fairly

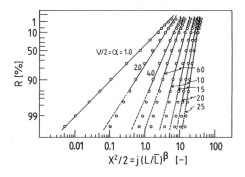

Figure 1. Some examples of CSD based on equation 2 on R–R chart. (Reproduced with permission from ref. 1. Copyright 1984 Elsevier Science Publishers B. V.)

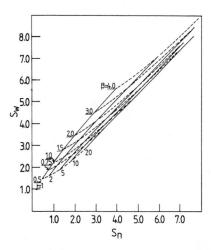

Figure 2. S_w vs S_n. (Reproduced with permission from ref. 1. Copyright 1984 Elsevier Science Publishers B. V.)

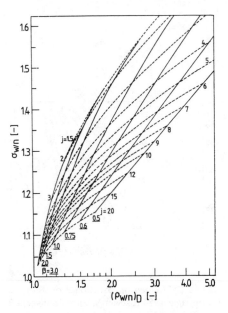

Figure 3. $\sigma_{w/n}$ vs $(\rho_{w/n})_D$. (Reproduced with permission from ref. 1. Copyright 1984 Elsevier Science Publishers B. V.)

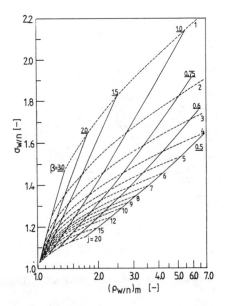

Figure 4. $\sigma_{w/n}$ vs $(\rho_{w/n})_m$. (Reproduced with permission from ref. 1. Copyright 1984 Elsevier Science Publishers B. V.)

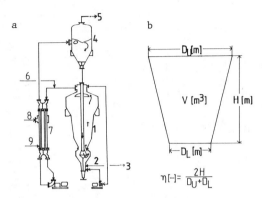

Fig.5   CEC style Krystal Oslo type crystallizer
a)Schematic diagram: 1.Crystallizer, 2.Exit crystal slurry, 3.slurry receiver,
4.Evaporator, 5.Condenser, 6.Feed solution, 7.Heat exchanger, 8.Steam inlet
9.Drain outlet
b)Conceptional diagram

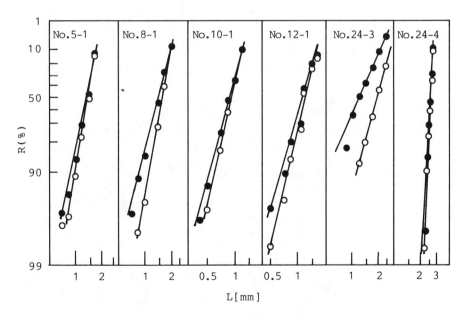

Fig.6   Some examples of CSD data on R-R chart
O: weight basis, ●: number basis

Table III CSD and j, β values estimated by different ways

No.		L_m [mm]	L_D [mm]	S		$S_n - S_w$	$\sigma_{w/n}$ $(\rho_{w/n})_m$	σ^2/L_{av}^2	$\rho_c f_v W_T/N_T$ / $W_T/N_T L_{D,n\&w}$
23-1	NB	1.07	1.03	6.03	j	4.9	4.0	5.0	
	WB	1.16	1.18	6.08	β	2.5	3.0	2.5	
23-2	NB	0.70	0.76	4.09	j	2.0	2.0	2.0	
	WB	0.82	0.85	5.17	β	2.6	3.0	2.8	
24-1	NB	1.21		2.28	j	1.1	1.1		
	WB	1.92		3.55	β	2.2	2.3		
24-2	NB	2.50		16.88	j	12.0			12.0
	WB	2.55		17.63	β	3.0			3.0
25-1	NB	1.31	1.35	5.98	j	3.9	3.8	4.0	4.0
	WB	1.47	1.55	6.86	β	2.8	2.5	2.8	2.5

needle-like crystal. The parameter, $\eta = [2H/(D_U + D_L)]$ is the shape parameter of crystallizer. The larger the η value, the crystallizer becomes more slim and approaches to an ideal classified tank ($j = \infty$), and the smaller, it is expected to approach to a well mixed tank ($j = 1$).

The samples were classified by f_v values to three groups: A 0.90-0.52, B 0.48-0.33, C 0.26-0.08. Figure 7 shows the relation between j and the shape parameter η, for each group. Number means the weight basis median size $L_{m,w}$[mm] and that with underline indicates the specific gravity of crystal ρ. The cross points on solid lines are the presumed j values of 1 mm size. As can be seen from Figure 7, with increase in L and ρ at constant η and in η at constant L and ρ, j increases and the crystallizer acts as more classified type. This tendencies might be reasonable, because that the large and heavy crystals are easily settled down with less mixing effect. Comparing the data between Figure 7-a and b, though it is impossible to compare at the same ρ and L, there might exist tendencies that the smaller the f_v value, j approaches to unity.

Since j values depends on at least ρ, L, f_v and η. Thus, another trial of arrangement was done, as shown in Figure 8. The ordinate η/j is the dimensionless height of one stage of crystallizer and $f_v^{1/2}$ in the abscissa corresponds to the ratio of breadth to length of crystal (b/l), assuming the crystal shape to be tetragonal form. Figure 8 might indicate a tendency, with some exceptions, that with increase in $f_v^{1/2}L^2$ (namely, in f_v and L) and ρ, the values of ordinate decreases and the crystallizer is expected to act as more classified type.

General speaking, the size dependency of growth rate, B ($= 1 - \beta$), in diffusion controlling region, is 0.5 for fluidized bed, -1.0 in mixed tank and, in reaction controlling region, is zero. As shown in Table IV, however, B values of about 40 % of whole samples ($= 76$) are fairly less than -1.5. Figure 9 shows B values plotted against crystal size, indicate that with increase in L, B value decrease below -1.5. Abnormal B values might come from the breakage for larger crystal and the agglomeration for smaller crystals.

Table IV Some Examples of CSD and j, β values

No[1]	Num[2]	Sample	f_v [-]	ρ_c [g/cm³]	V [m³]	η [-]	P [kg/h]	C[g/l] in	out	T [K]		L_m [mm]	L_D [mm]	L_{av} [mm]	S	J	β	B
1	1	KI	0.90	3.13	5.5	2.0	1060	2.6	1.0	303	NB	2.25	2.31	2.21	6.87	3.2	3.5	-2.5
											WB	2.41	2.48	2.37	7.85			
2	2	KF	0.87	2.48	3.9	1.8	533	2.6	0.9	335	NB	0.70	0.72	0.69	5.42	7.1	1.8	-0.8
											WB	0.72	0.74	0.71	6.05			
3	5	KBr	0.86	2.76	3.9	1.8	720	1.6	0.4	298	NB	1.60	1.64	1.58	4.90	1.9	3.3	-2.3
											WB	1.89	1.94	1.86	6.08			
4	2	NaCl	0.85	2.16	6.8	2.5	2170	3.5	1.1	360	NB	1.16	1.19	1.13	7.26	2.0	4.8	-3.8
											WB	1.23	1.26	1.20	8.48			
5	4	$NiSO_4 7H_2O$	0.81	1.95	4.8	1.7	944	1.8	0.6	298	NB	1.36	1.39	1.34	4.63	2.1	3.0	-2.0
											WB	1.41	1.45	1.39	5.85			
6	4	CrO_3	0.72	2.63	3.4	1.4	488	3.2	2.1	358	NB	0.82	0.78	0.84	2.55	8.0	0.8	0.2
											WB	0.92	0.91	0.92	3.14			
7	2	$Co(NH_2)_2$	0.68	1.34	6.8	2.5	1180	3.6	1.5	341	NB	0.72	0.74	0.71	5.08	1.7	3.7	-2.7
											WB	0.78	0.80	0.76	6.38			
8	2	$NH_4H_2PO_4$	0.62	1.79	4.2	2.3	922	3.1	1.0	296	NB	1.27	1.29	1.26	3.86	1.3	3.3	-2.3
											WB	1.43	1.47	1.41	5.26			
9	4	$La_2H(SO_4)_2 \cdot 9H_2O$	0.59	2.82	6.8	2.3	1330	1.8	0.7	315	NB	0.91	0.93	0.89	6.19	3.5	3.0	-2.0
											WB	0.94	0.97	0.92	7.10			
10	4	$NaHCO_3$	0.58	2.20	6.8	2.5	2826	4.8	1.2	302	NB	0.88	0.90	0.87	4.04	5.7	1.5	-0.5
											WB	0.92	0.94	0.91	4.73			
11	4	$KBrO_3$	0.57	3.27	3.9	1.8	518	2.3	0.6	298	NB	0.57	0.58	0.55	4.86	8.2	1.5	-0.5
											WB	0.60	0.61	0.58	5.45			
12	4	$MnSO_4 \cdot 4H_2O$	0.56	2.11	5.8	1.9	562	2.8	1.0	299	NB	1.14	1.15	1.12	4.01	2.2	2.5	-1.5
											WB	1.16	1.19	1.14	5.11			
13	4	$NiSO_4 \cdot 6H_2O$	0.52	2.03	5.8	1.3	1381	3.2	1.6	337	NB	0.87	0.82	0.89	2.51	7.7	0.8	0.2
											WB	0.96	0.94	0.96	3.07			
14	3	$Na_3H(SO_4)_3$	0.48	2.42	8.6	2.5	2289	3.0	0.9	304	NB	0.69	0.70	0.68	3.88	11.7	1.0	0.0
											WB	0.78	0.74	0.71	4.37			
15	2	CuS	0.47	4.64	4.8	2.1	1201	1.2	0.6	339	NB	0.71	0.73	0.70	6.02	2.5	3.5	-2.5
											WB	0.73	0.75	0.71	7.06			
16	4	$[Co(C_2H_8N_3)_3]Cl_3 \cdot 3H_2O$	0.46	1.54	3.9	1.6	322	2.1	0.5	300	NB	1.25	1.26	1.24	3.59	2.1	2.3	-1.3
											WB	1.34	1.38	1.32	4.71			
17	4	$NaSH \cdot 5H_2O$	0.42	1.79	12.0	2.3	1960	4.1	1.2	306	NB	3.34	3.39	3.31	3.97	3.3	2.0	-1.0
											WB	3.47	3.57	3.43	4.88			
18	2	$CuSO_4 \cdot 5H_2O$	0.38	2.28	4.8	1.7	766	2.8	0.6	298	NB	1.45	1.49	1.43	5.74	0.3	0.9	0.1
											WB	1.50	1.54	0.48	7.25			

*1)consecutive number of sample
*2)number of same substances with different production rate

Continued on next page

Table IV Some Examples of CSD and j, β values (Continued)

No[1]	Num[2]	Sample	f_v [-]	ρ_c [g/cm^3]	V [m^3]	η [-]	P [kg/h]	C[g/l] in	out	T [K]		L_m [mm]	L_D [mm]	L_{av} [mm]	S	J	β	B
19	3	Na$_3$AsO$_4$·2H$_2$O																
			0.38	1.76	8.9	2.4	1463	3.3	0.9	318	NB	0.74	0.73	0.74	3.15	2.6	1.8	-0.8
											WB	0.83	0.85	0.82	4.12			
20	4	(CH$_3$H$_6$)Cr(SO$_4$)$_3$·6H$_2$O																
			0.33	1.86	6.5	2.3	1850	3.2	1.6	307	NB	0.74	0.74	0.73	3.77	5.0	1.5	-0.5
											WB	0.80	0.81	0.78	4.47			
21	3	Ce(SO$_4$)$_3$·8H$_2$O																
			0.33	2.89	4.0	2.2	901	2.1	0.6	294	NB	0.85	0.87	0.83	5.17	2.1	3.3	-2.3
											WB	0.89	0.91	0.87	6.27			
22	2	Al$_2$O$_3$·3H$_2$O																
			0.26	2.32	4.0DTB		20	6.0	4.6	329	NB	0.13	0.13	--.--	2.31	5.0	0.9	0.1
											WB	0.16	0.17	--.--	2.98			
					6.2DTB		33	5.8	4.5	331	NB	0.08	0.08	--.--	2.23	3.5	1.2	-0.2
											WB	0.11	0.11	--.--	3.03			
23	2	Na$_2$SO$_4$ (irregular operation)																
			0.18	2.70	--.--	1.1	1580	0.5	0.1	354	NB	1.07	1.03	--.--	6.03	4.9	2.5	-1.5
											WB	1.16	1.18	--.--	6.08			
							1730	0.5	0.1	353	NB	0.70	0.76	--.--	4.09	2.0	2.6	-1.6
											WB	0.82	0.85	--.--	5.17			
24	4	(NH$_4$)Al(SO$_4$)$_2$·12H$_2$O																
			0.08	1.64	5.5	1.3	950	--.--	--.--	----	NB	1.21	--.--	--.--	2.28	1.1	2.3	-1.3
											WB	1.92	--.--	--.--	3.55			
					2.7		453	--.--	--.--	----	NB	2.50	--.--	--.--	16.9	12.0	3.0	-2.0
											WB	2.55	--.--	--.--	17.6			
25	1	L-alanine monohydride																
			0.08	1.40	5.8	2.3	-----	--.--	--.--	----	NB	1.31	1.35	1.34	5.98	3.9	2.8	-1.8
											WB	1.47	1.55	1.46	6.86			

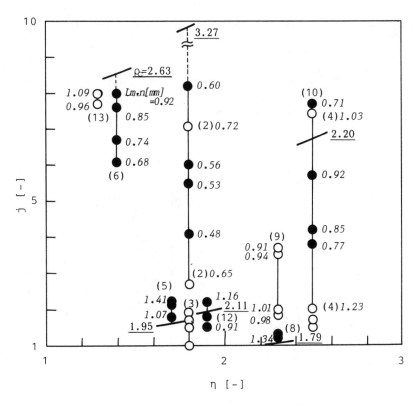

Fig.7-a   j vs. η : class A; ( ) sample no.

Continued on next page

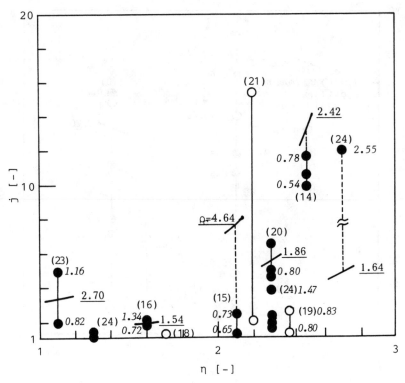

Fig.7-b   j vs. η : class B and C; ( ) sample no.

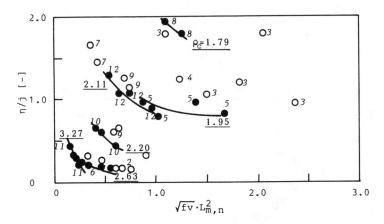

Fig.8-a η/j vs. $\sqrt{f_v} \cdot L_{m,n}^2$: class A
Numbers without underline mean sample no.

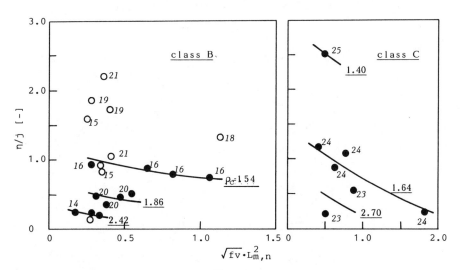

Fig.8-b η/j vs. $\sqrt{f_v} \cdot L_{m,n}^2$: class B , C
Numbers without underline mean sample no.

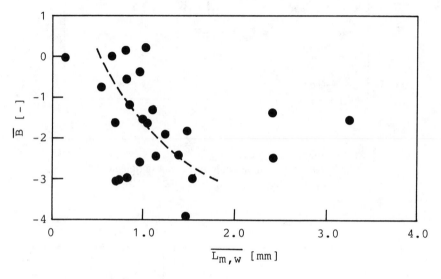

Fig.9 $\overline{B}$ vs. $\overline{L_{m,w}}$.

Conclusion.

At present, there remains complexity to make clear the relation between j values and its affecting factors, however, the procedures presented in this paper was found to be useful to inspect the characteristics of some industrial continuous crystallizers and to estimate the crystallization rate parameters from their CSD data.

Appendix 1.

The residence time distribution (RTD) function, E, of the tracer, which is injected into stream entering into a vessel as an instantaneous pulse, based on the tank-in-series (j-equal size) model is given by (2),

$$E(\theta) = \frac{C(\theta)}{C^0} = \frac{1}{(j-1)!} \theta^{j-1} e^{-j\theta}$$

(a-1)

or

$$E'(\theta) = \frac{C(\theta)}{C_0} = \frac{1}{(j-1)!} (j\theta)^{j-1} e^{-j\theta}$$

(a-2)

where θ is the dimensionless time ($\theta = t/\overline{\theta}$; $\overline{\theta}$=mean residence time, V/v; V=total volume of vessel=jV'). E or E' is the age-distribution frequency of the tracer leaving the vessel. C means the tracer concentration: C^0 is the initial (at $\theta=0$) average one, if uniformly distributed throughout the vessel, C_0 is the initial (at $\theta=0$) one in the 1st tank ($=jC^0$) and $C(\theta)$ is the concentration in exit stream from the j-th tank.

Considering the tracer entering the vessel at a given instant of time to be the nuclei formed at that time, C^0, C_0 and $C(\theta)$ can be converted to the number density of nuclei in the whole vessel, n^0, in the 1st tank, n_0 and that of crystals in the exit stream from the vessel, $n(\theta)$, respectively. The crystals having the residence time of θ grow up to the size L, which is given by Equation 1. Therefore, by using Equations a-1 or a-2 and 1, the number basis probability density function of final product crystals, $f_n(L)$ is obtained, as follows.

Denoting the number of crystals with size of L~L+dL per unit volume as N(L),

$$n_0 = \left(\frac{dN(L)}{dL}\right)_{L=0} = \left(\frac{dN(L)}{dt}\right)_{L=0} \left(\frac{dt}{dL}\right) = \frac{\dot{N}_0}{G}$$

(a-3)

where $\dot{N}_0$ is the nucleation rate per unit volume of the 1st tank, with dimension of [number/(volume•time)]. Replacing the variable $j\theta$ by y, the linear growth rate G is given by,

$$G = \frac{dL}{dt} = \frac{dL}{\overline{\theta} d\theta} = \frac{j}{\overline{\theta}} \frac{dL}{dy}$$

(a-4)

From Equations a-3 and a-4,

$$n_0 = \frac{\dot{N}_0 \overline{\theta}}{j} \frac{dy}{dL}$$

(a-5)

Then, from Equations a-2 and a-5,

$$n(L) dL = \frac{\dot{N}_0 \overline{\theta}}{j} \frac{1}{(j-1)!} y^{j-1} e^{-y} dy$$

(a-6)

When j is integer, $(j-1)!$ is the gamma function, $\Gamma(j)$. The total number of product crystals per unit volume of exit stream from vessel, N_T is given by

$$\int_0^\infty n(L)\,dL$$

and, from Equation a-6, is equal to $\dot{N}_0\bar{\theta}/j$. Thus,

$$f_n(L) \equiv \frac{n(L)}{n_0} = \frac{n(L)}{N_T}\cdot\frac{dL}{dy} = \frac{1}{\Gamma(\alpha_n)}y^{\alpha_n-1}e^{-y}$$ (a-7),[2n]

$$=2F_n(\chi^2/2,\nu_n)$$

The last term of Equation a-7 is the chi-square (χ^2) probability function and

$$\chi^2/2=y,\quad \nu_n=2\alpha_n=2j$$

$$y=j\theta=j\rho^\beta,\quad \rho=L/\bar{L},\quad \beta=1-B$$

$$\bar{L}=(\theta k_G\beta)^{1/\beta}$$ (a-8),[3]

$$n_0=\frac{\dot{N}_0}{G}=N_T\frac{dy}{dL}$$

$$N_T=\frac{\dot{N}_0\bar{\theta}}{j}\ (=\dot{N}^0\bar{\theta})=n_0 G\bar{\theta}/y\ (=n^0 G\bar{\theta})$$

where $\bar{L}$ is the size of crystal which has grown for a period of $\bar{\theta}$ and $\dot{N}^0$ is the nucleation rate per unit volume of the whole vessel.

From Equations a-7 and a-8, for the case of $j=1$, i.e., for MSMPR, the following well-known expressions,

$$f_n(L)_{j=1}=\frac{n(L)}{n_0}=\exp\left[-\frac{L^{1-B}}{G\theta(1-B)}\right]$$ (a-9$_1$)

and further, for $\beta=1$ (i.e. $B=0$, ΔL law applicable),

$$\frac{n(L)}{n_0}=\exp\left[-\frac{L}{G\theta}\right]$$ (a-9$_2$)

are obtained.

The weight basis probability density function of crystals in exit stream, $f_w(L)$ is derived from in a similar way. Multiplying both sides of Equation a-6 by $\rho_c f_v L^3$, where rc and f_v are the density and volumetric shape factor of crystals, respectively,

$$\rho_c f_v L^3 n(L)\,dL=W(L)\,dL$$

$$=\frac{\dot{N}_0\bar{\theta}}{j}\rho_c f_v\frac{L^3}{(j-1)!}y^{j-1}e^{-y}dy$$ (a-10)

Since $y=j\rho^\beta=j(L/\bar{L})^\beta$,

$$L=(y/j)^{1/\beta}\bar{L}$$ (a-11)

Substituting Equation a-11 for L in Equation a-10,

$$w(L)\,dL=\frac{\dot{N}_0\bar{\theta}}{j}\rho_c f_v\bar{L}^3 J\frac{1}{(j+(3/\beta)-1)!}y^{j+\frac{3}{\beta}-1}e^{-y}dy$$ (a-12)

where

$$J=j(j+1)\cdots(j+\frac{3}{\beta}-1)/j^{3/\beta}$$

(a-13)

Total weight of products crystals per unit volume of exit stream, W_T is given by

$$\int_0^\infty w(L)\,dL$$ and, form Equation a-13,

$$W_T=\overline{W}\dot{N}_0\theta J/\overline{j}=\overline{W}N_T\overline{J}$$ (a-14)

$$\overline{W}=\rho_c f_v\overline{L}^3$$

and then,

$$f_w(L)=\frac{w(L)}{W_T}\frac{dL}{dy}=\frac{1}{\Gamma(\alpha_w)}y^{\alpha_w-1}e^{-y}$$

$$=2F_w(\chi^2/2,\ \nu_w)$$

(a-15),[2w]

$$\nu_w=2\alpha_w=2(j+\frac{3}{\beta})$$

(a-16),[3]

Form MSMPR (j=1) and β=1 (B=0), from Equations a-14, a-13 and a-8, the following well-known equation is obtained,

$$W_T=\rho_c f_v 6n_0(G\theta)^4$$

(a-17)

Appendix 2.

From Equation 2 (or a-7 and a-15 in Appendix 1), the cumulative residue

distribution : $R_N(L)\ (=1-\frac{1}{N_T}\int_0^L n(L)\,dL)$ and $R_w(L)\ (=1-\frac{1}{W_T}\int_0^L w(L)\,dL)$ are

given by Equation 4, where, for convenience to use the numerical table (see Appendix 3), $\chi^2/2$ and ν are used as variables, instead of y and α.

From the characteristic of χ2-distribution function, for α=integer,

$$R=e^{-y}\sum_{n=0}^{\alpha-1}\frac{y^n}{n!},\qquad y=j\rho^\beta$$

(a-18)

From this relation, Equation 4_1, and 4_2 are given easily and, for j=∞, R=1. The latter corresponds to the CSD in complete classified crystallizer (piston flow pattern).

Appendix 3.

The average dimensionless size $y_{av}\ (=\int_0^\infty y^\alpha e^{-y}\Gamma(\alpha)^{-1}dy)$ the dimensionless dominant size y_D (y at $df/dy=0$), the variance $\sigma^2 (= \int_0^\infty (y-y_{av})^2 y^{\alpha-1} e^{-y} \Gamma(\alpha)^{-1} dy$

the slope $S \equiv dR/dy$ at the dimensionless median size (see Appendix 4) and W_T/N_T are derived from Equations 2, 4 and a-14, as shown in Table I. The median size y_m (y at $R=50$ %) are obtained by using the numerical table of probability integral of χ^2-distribution (3,4), as shown in Figure a-1. As the results, y_m was given by,

$$y_m = \alpha - 0.33 \tag{a-19}$$

$$\text{number basis } \alpha = j$$

$$\text{weight basis } \alpha = j + 3/\beta$$

$$\text{error (\%) : -7 (a=1), -2.6 (a=2), ngl.(a} \geq 3)$$

as shown in Table I.

The values at $\rho=1$ (at $\chi^2/2=y=\alpha$) in Table II are also obtained by using the numerical table mentioned above.

Appendix 4.

Since $y=j\rho^\beta$, the slope of Equation 4 on Rosin-Rammler chart, log(log(1/R)) vs. log ρ, is given by,

$$\frac{d[\log(\log(1/R))]}{d\log\rho} = \frac{\rho}{R\ln R} \cdot \frac{dR}{d\rho} = \frac{y}{R\ln R} \cdot \frac{dR}{dy}\beta \tag{a-20}$$

From Equation 4, dR/dy at median size ($y=y_m=\chi_m^2/2$) is given by,

$$\left(\frac{dR}{dy}\right)_{y_m} = -\frac{1}{\Gamma(v/2)} \cdot \left(\frac{\chi_m^2}{2}\right)^{v/2-1} \exp\left[-\frac{\chi_m^2}{2}\right] \tag{a-21},[5]$$

$$= -2F\left(\frac{\chi_m^2}{2}, v\right)$$

Substituting Equation a-21 (Equation 5 in Table 1) and $R=0.5$ to a-20,

$$S \equiv \left(\frac{d[\log(\log(1/R))]}{d\log\rho}\right)_{y=y_m} = 5.771 \left(\frac{\chi_m^2}{2}\right) F(\chi_m^2, v)\beta \tag{a-22},[6]$$

where

$$\chi_m^2/2 = y_m \doteq \alpha - 0.33 \tag{a-19}$$

Table a-1 and Figure a-2 show some examples of S values calculated by using the numerical table and from these data the following approximate relation was obtained.

$$S \doteq \alpha^{0.55}\beta \tag{a-23}, [7]$$

$$S_n \doteq j^{0.55}\beta, \quad S_w \doteq \left(j+\frac{3}{\beta}\right)^{0.55}\beta \tag{a-23'}, [7']$$

$$\frac{S_w}{S_n} \equiv \sigma_{w/n} = \left(1+\frac{1}{j\beta}\right)^{0.55} \tag{a-24}$$

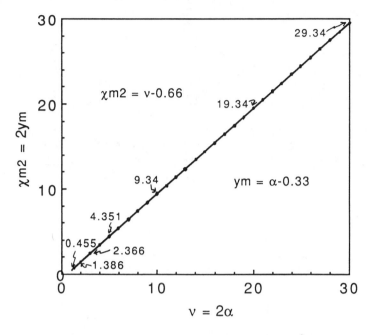

Figure a-1. Relation between median size y_m and α. $y_m = \chi_m^2/2$, $\alpha = \upsilon/2$.

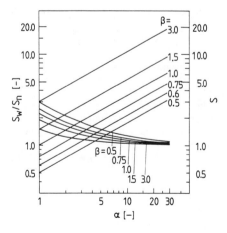

Figure a-2. log S and $\log(S_w/S_n)$ vs log α. (Reproduced with permission from ref. 1. Copyright 1984 Elsevier Science Publishers B. V.)

Table a-1 Some examples of S/β values

j	1	2	3	5	7	10	12	15	17	20
S/β,Eq.a-23	1.0	1.52	1.90	2.50	2.99	3.59	3.94	4.42	4.71	5.11
S/β,Eq.a-24	1.0	1.46	1.83	2.42	2.92	3.55	3.92	4.43	4.71	5.11
error(%)	0.0	-4.0	-3.7	-3.2	-2.3	-1.0	-0.5	-0.2	0.0	0.0

The approximate values of S/β calculated by Equation a-23' and their error are also shown in Table 1. Figure a-2 also show the slope ratio of weight basis S_w to number basis S_n ($=S_w/S_n=\sigma_{w/n}$) against α. The ratio vary from 3 to 1.25 for j=1~10, β=0.5~3.0, and is 2.22 for j=1 (MSMPR) and β=1 (ΔL law).

Legend of Symbols

B	size dependency of growth rate in power law	[-]
C	concentration	[g/l]
D_L,D_U	lower and upper diameter of crystallizer	[m]
f_n,f_w	number and weight basis probability density functions	[-]
fv	volumetric shape factor of crystals	[-]
G	growth rate	[mm/s]
H	height of crystallizer	[m]
j	tank number	[-]
k_g,k_G	constants defined in Equation 1	[-]
L	crystal size	[mm]
$\overline{L}$	size of crystal growth for the time q	[mm]
L_{av}	average size	[mm]
L_D	dominant size	[mm]
L_m	median size	[mm]
l	supersaturation dependency of growth rate in power law	[-]
$\dot{N}_0$	nucleation rate in 1st tank	[1/cm^{2}s]
N_T	total number of crystals in unit volume of exit stream from crystallizer	[-]
n_0	number density of nuclei (crystals at L=0), in 1st tank	[-]
n(L),w(L)	number and weight density of crystals in unit volume exit stream from crystallizer	[-]
P	production rate	[kg/h]
R	cumulative residue distribution function	[-]
S	slope of CSD on R-R chart at dimensionless median size	[-]
V	volume of crystallizer	[m^3]
W_T	total weight of crystals in unit volume of crystallizer	[g]
y	$=j\rho^\beta$	[-]
β	$=1$-B	[-]
η	shape parameter of crystallizer, $=2H/(D_U+D_L)$	[-]
ρ	dimensionless crystal size, $=L/\overline{L}$	[-]
ρ_c	gravimetric density of crystals	[g/cm^3]
$\overline{\theta}$	mean residence time	[s]
$\sigma_{w/n}$	$=S_w/S_n$	[-]

subscript
D   dominant size
m   median size
n   number basis
w   weight basis

Literature Cited

1. Harano, Y.; Douno, H.; Aoyama, Y. Industrial Crystallization 84; S.J.Jancic;
   E.J.de Jong, Ed.: Elsevier Science Publishers B.V.: Amsterdam, 1984; p.451.
2. Himmelblau, D.M.; Bishoff, K.B. Process Analysis and Simulation
   (Deterministic System): John Wiley and Sons, Inc., New York, 1968; p.70.
3. Handbook of Mathematical Functions With Formulas, Graphs and Mathematical
   Tables; Abramowitz, M; Stegun, I.A., Ed.: National Bureau of Standards, 1964;
   p.978.

RECEIVED August 6, 1990

CRYSTALLIZATION OF ORGANIC MOLECULES AND BIOMOLECULES

Chapter 14

Kinetics of the Ethanolic Crystallization of Fructose

M. R. Johns, R. A. Judge, and E. T. White

Department of Chemical Engineering, University of Queensland, Brisbane 4072, Australia

Batch crystallization studies of D-fructose from aqueous ethanolic solutions demonstrate that crystal growth rate is dependent on supersaturation (possibly to the 1.25 power), ethanol content and temperature. It appears that solution viscosity also has an effect. Growth rates of up to 1 μm/min were measured.

No nucleation occurs provided the supersaturation is kept below a value equivalent to 35°C of subcooling. There is a size spread effect, but it decreases with high ethanol contents. The results indicate that a practical process is feasible to grow large fructose crystals by the addition of ethanol to aqueous fructose solutions.

High fructose corn syrup (HFCS) has emerged in recent years as an alternative nutritional sweetener to sucrose. However, the use of HFCS has been confined to those applications suited to liquid syrups, in particular the beverage and canning sectors of the market. The manufacture of fructose as a crystalline product would open up further market opportunities for the sweetener. One company in the USA is producing crystalline fructose in commodity quantities (1), but at a higher price than sucrose.

Two methods are employed industrially to produce crystalline fructose, aqueous crystallization and alcoholic crystallization. Yields of fructose crystallized from water syrups are only of the order of 50%, due to the very high water solubility of the sugar, while the high viscosity of the concentrated solution results in long crystallization times, typically 50 hours or more (2). The second process requires the addition of lower alcohols (eg. ethanol) to a concentrated fructose syrup, generally 90% total solids or more, at temperatures of 50°C to 80°C and then cooling to cause crystallization. Fructose yields are from 70 to 80% and the total time involved is 8 to 12 hours (3). However, large quantities of

0097–6156/90/0438–0198$06.00/0
© 1990 American Chemical Society

alcohol are required and alcohol recovery must be included for economic operation.

Recently, Edye *et al.* (4) described a fermentation process which used a mutant strain of *Zymomonas mobilis* to produce high concentrations of fructose and ethanol when grown on a concentrated sucrose medium. Johns and Greenfield (5) proposed ethanolic crystallization as a means of recovering the fructose from the broth. The kinetic behaviour of fructose crystallization from ethanolic solution has not been previously reported, and this work investigates these crystallization kinetics.

Experimental Methods

The investigation was carried out using a seeded, batch crystallization in the absence of nucleation. Supersaturated solutions were prepared, seeded and maintained at a constant temperature while crystallization proceeded. Samples were taken periodically to give a solution for analysis and crystals for size analysis and crystal content determination.

The crystallizer was a stirred, 1-litre cylindrical glass vessel agitated with a central, 6-blade turbine impeller running at 500 RPM. The vessel was sealed to prevent alcohol evaporation. D-fructose (Boehringer-Mannheim GmbH, Mannheim) was dried at 70°C overnight in a vacuum oven and stored in a desiccator. Anhydrous ethanol (CSR Ltd., Sydney) was used. Weighed amounts of D-fructose, ethanol and distilled water were added to· the crystallizer, initially operated at 10 to 15°C above the crystallization saturation temperature by a heated water bath. When the fructose had completely dissolved, the temperature was reduced to give a supersaturated solution. A suitable quantity of seed crystals was then added.

The seed crystals of D-fructose were obtained by ball milling crystals produced by spontaneous nucleation from an aqueous ethanolic solution of fructose and allowing them to stand at room temperature in slightly supersaturated ethanolic solution until the desired crystal size (20-40 microns) was achieved. They were then stored at 30°C in saturated, anhydrous ethanol to prevent further crystal growth. A portion of this slurry was added to the crystallizer as the seeds.

At appropriate time intervals (several minutes), two known volume samples of the crystallizer contents were taken. One was used to determine the crystal content, where the crystals were recovered on a 0.45 μm membrane filter, washed, dried *in vacuo* at 60°C overnight and weighed. The other sample was also filtered on a membrane filter to give a solution for solute analysis and crystals for size analysis. The crystals were washed from the membrane by dipping it in an ethanol electrolyte solution containing 2% lithium chloride. Sizing was done using an electronic sensing zone sizer (Particle Data) with a 300 μm orifice. The solution was analysed for ethanol by gas chromatography and for fructose by both HPLC (5) and vacuum evaporation to dryness.

Results and Discussion

Properties of Ethanolic Fructose Solutions. Published information on the properties of aqueous ethanolic fructose solution is very limited. As a result, solubility data from 25 to 60 $^\circ$C was measured (Figure 1) and will be published separately. The equilibrium fructose/water mass ratio for zero alcohol, ranges from over 4 at 25°C to over 8 at 60°C (7,8). It can be seen that reasonable yields will only result with high alcohol additions (E/W at least 2). This is the range used in this study. Aqueous ethanolic solutions have a wide range of viscosities. These were measured at operating conditions using a Rheomat concentric cylinder viscometer.

Limiting Supersaturation for Nucleation. Like sucrose, D-fructose solutions can tolerate a high degree of supersaturation without nucleating, even in the presence of seed crystals. This is the metastable region on the Miers supersolubility diagram (9).

To measure this limiting supersaturation level before nucleation occurs, a batch of stirred, supersaturated solution containing large seed crystals was slowly cooled (0.5°C/min) and observed. When nucleation occurred, the degree of subcooling was noted. The results are shown on Figure 2. It can be seen that 35°C of subcooling is possible before nucleation occurs and this is independent of both the initial fructose concentration (saturation temperature) and ethanol content. Provided subcoolings are kept above this value, batch crystallization studies can be carried out without nucleation.

Crystal Growth Rate Studies. Nine runs were undertaken (Table I), three at each of three temperatures, 25, 40 and 55°C. Figure 3 shows size distributions of the samples in the cumulative number "greater than" form, for run 5 at 40°C with an E/W content of 4.42. The total crystal number counts (> 10 μm) were substantially constant, showing that there were no nuclei growing into the size range. For the plot, all the distributions were scaled slightly to give the same value at 10 μm, to increase the accuracy in evaluating the growth rate. The cumulative size distributions are seen to broaden (covering an increasingly wider range of sizes) as they grow, showing either size dependent growth or growth dispersion.

From the change in the mean size with time, an average growth rate can be calculated for each time interval. Figure 4 shows a plot of the average growth rate and the fructose concentration during run 5. From this plot, the average concentration corresponding to each growth rate was evaluated.

The growth rates for the nine experiments plotted against solution concentration, C are shown in Figure 5. The growth rates are quite substantial (up to 1 μm/min). These could be increased further (without nucleation) by operating at a higher supersaturation. As the E/W ratio increases, the value of C is smaller and the dependence on C becomes steeper. Thus a process to grow large single crystals of D-fructose is feasible.

The order of the dependence of growth on supersaturation could not be evaluated. The solutions take a considerable time (up to a day) to reach equilibrium and final equilibrium solubility values were not determind. As it eventuated, solubility values given in

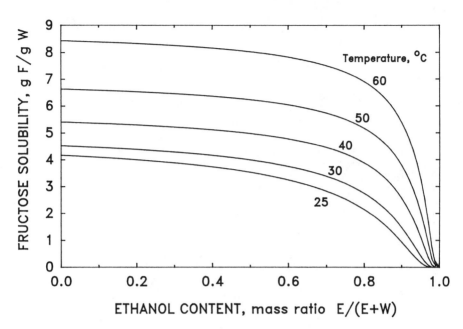

Figure 1. Solubility of fructose in aqueous alcoholic solutions, 25–60°C.

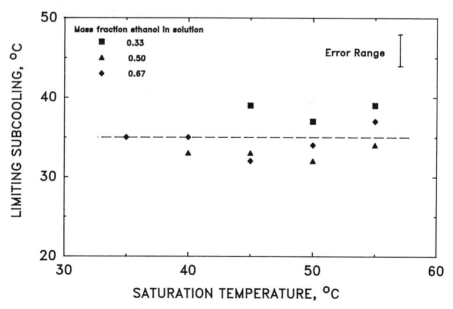

Figure 2. Limiting subcooling before aqueous fructose solutions nucleate.

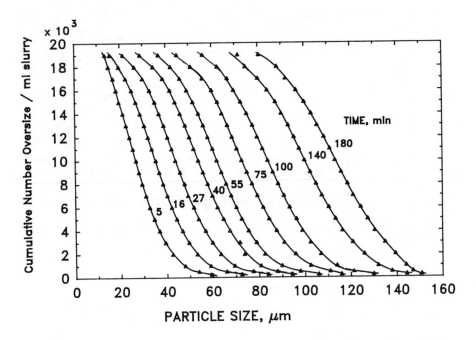

Figure 3. Cumulative number size distributions of samples from fructose crystallization for Run 5.

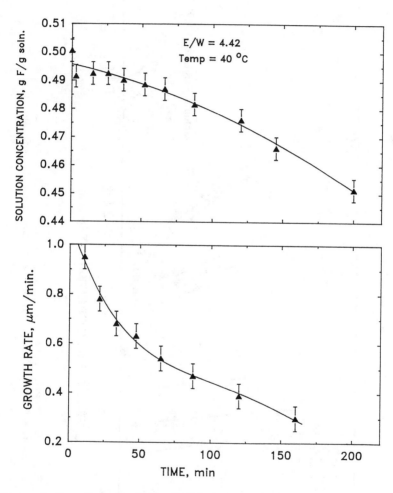

Figure 4. Growth rate and solution concentration as a function of
time for Run 5.

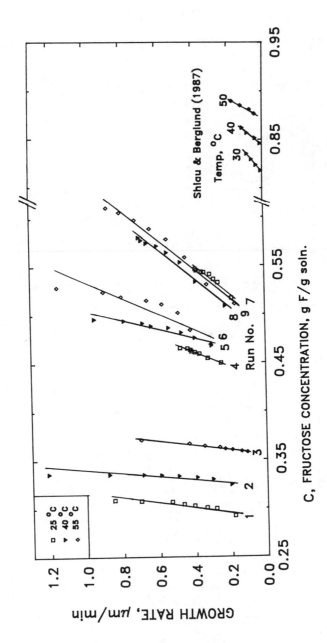

Figure 5. Growth rate vs C for fructose crystallizing from aqueous ethanol. Also shown are the results of Shiau and Berglund (10) for aqueous solution.

Figure 1 were not known with sufficient certainty to pinpoint a value. Further investigations, using an accurate evaluation of the final equilibrium fructose concentration, are in progress to determine this dependence.

Table I.   Conditions for the Crystallization Experiments

Run No.	Temp. (°C)	$\dfrac{E}{W}$	Typical F/W	κ_G^*	$q^\#$
1	25	4.62	2.2	92	0.07
2	40	6.09	3.2	130	0.05
3	55	8.80	4.6	112	0.05
4	25	3.29	2.4	49	0.14
5	40	4.42	3.7	37	0.10
6	55	7.40	4.9	21	0.12
7	25	2.48	2.7	11.4	0.19
8	40	3.48	4.0	9.8	0.14
9	55	5.94	5.4	10.4	0.16

* κ_G from equation 1, $\mu m/min$ (g fructose /g soln)$^{1.25}$

\# q from equation 2

Shiau and Berglund (10) investigated the crystallization of fructose from aqueous solutions (no alcohol) at 30, 40 and 50°C. They determined a dependence of growth rate on supersaturation (in the same units as Figure 5) at the 1.25 power, although there is uncertainty on this value. Their results have also been plotted on Figure 5. For further analysis of the data, their dependence was taken and fitted to the data of Figure 5 to give the best growth rate constant, κ_G where

$$\kappa_G = G/(C-C^*)^{1.25} \qquad (1)$$

Values of κ_G are plotted against E/W in Figure 6 with temperature as a parameter. It might be tempting to fit the effect of temperature at a given E/W with an Arrhenius plot, but this would show the surprising result of an inverse dependence, ie. the growth rate is higher at the lower temperatures. This is no doubt due to the effect of viscosity. At higher temperatures the increase in fructose solubility increases solution viscosity, and thus the film thickness increases and the diffusivity decreases. Further experiments are in progress to evaluate these effects.

It is likely that the temperature dependence found by Shiau and Berglund (10) (for E/W = 0) could also be influenced by this viscosity effect, although they found a normal Arrhenius dependence on temperature.

If the solution viscosity has a marked effect on the growth rate, then mass transfer must play an important part in the growth mechanism.

Crystal Size Spread.   Figure 3 shows that the size range of the crystals increases as the crystals grow. In Figure 3 the standard deviation of the distribution, σ_L is plotted against the mean size, $\bar{L}$, for run 5. This plot is approximately linear, giving a slope q,

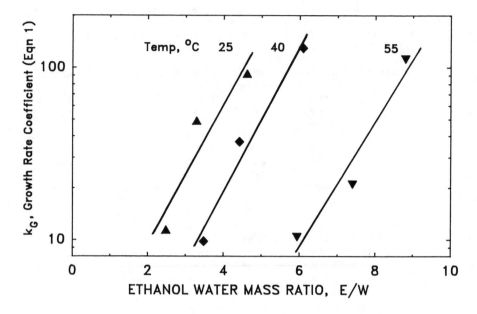

Figure 6. Growth rate coefficient κ_G vs E/W ratio.

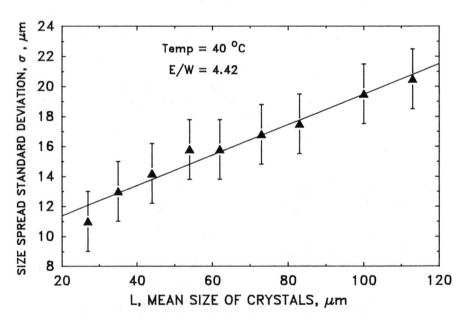

Figure 7. Size spread standard deviation vs mean size for Run 5.

termed the growth spread coefficient, and defined by,

$$q = \Delta\sigma_L/\Delta\bar{L} \qquad (2)$$

Values of q for the nine runs are shown in Table I and are plotted against E/W in Figure 8. Shiau and Berglund (10) measured growth dispersion following individual fructose crystals in a microscope cell. Figure 9 replots their data as σ_G, the standard deviation of the spread of growth rates vs $\bar{G}$, the mean growth rate. This plot appears to give a better linear fit to the data than their plot of σ_G^2. This would be expected from the constant crystal growth (CCG) model (11). The slope of the straight line from Figure 9 has also been plotted on Figure 8, although it is a growth spread coefficient, whereas the data from this work is a size spread coefficient. None the less, the values are comparable.

An increase in the spread of sizes in a batch crystallization could be due to size dependent growth or also growth dispersion. Whatever the cause, the spreading decreases substantially for the high alcohol content crystallizations. There seems to be no current explanation for this effect. In practical terms, this means that batch growth from alcoholic solutions should give a more uniform product than from aqueous solution.

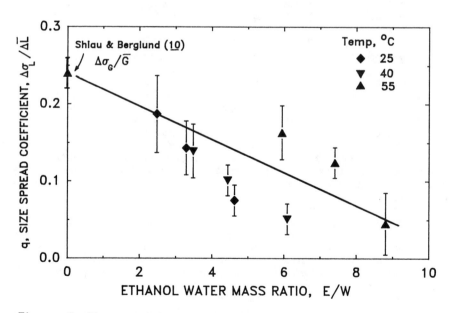

Figure 8. Size spread coefficient, q vs E/W for fructose crystallization from aqueous ethanol solutions.

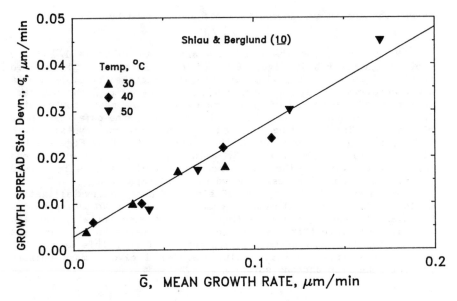

Figure 9. Growth spread standard deviation vs mean growth rate from Shiau and Berglund (10) for fructose grown from aqueous solution.

Conclusions

The kinetic behaviour of fructose crystallization from aqueous ethanolic solutions, typical in composition to those operated on an industrial scale, is strongly dependent on supersaturation, solvent composition and temperature. Provided the supersaturation is kept below 35°C of subcooling, nucleation does not occur.

High growth rates (µm/min) can be achieved. The super-saturation required is less for the higher alcohol contents. The high growth rates in the absence of nucleation means that a practical industrial crystallization process can be developed to grow large fructose crystals.

The limited results obtained are not inconsistent with a 1.25 dependence on supersaturation as proposed by Shiau and Berglund (10). The dependence on solvent composition and temperature appears complex and is probably related to the solution viscosity which varies considerably over this range. It thus appears that mass transfer plays a significant part in the crystallization process. Further work needs to be done to clarify these effects.

A size spread effect was observed. It is comparable with the growth dispersion effects found by Shiau and Berglund (10). The effect decreased with increasing ethanol content.

Acknowledgements

    The authors acknowledge financial support from Queensland Science & Technology Ltd., Australia for this work.

Legend of Symbols

C   —   Fructose concentration, g F/g soln.
C*   —   Fructose concentration at equilibrium, g F/g soln.
E   —   Mass ethanol
F   —   Mass fructose
G   —   Crystal growth rate, μm/min
$\overline{G}$ —   Mean crystal growth rate, μm/min
$\overline{L}$ —   Mean crystal size, μm
q   —   Growth spread coefficient, dimensionless
W   —   Mass water
κ_G —   Growth rate constant, μm/min (g F/g soln)$^{1.25}$
σ_G —   Growth spread standard deviation, μm/min
σ_L —   Size spread standard deviation, μm.

Literature Cited

1. Dziezak, J.D. Food Technol. 1987, 41, 66-7, 72.
2. Forsberg, K.H.; Hamalainen, L.; Melaja, A.J.; Vistanen, J.J. U.S. Patent 3 883 365, 1975.
3. Anon. French Patent 2 154 951, 1973.
4. Edye, L.A.; Johns, M.R.; Ewings, K.N. Appl. Microbiol. Biotechnol. 1989, 31, 129-33.
5. Johns, M.R.; Greenfield, P.F. Proc. 21st Convention Aust. Inst. Food Sci. Technol. 1988, p 36.
6. Doelle, H.W.; Greenfield, P.F. Appl. Microbiol. Biotechnol. 1985, 22, 411-15.
7. Bates, F.J. and associates Polarimetry Saccharimetry and the Sugars; Circ. C-440, Natl. Bur. Standards: US Dept Commerce, Washington, DC, 1942.
8. Young, F.E.; Jones, F.T.; Lewis, H.J. J. Phys. Chem. 1952, 56, 1093-6.
9. Mullin, J.W. Crystallization; Butterworth: London, 1972;
10. Shiau, L.D.; Berglund, K.A. AIChE J. 1987, 33, 1028-33.
11. Berglund, K.A.; Murphy, V.G. Ind. Eng. Chem. Fundam. 1986, 25, 174-6.

RECEIVED May 15, 1990

Chapter 15

Developments in Controlling Directed Crystallization and Sweating

J. Ulrich and Y. Özoguz

Verfahrenstechnik, FB 4, Universität Bremen, Postfach 330 440, D–2800 Bremen 33, Federal Republic of Germany

Directed crystallization of multi-component mixtures is a more and more important purification process for organic components. In order to understand, optimize and control this already industrially applied technology, it is important, to control, and that means also to measure the characteristic parameters of the process. Different new on-line-measuring methods of the growth rate of a crystal layer on a cooled tube give a possibility to predict the achievable purity of the directed crystallization process. After the process step of the crystallization the crystal product can be furthermore purified by the so called sweating step. A first approach to predict the purity during the crystal growth through an on-line-measurement of the growth rate are introduced. Additionally experimental results of the sweating process by the example of the purification of the binary mixture of dodecanole and decanole are discussed.

"Following years of glacial slow development, freeze concentration processing is suddenly heating up" says J. Chowdhury in 1988 (1). He underlines that point by finding more than ten new freeze concentration plants installed in the USA in that year. About the same number of companies offer new or improved processes in the field of directed crystallization as a separation technology.
Compared to the evaporation processes crystallization has a lot of advantages like the energy advantage of about 7 to 1 (for water) and other advantages like a smaller plant volume, less corrosion problems and easier treatment of heat sensitive materials due to lower temperature level, advantages in the field of environment safety due to no possible gas leakages. However it is still the case, that freeze concentration has to look for the niches and is not competing in fields where evaporation or distillation is estab-

0097–6156/90/0438–0210$06.00/0
© 1990 American Chemical Society

lished. Reasons for this are, that directed crystallization as a separation technology is yet not enough explored e.g. problems of transport and treatment of solids are not yet solved as they are for liquids. Besides that an on-line control of the purity as a key value of the process is still not realized. An approach to solve some of these problems are introduced here.

Directed Crystallization as a Separation Process

The main fields of application for directed crystallization as a purification (separation) technology are: organic-chemical separation and purification (2), fruit juice or liquid food concentration (3,4), waste water treatment (1).

The principle of separation by directed crystallization is based on the phase change liquid/solid. A binary or a multi-component mixture is fed to a cooled surface, in most cases as a falling film on the in- or outside of tubes. The crystallization takes place on the cooled surface of the tube and is "directed" horizontal to the surface, therefore results the name "directed crystallization". The crystal layer has a different concentration from the multi-component feed mixture. This effect is used to purify the liquid mixture (melt). In theory complete purity can be reached in one step for the case of eutectic mixtures, if the process is just operated slow enough. In reality however the component with the highest melting point of a multi-component mixture crystallizes (in our case on the cooled surface) incorporating some impurities. The amount of incorporated impurities depends mainly on the growth rate of the solid layer. The residue leaves the process in liquid form. The purified product has to leave the apparatus in solid (crystallized) form or in remolten form also as a liquid. Investigations to improve understanding, performance and control of such a process can be found in earlier investigations (5, 6).

In order to control the separation process it is necessary to control the dominant parameters of the process. Their dependence on each other is demonstrated in Figure 1.

Besides the physical properties of the material the growth rate of the crystal layer is the dominant parameter influencing the purity and the main parameter influencing the yield of product. The growth rate itself is influenced by several parameters such as supercooling, physical properties of the substances, Reynold's number of flow regime, concentration and starting temperature of feed stream. If parameters like volume flow rate, starting conditions of temperature and concentration are constant, the supercooling is directly proportional to the cooling rate of the crystal layer.

The Growth rate of the layer. The growth rate of the layer is rather different of the growth rates of a single crystal. The cumulative growth rate in one dimension of a group of single crystals will add up to the layer growth rate. There is a linear relation between cooling rate of the cooled surface and the growth rate of the crystal layer v_w (as shown in (7)). The equation parameters of this linear relation depend on the above mentioned constant parameters and the physical properties of the material.

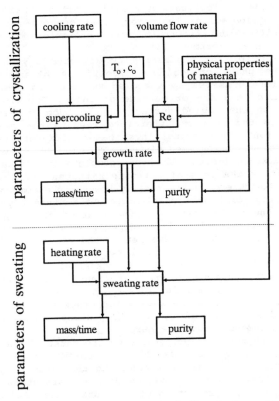

Figure 1. Parameters of the process of directed crystallization
and sweating

Starting from this relation it is possible to define a factor, which describes irregularities in the growth rate occurring by real crystal processes. These irregularities have an important, mostly negative influence on the purity of the product. Among other reasons the metastable zone of the fluid and unknown impurities in ppm concentration ranges are their explanation (the broader the metastable zone the greater probability of irregularities). The irregularities of the growth rate can be qualified by a growth rate deviation defined to:

$$d_v = \frac{v_{W,real}}{v_{W,ideal}} \tag{1}$$

Here the so called ideal growth rate of the crystal layer $v_{W,ideal}$ is a calculated one. It is calculated from the linear dependence between the cooling rate of the surface and the resulting growth rate. The measured growth rate of crystal layer $v_{W,real}$ includes all possible irregularities. The growth rate deviation d_v defined in that way has in most cases a value greater than one, due to growth rate irregularities which are normally leading to a higher growth rate. For example irregularities in growth can lead to more liquid inclusions in the crystal coat and therefore to a larger volume but a lesser purity of the coat.

Describing and Calculating the Quality of the Product

Aim of the directed crystallization is in most cases to reach a crystal coat of highest possible purity. The most used measure to qualify the purification effect of the crystallization is the effective distribution coefficient k_{eff}. For the case of crystallization processes in which the concentration of the feed remains constant the effective distribution coefficient is defined as the ratio of the impurity concentration in the crystal product c_k to the impurity concentration in the feed c_0 :

$$k_{eff} = \frac{c_k}{c_0} \tag{2}$$

k_{eff} is zero for the unrealistic case of theoretical purification to 100% purity of the product and k_{eff} is one for no purification of the product.

Several investigations (8,9) show, that the effective distribution coefficient can be described as a function of the growth rate. Own experiments show that a model to calculate the purity in dependence of the real growth rate is rather realistic. To considerate irregularities by the calculation of k_{eff}, the effective distribution coefficient is presented as a function of the ideal growth rate of the crystal layer $v_{W,ideal}$ and the growth rate deviation d_v .

$$k_{eff} = a \cdot v_{W,ideal}^{b} \cdot d_v^{e} \tag{3}$$

In this equation a, b and e are constants depending on the physical properties of the material to be separated. In Figure 2 the effective distribution coefficient k_{eff} is plotted over the growth rate on the example of a binary mixture of dodecanole with 3,3% impurity of decanole. One curve represents the measured purities ($k_{eff,real}$). It is compared to the curve of $k_{eff,ideal}$ for the ideal growth rate and the curve of $k_{eff,calc}$ calculated from Equation 3. The growth rate deviation used in Equation 3 is calculated according to Equation 1. As to be seen in Figure 2 the calculated effective distribution coefficient fits the measured one rather well. This correlation gives a possibility to predict the purity of the crystal coat during its build up. Supposition for the prediction is the on-line measurement of the growth rate of the crystal coat.

Influence of the growth rate on the purity of product. Besides the physical properties of the substances the growth rate is the dominant parameter influencing the purity of product. The most important effect hereby is, that higher growth rates cause more liquid inclusions. The variation of growth rate can be reached by a variation of the volume flow rate of the falling film or by a variation of the cooling rate of the cooling medium, that means a faster temperature change of the cooled surface. Earlier investigations (10) with the above mentioned binary mixture show, that results of the purity of the crystal layer reached by a variation of cooling rate are nearly the same as those results reached by a variation of volume flow rate. Therefore it is enough for the prediction of the purity to measure the growth rate. It is not important however to know the reason of change of the growth rate.

How the growth rate influences the purity of a product is demonstrated in a simple experiment with a microscope. Under a microscope with a Zeiss Luminar Objective (40mm 1:4/A 0,13) the crystallization was carried out with different cooling rates of a cooled surface. It results in different growth rates of the crystal layer. The different structures of crystal layers underline the great influence of growth rate on the structure as shown in an earlier investigation (10). While the structure by a cooling rate of 0,5°C/min is nearly homogeneous, only dendritic growth (rather rough surfaces) can be found by a cooling rate of 4°C/min. Dendritic growth leads to lots of liquid inclusions, and therefore the resulting separation effect will be rather small.

Measurement of the Growth Rate of Crystal Layer

The conventional ways to measure the growth rates of the crystal layers has always been combined with a stop or interruption of the crystallization process. The most used way of growth rate measurement was to determine the loss of in the feed tank after crystallization or to weigh the mass of crystallized material. With the assumption of a constant growth of the layer thickness, it is possible to calculate an average growth rate. Methods measuring the thickness of the crystal layer after interrupting the process are of the same type as methods measuring the loss of volume in the feed tank. The last mentioned technique is often used in industrial applications in order to find the point to stop the process. All these measuring

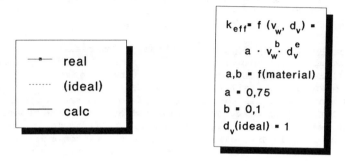

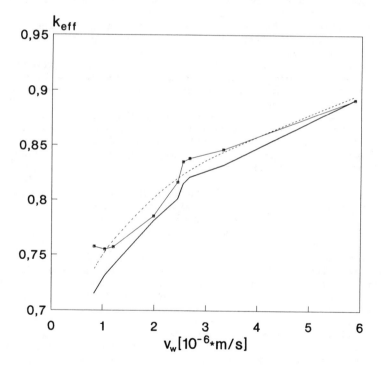

Figure 2. Effective distribution coefficient k_{eff} as a function of the growth rate v_w

methods involve an interruption of the crystallization process and
therefore do not allow an on-line control of the crystallization
process. The only on-line measuring technique so far has been the
feed tank level control. For an on-line purity control however an
on-line crystal layer growth control is necessary.

On-Line Measurement of the Growth Rate. Four examples of on-line
measurement of the growth rate of a crystal layer for the most used
crystallization applications on or in tubes are shown in Figure 3.
The already industrially used method is the liquid level control.
But as mentioned it is so far only used to find the point to end the
crystallization process. In an expanded way it could be used as an
on-line measurement for the growth rate. If irregularities of crystal growth could be measured to correct the cooling rate of the
crystal surface in a servo-loop it would be possible to correct the
growth rate of crystal layer. By this way a control of the purity of
the crystal product is possible. The liquid level measurement can be
used for inside tube crystallization as well as for outside tube
crystallization and has no impact on the crystal coat. Therefore it
is a useful measuring technique for industrial application. The only
disadvantage of this technique is the high difference in liquid
level necessary in order to receive good enough informations about
growth rate. Therefore this method is only applicable in cases of
large feed tanks and therefore not useful for laboratory experiments.
    All advantages of the liquid level measurements can also be
found by the on-line-weighing. Since mass differences are more easy
measurable also in cases of little changes, this technique can be
used successfully in laboratory applications. Some technical
problems in the realization in case of an industrial application
seem however solvable.
    A modern and very exact method of growth rate measurements is
the laser-scanner. The disadvantage of this method is, that it can
only be used on the outside of tubes. In the industrial application
the disadvantage is that by this method only one tube of a bundle of
tubes can be measured at a time.
    The same accuracy as with the laser-scanner can be reached
with thermocouples. However this is only a quasi-on-line measuring
technique. The number of measuring points is limited by the number
of thermocouples. The greatest disadvantage of this method is its
interference during the build up of the crystal coat.
    While the process parameters such as the cooling rate of the
cooling fluid or the volume flow rate of the feed can be measured,
regulated and kept constant, the control of other parameters is not
that easy. One of those other parameters is for instance the supercooling of the mixture. In case of the crystallization it does not
only depend on the cooling rate of the cooled surface, but it varies
also over the length of the tube. It could lead to a non constant
growth rate of the crystal layer over the length of the tube. This
is also an important point in deciding a measuring technique. While
the on-line-weighing and the liquid measuring give a cumulative
growth rate of the whole apparatus, the laser scanner and the thermocouples measurement give only a local growth rate of the layer.

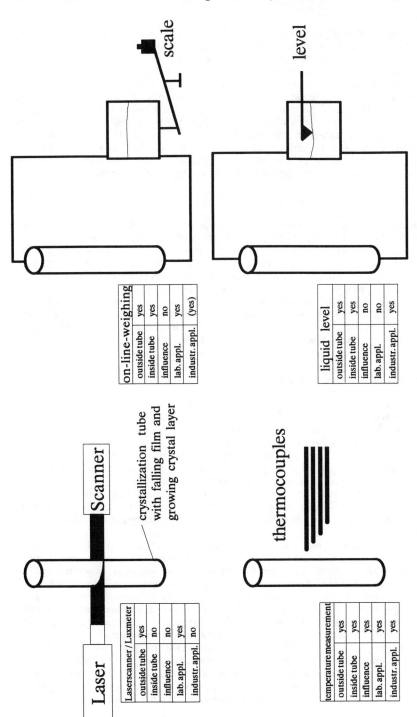

on-line-weighing	
outside tube	yes
inside tube	yes
influence	no
lab. appl.	yes
industr. appl.	(yes)

liquid level	
outside tube	yes
inside tube	yes
influence	no
lab. appl.	no
industr. appl.	yes

Laserscanner / Luxmeter	
outside tube	yes
inside tube	no
influence	no
lab. appl.	yes
industr. appl.	no

temperature measurement	
outside tube	yes
inside tube	yes
influence	yes
lab. appl.	yes
industr. appl.	yes

Figure 3. On-line measurement techniques of the growth rate of the crystal layer

All the mentioned methods have their advantages and disadvantages, therefore every special application will ask for an individual decision of the most useful method.

When discussing growth rate measurements also nucleation has to be taken into consideration. There is an important influence on the consistency of growth rate through nucleation at the beginning of crystallization. To have a good calculable separation efficiency by crystallization it is not only important to have a controlled constant growth rate of the crystal layer, but it is also important to have a constant growth rate over the whole time of crystallization, especially in the unsteady phase of the nucleation, before having a first complete crystal layer on the tube surface.

Apparatus and measuring technique.

The basis of the experimental investigations in this paper is an apparatus where crystallization is performed on the outside of a vertical tube out of a falling film. The tube is temperature controlled from the inside with a constant temperature over the whole length of tube. This tube is representing one out of a bundle of tubes and one stage of a number of stages of a process, which is already industrially applied. The diameter of the tube is 21 mm. The construction of this tube allows the variation of the length of the tube between 200 mm and 400 mm. All growth rate measuring techniques, which have been explained above, are attachable. A more detailed description of the apparatus can be found in earlier works (<u>10</u>).

Sweating for Higher Purity

Additionally to the main crystallization process higher purity can be obtained with the additional steps sweating and washing, whereby the sweating procedure as a partial melting by means of a controlled heating of the crystal layer is used to separate included impurities. The washing procedure is a rinsing of the crystals by a reflux of liquid of higher purity in order to push away the highly contaminated liquid sticking to the crystal surface.

As to be seen in Figure 1 the the purification effect by sweating does not only depend on the heating rate, it depends also on the history (the crystallization) of the crystal coat to be sweated.

The sweating was carried out by means of a controlled heating of the tube from the inside with different heating rates. The investigations show, that the purification of the crystal coat by sweating is independent from the heating rate but the yield of pure material can be enlarged with higher heating rates. The sweating is more effective: 1. when the crystal layer has a higher impurity content. This is analogous to the purification by crystallization, and 2. when the thickness of the crystal layer is greater.

A comparison of the steps crystallization and sweating shows, that the purification effect by sweating can be as high as the purification by crystallization (see also (<u>6</u>)). The loss of mass by sweating (7%-10%) is even less than the loss of mass by crystallization (20%-30%) (<u>7</u>) as to be found in industrial applications.

This shows the possibilities to save energy and time by a good introduction of the sweating procedure and the co-ordination of the two steps crystallization and sweating.

Acknowledgment

The support of this project by the Deutsche Forschungsgemeinschaft (DFG) is gratefully acknowledged.

Legend of Symbols

a,b,e constants depending on material
c   concentration
d_v growth rate deviation
k_{eff} effective distribution coefficient
v_w growth rate of crystal layer

Indices
0   feed
calc   calculated
ideal by ideal conditions
k   crystal product
real   by real conditions including irregularities

Literature Cited

1.   Chowdhury, J. Chem.Engn., 1988, 25, 24-31
2.   Rittner, S.; Steiner, R. Chem.-Ing.-Tech. , 1985, 2, 91-102
3.   Almasi, E. Die Kälte, 1986, 21, 464-7
4.   Harz, P. Chem.-Ing.-Tech., 1989, 2, 155-7
5.   Ulrich, J. In Industrial Crystallization 87, Nyvlt, J., Ed.; Elsevier: Amsterdam, 1989; pp 585-8
6.   Ulrich, J.; Özoğuz, Y. Chem.-Ing.-Tech., 1989, 1, 76-7
7.   Ulrich, J.; Özoğuz, Y. Proc. The 9th International Conference on Crystal Growth (ICCG-9) in Sendai (Japan), 1989, to be publ. in J. Crystal Growth
8.   Myasnikov, S. K.; Murav'ev, Yu. N.; Malyusov, V. A. Theor. found. of chem. Eng., 1988, 1, 23-30
9.   Kasaymbekov, B. A.; Myasnikov, S. K., Malyusov, V. A. Theor. found. of chem. Eng., 1985, 1, 17-23
10.   Ulrich, J.; Özoğuz, Y. Proc. 5th Int. Conf. on Eng. and Food in Köln, 1989, to be published by Elsevier Sciences Pub. Comp.

RECEIVED March 29, 1990

Chapter 16

In Situ Observation of *p*-Xylene Crystal Growth by Controlling Pressure

Masato Moritoki and Nobuhiko Nishiguchi

Kobe Steel, Ltd., 1–3–18, Wakinohama-cho, Chuo-ku, Kobe 651, Japan

By means of in-situ observation of crystal growth
under very high pressure, a defect healing process was
studied with p-xylene crystals in p-, m-xylene
mixture. Crystals, partially melted by decrease in
pressure, were repressurized to grow and to heal
their melted surfaces. It was found that the
half-melted crystals rapidly grow and recover their
rectangular growth shape within a few minutes.

High pressure crystallization process has already been applied in
industry to separate a pure substance from a mixture. In this
process, pressure is used for crystallization, instead of cooling in
the conventional method. Although the pressure is usually applied
in industry so quickly that it reaches 200 MPa in 10 sec, a
sufficient amount of crystals can be grown, which results in
attaining the separation and refining of a short cycle within a few
minutes (1).
It was interesting to study the nucleation and growth behaviour
of crystals during pressure application, or under a steady
pressurized state in a supersaturated condition. So, they were
studied with p-xylene crystals from m-, p-xylene mixture by means of
in-situ observation. And some results have been presented both on
nucleation (2) and on growth behaviour (3) elsewhere, which were
related to supersaturation pressure and p-xylene concentration.
In this paper, observation results of defect healing of a
p-xylene crystal are discussed. The crystal had been partially
melted by a pressure decrease, and healing was caused by a slight
re-application of pressure.

Experimental apparatus and procedure

Figure 1 shows a schematic diagram of the experimental apparatus. A

0097–6156/90/0438–0220$06.00/0

high pressure optical cell was kept in an oil bath. A sample space
inside the cell was surrounded by an cylindrical thermal insulator,
and the sample was pressurized with an intensifier. Nucleation and
growth behaviour were observed with a microscope through a sapphire
window, and recorded by a VTR. The inside temperature, measured by
a thermo-couple, and the pressure were recorded together with the
elapsed time. The apparatus can be used upto 500 MPa, and the
visible diameter is 10 mm. The length of the sample space is
adjustable, and was 10 mm in these experiments.

The experiments were carried out with samples of m- and
p-xylene mixtures at around 20°C. P-xylene concentration was 70, 80
and 90 mol %.

A fundamental pressure procedure used in this experiment is
shown in Figure 2. First, pressure was applied to the sample to
crystallize some portion of it. After a few seconds, pressure was
decreased rather rapidly so as to allow most of the crystals to
melt, but some crystals still remained. Soon after, the pressure
was increased again to prevent further melting, and adjusted to the
pressure Po, where the remaining crystals did not grow or melt
further. It was maintained there for five or ten minutes. After
the temperature stabilized, a slight additional supersaturation
pressure   P was applied. And then, the growth behaviour of the
partially melted crystal surfaces was observed continuously.

When the pressure was adjusted to Po, the phase diagram, Figure
3, presented by us before (4), was referred to. However, in
principle, Po is not always the same as the melting pressure in this
figure, because there were some crystals remaining in the cell.
Moreover,   P is not exactly the same as the supersaturation
pressure, because there was a slight change of the equilibrium
pressure caused by a temperature change due to the application of
  P. However, these errors were not so serious that they needed
correction in these experiments.

Results

Figure 4 shows an example of the healing process of a partially
melted p-xylene crystal in a 90% p-xylene mixture. The crystal
surface on the far side was close to the innersurface of the window.
The smooth line shows the half-melted surface. A   P of 4 MPa was
applied, changing the pressure from 37 MPa to 41 MPa in five
seconds. Soon afterwards, numerous very fine steps appeared all
over the half-melted surface. They grew and grew, and the adjacent
steps bunched together. In the time of 2.6 minutes, the crystal
almost formed a rectangular plate-like shape, which is the growth
shape of this material, though the thickness was still gradually
increasing and healing to form a plane surface at the near side of
the crystal surface.

This process was schematically illustrated in Figure 5. On the
melted surface, first, numerous steps appear and grow accompanied by
bunching. When the height or width of steps reaches about 0.1 or
0.2 mm, all the steps begin to grow independently without causing
further bunching. As a result, every corner and every edge moves in
a parallel fashion as shown by the arrows in this figure. Each step

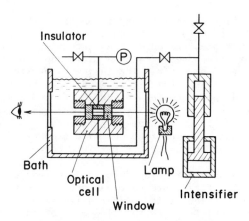

Figure 1.   Schematic diagram of the optical cell.

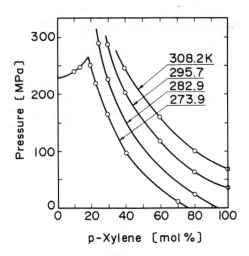

Figure 2.   Typical pressure procedure in the experiments.

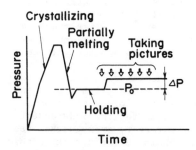

Figure 3.   Isothermic phase diagram of m-, p-xylene mixture.

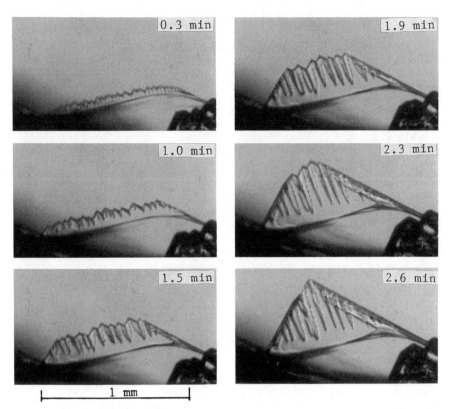

Figure 4.   Healing behaviour of a p-xylene crystal in a 90%
p-xylene mixture with elapsed time (Po=37MPa,   P=4MPa, T=21°C).

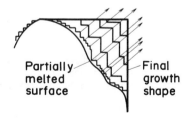

Figure 5.   Schematic diagram showing the step growth.

stops growing when it reaches an outer surface to form the final growth shape.

We have already stated in the previous paper (3) that, under the supersaturation pressure of 5 to 8 MPa, the longitudinal growth rate of a rectangular pillar-shaped crystal attained 1.0 mm/min.MPa in the same concentration, whereas the transverse growth rate was too low to detect.

In these pictures, the growth rates of the step surfaces in both directions are quite similar to each other, and they are roughly evaluated to be 0.05 mm/min.MPa under a supersaturation of 4 MPa. On the other hand, there is scarcely any growth in the maximum length and width of the parially melted crystal during the shape healing.

Figure 6 shows an example of the healing behaviour in a mixture of 70% p-xylene. Po was 62.9 MPa and   P was 12 MPa. Fine step appearance, bunching, parallel growth of bigger steps were all similar to the above. However, when the steps became bigger, as seen in the pictures at 1.1 and 2.0 min, the growth rate at the corner on the near surface began to slow down, resulting in the formation of a groove, and the groove deepened.

But still, as seen in the pictures at 2.0, 3.2, and 5.5 min, the corners at the far surface were at right angles, and the bottoms of the grooves gradually grew, following to the far side corners. During these times, the thickness also increased very gradually and steadily. The same phenomena as Figure 6 can be seen in Figure 7. It was found that this tendency increases with increasing supersaturation pressure and with decreasing p-xylene concentration in the liquid.

Through all the experiments, it was interesting to see that the half-melted crystal seemed to have known by itself the final rectangular growth shape whose surfaces are geometrically touching the initial half-melted crystal surfaces. Thus, these surfaces seemed to appear first and to grow most easy.

In the first picture of Figure 4, the surface shown by the arrow appeared very soon after the superpressurization, and it formed a portion of the outer surface of the stable growth shape.

Figure 8 shows the healing process in a 70% p-xylene mixture. In the first picture, numerous stripes seen on the slope are the fine steps. The top of the crystal soon became flat, and then became like the top of a hammer. As time passed by, the top became bigger, and then, a rectangular pillar formed.

A similar phenomenon is seen in Figure 9, which was in the 90% p-xylene. The top first became flat, and grew increasing in width. And this part grew to form an independent crystal in the last picture. However, in this case, the both crystals combined after a short time.

Considerations

Based on the phenomena mentioned above, the growth mechanisms were considered to be as follows.

(1) By the application of   P, numerous invisible and microscopic steps will soon appear, and will grow and bunch together to become

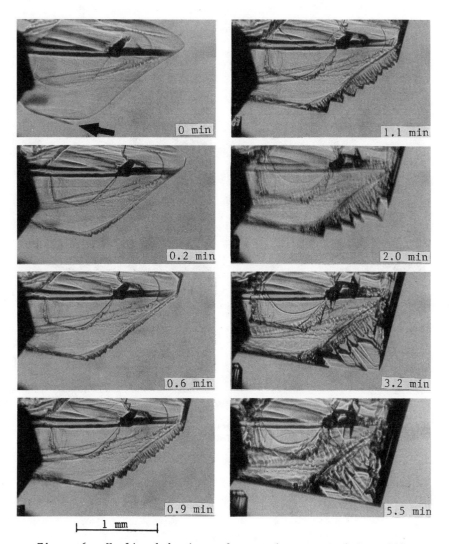

Figure 6.   Healing behaviour of a p-xylene crystal in a 70%
p-xylene mixture (Po=62.9MPa,   P=12MPa, T=19.8°C).

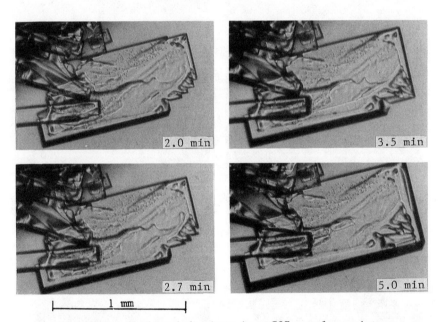

Figure 7. Healing behaviour in a 70% p-xylene mixture
(P=63.9MPa,   P=5.0MPa, T=20°C).

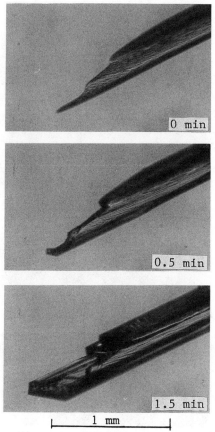

Figure 8. Healing at the top in a 70% p-xylene mixture
(Po=65MPa,   P=12MPa, T=20.6°C).

visible. As the half-melted surface is thermally roughened, there
must be a lot of molecular size defects, which will play the role of
kinks in Kossel's Model. The steps also grow in the same manner,
when they are small.

(2) When the steps become somewhat large, there must be some
concentration gradient in the liquid phase near the steps. The
edges should contact the liquid containing a higher concentration of
p-xylene than the corners as illustrated in Figure 10. Still, the
steps grow in a parallel fashion. In this stage of growth, Berg's
Effect plays an important role in forming the plane surface of the
step, where the surface integration will be the predominant growth
rate determining factor.

(3) When the steps become bigger, or p-xylene concentration is
lower, the corner in the steps forms a groove. In this tage, the
diffusion effect is increasing in rate determining factors instead
of the surface integration.

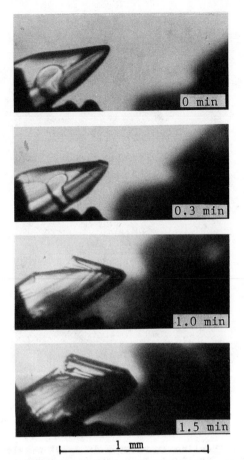

Figure 9. Healing at the top in a 90% p-xylene mixture
(Po=37MPa,   P=5MPa, T=20.5°C).

(4) The growth rate of the total length or the total width of the
partially melted crystal, or of the outer surface of the growth
shape was negligiblly small at under the experimental conditions,
but all the steps grew rapidly and similarly in a parallel fashion.
Besides, the outer surface of the growth shape could be formed
rather easily. Judging from these facts, it might be possible that
the edge of the steps plays an important role in the growth.

Conclusions

The healing phenomena of the partially melted p-xylene crystals
could be successfully studied by means of in-situ observation under
pressure. This was due to the fact that the supersaturation was
applied by pressure without causing movements of the crystals or the
liquid during the observation, and without introducing much of a
temperature gradient.

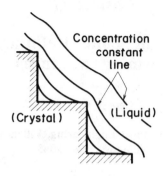

Figure 10. Concept showing relation between the steps and the concentration gradient in the liquid phase.

Defects of crystal surfaces were easily and rapidly healed to form the simple growth-shaped crystals. This suggests that the method used in these experiments can be usefully applied to form a single crystal in the liquid phase.

The growth mechanisms of healing were also discussed. It is believed that these mechanisms discussed above will lead to important suggestions to study the growth behaviour caused by such a rapid pressure application as used in the practical high pressure crystallization.

Literature cited

1. Moritoki, M., Kitagaw, K., Onoe, K., Kaneco, K. In Industrial Crystallization 84; Jancic, S.J., de Jong, S.J., Ed.; Elsevier; Amsterdam, 1984; p 377.
2. Nishiguchi, N., Moritoki, M. In Proc. Int. Sym. on Preparation of Functional Materials and Industrial Crystallization '89 Osaka; Harano, Y., Ed.; The Kansai Branch of Soc. Chem. Eng. Jap., 1989, p 189.
3. Nishiguchi, N., Moritoki, M., Tanabe, H., In Abstr. of The Ninth Int. Conf. on Crys. Growth (Sendai), 1989, p 132.
4. Moritoki, M., Kagaku Kogaku Ronbunshu, (Soc. Chem. Eng. Jap.), 1979, 5, 79 - 84.

RECEIVED May 15, 1990

Chapter 17

Solute Transfer in Zone Refining of Eutectic-Forming Mixtures

George C. Yeh

Department of Chemical Engineering, Villanova University,
Villanova, PA 19085

A simple physical model called "Filtration Model"
has been proposed to describe the separation process
of eutectic—forming mixtures by zone refining, in
which the solute transfer through the partially
solidified zone (mushy region) is considered to be
the rate—controlling step of the overall solute
transfer process. Two generalized correlations to
account for the effects of free convective mixing
upon the macrosegregation rate have been developed.
The applicabilities of the proposed model and of
these correlations have been demonstrated for a
wide range of experimental conditions; and the
criteria for the maximum and zero separation were
also established from the correlations.

Following the earlier work by Pfann (10) in 1952 on zone refining many
publications have appeared in the literature. In order to describe the
longituidinal solute transfer in zone refining, various theories
assuming complete mixing (5 & 11) or pure molecular diffusion mechanism
in the melted zone using the boundary—layer treatment have been
proposed (19 through 23). Most of these early works have assumed the
existence of a constant distribution coefficient and an equilibrium
between the solid and melt with a few exceptions, which have devoted
to eutectic—forming mixtures which do not have constant distribution
coefficients (5, 20 through 23).

In all early works, it has been assumed that there exists two
well—defined planner interfaces. However, in the actual experiments
a partially solidified zone (P.S.Z.) or 'mushy region' and a partially
melted zone (P.M.Z.) always exist between the completely resolidified
and fresh solids. The existence of a P.S.Z. (mushy region) behind the
freezing front is especially important since it offers the greatest
resistance to the forward solute transfer and can trap the solute
flowing backward and can affect the nature and the extent of 'macro-
segregation'.

The existence of a P.S.Z. (mushy region) and its importance in
solute transfer during solidification of a binary mixture have been
recognized by more recent workers (24 through 29). Simple models

0097–6156/90/0438–0230$06.25/0
© 1990 American Chemical Society

relating the temperature and the liquid mass fraction for mushy region
(30, 31), and more comprehensive models taking into account the
interaction of momentum, heat and mass transfer processes inside the
mushy region in order to predict macro-segregation patterns were
presented recently (32 through 38). All these works dealt with
solidification in a stationary system and have not been applied to zone
refining, which takes place in a nonstationary system in which the
P.S.Z. (mushy region) and the melt move in a spacial relation.
    This study has been conducted with the objectives: 1) to analyze
and determine the effect of the free convective mixing in the melted
zone upon the over-all solute transfer process in zone refining under
various conditions and 2) to take into account the solute transfer both
in the melted zone and in the adjacent P.S.Z. (mushy region) in
predicting the rate of over-all solute transfer. The temperature
distribution, composition, zone travel speed, zone dimension and other
important parameters are considered in this study.

THEORETICAL

The conditions of freezing and of mixing in the molten solid play
decisive and delicate roles in determining the separation results. In
normal operation, the molten solid is cooled externally; as a result
heat transfer in both radial and longitudinal directions, and the
crystals of solid grow in the directions opposite to that of heat
transfer. The radial mass transfer becomes important when rapid
cooling rates and high zone speeds are employed; under these conditions, the simple assumption of unidirectional mass transfer across the
planner freezing interface deviates greatly from the real conditions.
    Incomplete separation is resulted not only by the radial and the
anisotropic segregation due to "coring" (1, 2, 3, 4, & 8) or "constit-
utional subcooling" (2, 3, 8, 12, 13 through 18) but also by the
incomplete mixing of the molten solid. The melt may be mixed by free
convection. In fluid mechanical sense, complete mixing of a fluid
requires the conditions for locally isotropic turbulence to exist, and
consequently some of the solute which have once been removed and
carried away from the freezing interface may be transported back by the
convective currents. Such a redistribution of solute in the P.S.Z.
(mushy region) is referred to as micro-segregation. An increase in the
intensity of free convection always results in the corresponding
decrease of the thickness of the liquid layer or the resistence to the
forward solute transfer; as a result, the degree of back-mixing of
solutes and the rate of resultant contamination in the P.S.Z. (mushy
region) are also increased.
    Since the results of zone refining depend on the interaction of
momentum, heat and mass transfer in the system, all the basic factors
affecting these three processes, both molecular and convective, have
to be taken into consideration. These basic factors are: concentration W, Density ρ , viscosity μ, heat capacity Cp, temperature densification coefficient β , thermal conductivity k, molecular diffusivity D, zone diameter d_t, zone length L, zone travel speed u,
temperature difference in zone Δ T and acceleration g. The concentration W may affect ρ , μ , Cp, β , k, and D as well as the properties
of the P.S.Z. (mushy region). Aside from the concentration W, all

other factors may be combined into the following dimensionless groups:

D/uL, D_e/uL ($=D/uL_e$)   :   Parameter for molecular diffusion model in a moving zone, equivalent to the reciprocal of Peclet number, dispersion number

$d_t u \rho / \mu$:   Reynolds number, Re

$Cp\mu/k$:   Prandtle number, Pr

$\mu/\rho D$:   Schmidt number Sc

$h^3 \beta \Delta Tg(\rho/\mu)^2$:   Grashof number, Gr, where $h = (L' + d_t)/2$ for melted zone.

The product of Pr and Gr numbers may be used as the parameter to account for free convection effect.

The boundary layer treatment requires the knowledge of the effective thickness of the layer, which cannot be determined experimentally. Wilcox (19) and Thomas (13) applied to zone melting the prediction method for vertical flat plat originally proposed by Holmes et al (6); but its applicability has not been proven. Kraussold (7) has developed a generalized correlation between the ratio of effective thermal conductivity for free convection to the molecular thermal conductivity k_e/k and the product of Pr Gr for a fluid enclosed between two plain walls. By the analogy between heat and mass transfer, the values of K_e/k should be identical to that of D_e/D, the ratio of effective diffusivity for free convection to molecular diffusivity in a given system. By the definition, the effective boundary layer thickness L_e may be calculated by the relationship $L_e=L(D/D_e)$. The values of L_e so obtained depend on the degree of free convective mixing, which may be represented by the Pr Gr value of the system. It may be suggested that either molecular diffusivity D be replaced by the effective diffusivity D_e or the total zone length L be replaced by the effective boundary layer thickness L_e to account for the effect of free convection. To do this, a new parameter uL/D_e may be used instead of uL/D in correlating the separation results of zone refining.

The solute transfer in zone refining is an extremely complicated, unsteady state process involving numerous steps and phenomena. In the direction of zone travel, the solute transfers through the P.S.Z. (mushy region) mainly by molecular diffusion, and partly by eddy diffusion near the freezing front. In the completely melted zone (C.M.Z.), the solute transfers by both molecular and eddy diffusion from the freezing front to the plane of complete melting. Near the front of the outer boundaries of both the P.S.Z. and the P.M.Z., molecular diffusion is predominant, but near their inner boundaries eddy diffusion plays a major role. In the direction opposite to the zone travel, there is the transfer of solute from the C.M.Z. into the P.S.Z. as a result of the relative movement of the P.S.Z. into the melt and the back-mixing of the solute by convective currents. The rate of backward solute transfer depends not only on the degree of back mixing and the zone travel speed, but also on the nature of the P.S.Z., the freezing rate, and the concentration difference across the P.S.Z., as may be obvious.

Each of the above mentioned steps is at a highly unsteady state; the individual transfer coefficients, the driving forces and the resistances vary with distance as well as time. The theoretical prediction of all of these values is impossible, and their experimental determination would be impractical. Therefore, the solution of any rigorous mathematical expressions for the rates of both forward and backward solute transfer processes is impossible. However, the rate of the overall process (macro-segregation) may be given by using a simple model which represents the overall effect of the solute transfer in zone refining.

Consider a physical model with the following simple mechanisms:

1. A molten solid containing a solute of increasing concentration is flowing at a constant velocity through a porous medium whose porosity decreases with the distance; or one may visualize the porous medium as moving through the molten solid at the same relative velocity. The effect would be the same. In this case, the porous medium is the P.S.Z. (mushy region).

2. The porous medium, namely the P.S.Z., possesses a certain filtration capacity to filter out the solute. This corresponds to the rate of forward mass transfer across the P.S.Z. in zone refining.

3. The unfiltered solute in the molten solid transfers through the porous medium, viz. the P.S.Z., and finally reaches the plane of complete freezing, where it finally solidifies completely. This determines the solute concentration in the treated solid.

In Step 1, one visualizes the increasing solute concentration in the complete melted zone with time, the varying degree of solidification along the thickness of the P.S.Z., or mushy region and the constant relative velocity between the two zones (or the heater and the cooler). In Step 2, one visualizes the rate of forward mass transfer through the P.S.Z., which is the rate-controlling step in the overall forward mass transfer process. In Step 3, one visualizes the rate at which the unfiltered solute is trapped inside the treated solid. If in Step 1, the relative velocity of the two zones approaches zero then the rate of Step 2 becomes indefinitely large, and according to the phase diagram the separation should be complete, or the rate of Step 3 would be zero.

The model proposed above is analogous to a continuous, unsteady state filtration process, and therefore may be called "Filtration Model". In this model, the concentration of the filtrate, viz. the concentration of the solute remained in the treated solid is one's major concern. This is given by the rate of Step 3, which may be expressed by an equation similar to Fick's Law including a transmission coefficient D_m for the porous medium, viz. the P.S.Z. and the concentration difference ΔW across the P.S.Z. as the driving force, and the thickness of the P.S.Z. as the distance Δx.

$$j = \frac{dm}{dtA} = -\rho_l D_m \frac{dW}{dx} \text{ , or}$$

$$\rho_l D_m \frac{\Delta W}{\Delta X} = \rho_l D_m \frac{W_l - W_s}{\Delta X} \tag{1}$$

The following assumptions may be made in order to solve Equation 1:
1) Constant A and ρ
2) $\rho_s = \rho_1 = \rho$, or $\Delta\rho = 0$
3) The amount of solute in the P.S.Z. having the thickness Δx is negligible compared to that in the C.M.Z., and may be neglected in the material balance.

To determine the value of D_m from the experimental results, the data of W_s as a function of z are needed. Thus, from the mass balance the amount of solute in the treated solid between 0 and z is

$$A\rho\int_0^z W_s(z)\,dz$$

the amount of solute in the untreated solid between 0 and z+L is

$$A\rho(z+L)W_0$$

and therefore the amount of solute in the C.M.Z. at t=t, or z=z is

$$A\rho\left[(z+L)W_0 - \int_0^z W_s(z)\,dz\right]$$

The total mass in the C.M.Z. is equal to $A\rho L$; therefore the average solute concentration in the C.M.Z., $\overline{W}_1$ is

$$\overline{W}_1 = \frac{z+L}{L}W_0 - \frac{1}{L}\int_0^z W_s(z)\,dz \tag{2}$$

On the other hand, dm/dt may be evaluated as follows:

$$(m_s)_{0\to z} = A\rho\int_0^z W_s(z)\,dz$$

$$(m_s)_{0\to z+\Delta z} = A\rho\int_0^{z+\Delta z} W_s(z)\,dz$$

$$(\Delta m_s)_{z\to z+\Delta z} = A\rho\int_z^{z+\Delta z} W_s(z)\,dz$$

$$\frac{dm}{dz} = \lim_{\Delta z\to 0}\frac{\Delta m}{\Delta z} = \lim_{\Delta z\to 0}\frac{A\rho\int_z^{z+\Delta z} W_s(z)\,dz}{\Delta z}$$

$$= \lim_{\Delta z\to 0} A\rho W_s(z+\xi\Delta z)$$

$$= A\rho W_s(z) \tag{3}$$

where $0 < \xi < 1$. Since $dm/dt = u(dm/dz)$, Equation 3 may be combined with Equation 1 to give

$$\frac{dm}{dt} = u A \rho W_s(z) = A \rho \frac{D_m}{\Delta x} (W_1 - W_s) \text{, or} \tag{4}$$

$$A \rho \frac{D_m}{\Delta x} = A \rho \frac{u W_s(z)}{W_1 - W_s(z)} \tag{5}$$

$$\therefore u W_s(z) = \frac{D_m}{\Delta x} \left[\left(\frac{z+L}{L}\right) W_0 - \frac{1}{L} \int_0^z W_s(z) dz - W_s(z) \right] \tag{6}$$

D_m may be evaluated by Equation 6 by using the experimental data of W_s as a function of z, $w_s(z)$. D_m as determined by Equation 6 is the differental D_m which, according to the Filtration Model proposed, should vary almost linearly with the variation of W_s. The rate of solute contamination by the surface adsorption usually increases as the solute concentration in the completely melted zone $\overline{W}_1$ increases. The integral coefficient $\overline{D}_m$ may be evaluated by

$$\overline{D}_m = \frac{1}{W_s(z_2) - W_s(z_1)} \int_{W_s(z_1)}^{W_s(z_2)} D_m(W_s) dW_s \tag{7}$$

in which D_m is given as a function of W_s, $D_m(W_s)$. The value of $\overline{D}_m$ thus obtained is a mean value of D_m for a run carried out between W_{s1}, and W_{s2}. To develop any generalized correlations between $\overline{D}_m$ values of experimental runs caried out at various W_s and any experimental variables, the ratio $\overline{D}_m/\overline{W}_s$ may be used.

Solute Redistribution in a Single Zone Pass:

In order to predict the solute redistribution, $W_s(z)$ by Equation 6 the values of D_m and Δ x are needed. As will be shown below, these values may be obtained from the generalized correlations developed in this study (see Figure 5 and Figure 6). Thus, if D_m may be given by the simple expression, $D_m = \alpha W_s$ (see Results section) where α is a constant, then Equation 6 may be rewritten as

$$\frac{u \Delta x - \alpha W_0}{\alpha} = \frac{W_0}{L} z - \frac{1}{L} \int_0^z W_{s1}(z) dz - W_{s1}(z) \tag{8}$$

After obtaining derivatives for both sides, separating the variables, and integrating one obtains

$$\int_0^z \frac{d W_{s1}(z)}{W_0 - W_{s1}(z)} = \int_0^z \frac{dz}{L} + C_1$$

$$\therefore W_{s1}(z) = W_0 - e^{-(C_1 + \frac{z}{L})} \tag{9}$$

where C is a constant.

By substituting Equation 9 into Equation 8 and simplifying, one obtains

$$e^{-C_1} = \frac{u \, \Delta X}{\alpha} \tag{10}$$

$$\therefore W_{s1}(z) = W_0 - \frac{u \, \Delta X}{\alpha} e^{-z/L} \tag{11}$$

Equation 11 applies to the whole refrozen solid except the last melted zone to freeze, for a single zone pass.

Solute Distribution in Multi-pass of Zone:

For ith zone pass, the corresponding mass balance equation similar to Equation 6 would be

$$u W_{si}(z) = \frac{D_{mi}}{\Delta X} \left[\frac{1}{L} \int_0^{z+L} W_{si-1}(z) dz - \frac{1}{L} \int_0^z W_{si}(z) dz - W_{si}(z) \right] \tag{12}$$

where $D_{mi} = \alpha \, W_{si}(z)$. This equation applies to all the zones where the ultimate distribution has not been reached. Solving for $W_{si}(z)$, one obtains

$$W_{si}(z) = W_{si-1}(z) - e^{-\left(C_i + \frac{z}{L}\right)} \tag{13}$$

where C_i is a constant. To evaluate the constant C_i, Equation 13 is substituted into Equation 12, resulting

$$\frac{u \, \Delta X}{\alpha} = \frac{1}{L} \int_0^{z+L} W_{si-1}(z) dz - \frac{1}{L} \int_0^z \left[W_{si-1}(z) - e^{-\left(C_i + \frac{z}{L}\right)} \right] dz$$

$$- \left[W_{si-1}(z) - e^{-\left(C_i + \frac{z}{L}\right)} \right] \tag{14}$$

In order to evaluate each integral in Equation 14, the values of $W_{si-1}(z)$ must be known as a function of z, in the form of Equation 13. Therefore, the general analytical solution of Equation 14 is impossible. Instead, the computer solutions may be obtained by first substituting Equation 9 into Equation 12 to obtain W_{s2} as a function of z, $W_{s2}(z)$ as Equation 13 and then substituting $W_{s2}(z)$ into Equation 12 again to obtain $W_{s3}(z)$. This procedure is repeated until $W_{si}(z)$ is obtained. In this method, it is assumed that the values of u , L,Δ X, and α are constant. The deviation of the predicted results from the actual distribution for each zone pass depends mainly on the adequacy of the expression, $D_m = \alpha W_s$, and would be magnified as the number of zone passes is increased.

EXPERIMENTAL

A. Apparatus and System

A modification of the zone refiner made by the Fisher Scientific Co. was used, and its working mechanism is shown in Figure 1. The mixture to be separated is casted and sealed in a heavy-wall glass tube, and the melted zone is formed by heat from a loop of Nichrome ribbon. The power supplied to the heater and hence the heat generated, is regulated by a powerstat. The heat input throughout each run was kept constant. Cool air of a constant temperature is blasted from a hollow ring mounted above the heater; the distance between the two can be varied in order to vary the melted zone length. An atmosphere of constant temperature around the freezing interface was maintained by closing both upper and lower sides of the cooling ring with plastic disks to make an annular enclosure around the glass tube. The carriages carrying the heater and the annular colling ring are moved together along the sample tube by a variable-speed motor. To measure the zone travel speed a needle attached to the cooling ring is pointed to a scale located in parallel with the sample tube, and the distance moved is read directly to give the zone travel speed, which can be varied from 0.77 to 7.24 cm/hr. The sample tube is held securely by a collar with three positioning screws to permit the use of a sample tube as large as 2.0 cm O.D. The zone refiner can be used in any desired positions although the vertical position with the zone traveling downward was chosen in this study for better separation. The zone refiner can be used for a single zone pass or automatic multipass refining. The temperature of cooling air is controlled using a cold bath. The air rate can be varied from zero to 430 in³/min at 14 psi. A constant air temperature between 13° and 30°C can be maintained, and is measured in the cooling ring by a thermocouple. The eutectic-forming mixtures of benzoic acid (9.2 to 17.5 wt %) in naphthalene were used.

B. Procedure

To prepare samples for separation experiments, mixtures of naphthalene and benzonic acid of known concentrations are melted and thoroughly mixed; and the melt is poured into a glass tube immersed in a hot bath to obtain a compact, homogeneous casted sample containing neither air

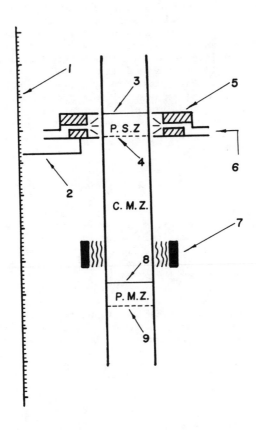

1 . Distance scale	6 . Cooling air inlet
2 . Indicator needle	7 . Heating ribbon
3 . Completely solidified plane	8 . Completely melted plane
4 . Freezing front	9 . Melting front
5 . Cooling ring	

Figure 1.   Experimental Zone Refining Apparatus

bubbles nor segregated parts. To analyze the solute redistribution, the treated sample is removed from the glass tube by breaking the glass; and then cut into piece by piece 1.0 cm long. Each sample piece is then weighed and dissolved into 95% ethanol for the titration by a standard sodium hydroxide solution. Except the results from the runs using a glass tube of 0.3 cm O.D. (or 0.10 cm I.D.) the titrated results were reproduceable within $\pm$ 5%.

In order to determine the temperature profile across the melted zone of each run, a separate, identical run is carried out under the exactly same conditions. The reproduceability of the experimental runs was checked by comparing the separation results of both runs. A copper–constantan thermocouple is situated in the casted sample by inserting it during the casting, to measure the temperature of the melted zone while it is passing the thermocouple.

RESULTS AND DISCUSSION

All the experimental runs are listed in Table I. The thirteen double runs identified by the letters BT in their run numbers were made for determining both solute redistribution and temperature profiles by carrying two identical runs under the same conditions. These thirteen double runs were designed to cover the maximum range of Sc·Re (4.7 to 74.0) and of Pr·Gr (6.6×10^{-7} to 1.0×10^{-5}) possible in this study. The temperature profile across the melted zone of each run has been plotted in Figure 2 from which the lengths of the P.M.Z., of the C.M.Z. and of the P.S.Z. (mushy region) were measured. To determine the lengths of the P.S.Z. from the temperature profile, the location of the cooling ring is taken as the freezing front and plane of complete solidification is considered to correspond the eutectic point. The region between these two temperature points is defined as the P.S.Z. The plane of complete melting is found by noting the temperature corresponding to the initial composition W_o from the phase diagram, and the region between this temperature point and the freezing front is defined as the C.M.Z. Accordingly, the region between the plane of complete melting and the melting front corresponding to the eutectic point is considered to be the P.M.Z. The maximum temperature difference ΔT, the distance between the hottest and the coldest points L' and the average temperature $\overline{T}$ in the C.M.Z. also are determined from the same plots.

The solute concentration W_s from each run has been expressed as a step function of the distance z, from which ϕ_s ($=W_s/W_o$) is calculated. The values of all the experimental parameters, D/uL, D_e/uL, $Lu\rho$ $/\mu$, μ/LD, $C_p\mu/k$ and ($h^3\rho^2 g/\mu^2$) ($\beta\Delta T$) etc. are calculated from the values of the related physical properties in the literature (19). In Figure 3, ϕ_s is correlated with z/L for all the thirteen runs, for which the values of effective diffusivity D_e in the melted zone have been predicted from the Kraussold correlation (7) using the experimental values of Pr·Gr number; and the values of the parameter Pe calculated show that little improvement has been made by using Pe instead of P in the correlation.

The experimental values of the differential transmission coefficient D_m and of the integral transmission coefficient $\overline{D}_m$ for thirteen runs have been calculated by Equation 6 and Equation 7

Table I. Experimental Conditions and Measured Values

Run No.	dt (cm)	u (cm/hr)	Wo (-)	b×10⁴ (-)	Ta (±0.5°C)	ΔT (C°)	L (cm)	L' (cm)	ΔX (cm)	T̄ (C°)
BT-22	1.76	3.98	0.175		15.5	2.0	0.95	0.95	1.20	353
BT-21	1.76	2.49	0.146	0.240	15.0	5.6	1.50	1.05	0.44	351
BT-19	1.76	1.70	0.132	0.035	15.4	8.1	1.70	1.60	0.11	352
B-23	1.76	1.70	0.172		15.5		1.70			
BT-45	1.76	0.77	0.099	0.029	17.3	12.0	2.00	1.55	0.10	353
BT-36	1.26	3.98	0.096	0.60	15.3	3.5	1.85	1.70	0.75	354
B-38	1.26	3.98	0.095		15.4		1.85			
BT-40	0.80	7.24	0.108	1.75	17.8	7.0	1.60	1.20	0.15	349
BT-28	0.80	3.98	0.093	0.12	15.3	16.5	1.90	1.30	0.10	354
BT-39	0.80	3.98	0.103	0.10	18.5	8.5	1.95	1.15	0.10	350
BT-29	0.80	1.70	0.098	0.18	15.4	16.5	2.30	1.65	0.10	354
BT-37	0.80	0.77	0.097	0.01	23.0	23.0	2.57	1.85	0.10	356
BT-26	0.34	3.98	0.156	0.014	15.6	31.3	2.85	1.85	0.04	359
BT-24	0.34	1.70	0.162	0.016	16.6	34.5	3.10	1.70	0.05	363
BT-41	0.14	3.98	0.103	0.02	17.3	88.2	3.55	1.70	0.05	387

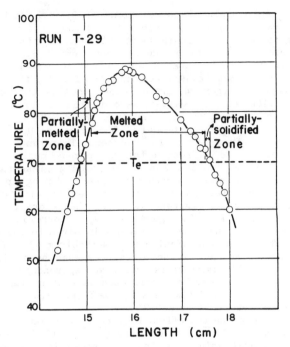

Figure 2. Temperature Profile Across Melted Zone

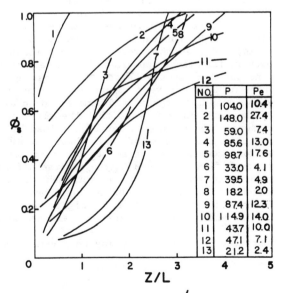

Figure 3. Separation Factor, ϕ_s vs. Dimensionless Travel Distance, Z/L

respectively. An example of correlations of D_m with W_s is shown in Figure 4. As may be expected, a plot of D_m versus W_s is a perfect straight line for each run within about 2.5 zone lengths during which the surface adsorption rate is nearly constant. Since the total backward solute transfer is the sum of the solute adsorbed on the freezing front plus that transmitted into the P.S.Z., the value of D_m at the intercept of Figure 4 may be considered as that due to the surface adsorption and D_m values beyond this intercept is that contributed by the solute transmitted into the P.S.Z. as a result of its relative movement through the melt. The values of D_m found for the thirteen runs range from 10^{-6} to 10^{-5} cm^2/sec. This range corresponds to that for the liquid diffusion through porous media. The fact suggests that the "Filtration Model" proposed represents well the overall effect of the mass transfer in zone refining in which the P.S.Z. plays the key role. To account for the effect of W_s differences between runs, $\overline{D}_m/\overline{W}_s$ is used instead of $\overline{D}_m$ for the generalized correlations with Pr·Gr and Sc·Re. In Figure 5, $\overline{D}_m/\overline{W}_s$ is correlated with Sc·Re ($=ud_t/D$). Considering the great many steps involved in the experiment and the calculation, the correlation is remarkably good. The value of $\overline{D}_m/\overline{W}_s$ approaches a constant as Sc·Re is decreased below 10^0, and it increases geometrically as Sc·Re is increased, approaching infinity as Sc·Re exceeds 10^2. This may be interpreted as that no improvement in separation may be expected by decreasing the Sc·Re value below 10^0, and no separation may result is Sc·Re$>10^2$.

   In order to examine the effect of free convection, the parameter for the filtration model $(\overline{D}_m/\overline{W}_s)/(ud_t)$ and $(\overline{D}_m/\overline{W}_s)/(ud_t')$ are correlated with the experimental values of Pr·Gr as shown in Figure 6, and a linear plot is obtained in each case. d_t' has been included in one of the parameters in order to account for the effect of Δx. To determine the dimensional length for free convection in the melted zone both L' and d_t have been included to reflect their effects since the zone was oriented vertically for all the runs. As may be seen from Figure 6, the transmission rate of solute into the P.S.Z. decreases as the value of Pr·Gr is increased under constant zone speed u. In other words, the separation is improved when the free convective mixing becomes more intensive. However, its effect appears to be far less significant than that of the zone speed u, as may be seen by comparing Figure 6 with Figure 5. Furthermore, it is very important to note that the effect of free convective mixing in the melted zone is advantageous only if $10^0 <$Sc·Re$<10^2$. The fact suggests also that the ultimate distribution or the maximum separation of solute may be obtained by a single zone pass if the experimental value of Sc·Re($=ud_t/D$) $\langle 10^0$; under such conditions the forward transfer of solute by molecular diffusion is balanced by an equal flow of solute in the reverse direction. From a practical viewpoint it would be more advantageous to carry out zone refining process in single zone pass under conditions of very low Sc·Re values than using multi-pass of zone at higher Sc·Re values. The values of D, D_e, $\overline{D}_m$ and of all the experimental parameters are given in Table II.

   As demonstrated in Figure 4, D_m may be given by an equation of the form,

$$D_m = aW_s + b \qquad (15)$$

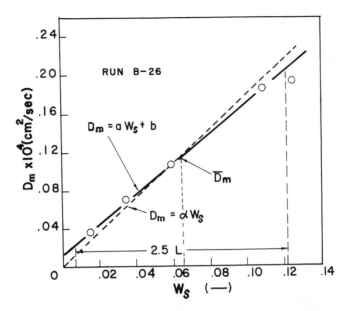

Figure 4. Solute transmission coefficient, D_m vs. solute concentration, W_s.

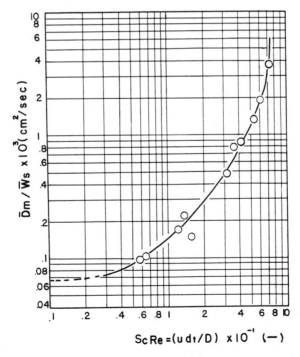

Figure 5. Filtration model parameter, $(D_m/W_s)/udt$ vs. Pr · Gr.

Table II. Values of the Experimental Parameters

Run No. (-)	$DX10^5$ (cm²/sec)	$ScX10^{-2}$ (-)	$ReX10^2$ (-)	Pr (-)	$GrX10^5$ (-)	D_eX10^5 (cm²/sec)	L_e (cm)	P (-)	Pe (-)	$\bar{D}_mX10^5$ (cm²/sec)	$\alpha X10^4$ (cm²/sec)
BT-22	2.4			15.9	0.27	10.5	0.11	43.7	10.0		
BT-21	2.2	5.22	10.40	15.5	0.78	14.9	0.11	47.1	7.1	6.54	13.15
BT-19	2.4	4.83	7.09	16.1	2.09	19.3	0.11	33.0	4.1	2.58	4.77
BT-45	2.4	4.58	3.40	14.8	3.68	21.1	0.11	18.2	2.0	0.59	1.44
BT-36	2.4	4.36	1.38	14.5	0.85	15.8	0.14	85.6	13.0	18.50	17.50
BT-40	2.2	5.23	14.15	16.5	0.41	12.3	0.15	148.0	27.4	29.00	37.20
BT-28	2.4	4.33	8.51	14.5	1.37	17.8	0.14	87.4	12.3	3.54	8.05
BT-39	2.2	5.21	7.79	16.5	0.46	13.6	0.17	98.7	17.6	4.93	8.96
BT-29	2.8	3.79	3.62	14.5	2.19	21.9	0.14	39.5	4.9	0.57	2.22
BT-37	2.6	3.86	1.72	13.7	4.23	23.5	0.14	21.2	2.4	0.19	1.05
BT-26	2.7	3.76	3.22	14.2	2.86	12.6	0.18	114.9	14.0	1.09	1.70
BT-24	2.5	4.62	1.24	15.9	2.00	19.0	0.19	59.0	7.4	0.64	1.00
BT-41	3.8	1.99	2.36	10.2	9.91	27.2	0.18	104.0	10.4	9.05	11.30

However, in Figure 5 and Figure 6 the intercept b is not included since the values of b cannot be correlated easily with the surface adsorption effect or any known effects of the experimental variables. Therefore, the following expression has been chosen for D_m to replace Equation 15 in order to use the generalized correlations and to examine their applicabilities and the predictability of the theory proposed in this study.

$$D_m = \alpha\, W_s \qquad (16)$$

The value of D_m predicted from W_s by Equation 16, would deviate slightly from the actual value given by Equation 15, as shown in Figure 4. The deviation depends mainly on the intercept b and the change in the slope from a to α. The deviation of D_m predicted by Equation 16 using the values $\overline{D}_m/\overline{W}_s (=D_m/W_s)$ from Figure 5 and Figure 6 would reflect on the accuracy of the prediction by Equation 11. An example is shown in Figure 7 for the conditions used in Run B–26; the prediction deviates less than 10% from the experimental results for the entire run. The theoretical curve intercepts the experimental curve in the same manner the line of $D_m = \alpha\, W_s$ intercepts the line of $D_m = aW_s + b$ in Figure 4, indicating that the accuracy of the prediction by Equation 11 depends primarily on the deviation of Equation 16 from Equation 15. It may also be noted that the prediction by the molecular diffusion theory as proposed by Wilcox (20 & 21) is completely uncomparable with the experimental results. It should be pointed out that Equation 6 may be solved by computer readily using Equation 15 for D_m without approximating by Equation 16, in order to obtain the correct results, as needed. Since the correct values of D_m cannot be calculated by the generalized correlations, such as Figure 5 and Figure 6, the above approximation method for D_m was used in solving for Equation 6. Calculation of Equation 11 for predicting $W_{s1}(z)$ for a single zone pass requires first predicting the values of $\alpha (=D_m/W_s)$ and of Δx from Figure 5 and Figure 6. The values of u, L and W_0 are from the experimental conditions.

CONCLUSION

From the results of this study the following conclusions may be drawn:

1. The thickness of a partially solidified zone (mushy region) in zone refining under most experimental conditions has been determined. The transfer of solute within this zone is, no doubt, most important in determining the overall solute transfer. All the zone refining theories have failed to take this most important factor into account.
2. The "Filtration Model" as proposed in this study represents the overall rate of the solute transfer (macro–segregation) in zone refining very well for a wide range of experimental conditions.
3. Though free convective mixing in the melted zone increases the separation rate, its effect is far less important than the opposed effect of the Prameter, D/ud_t. The theoretical maximum separation rate from a single zone pass may be expected if the D/ud_t value is less than 10°; and no separation may result if the experimental value of D/ud_t is greater than about 10^2.

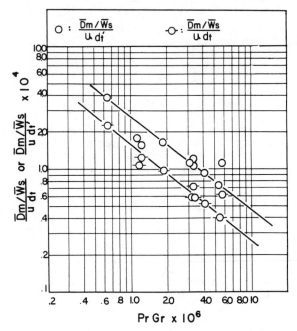

Figure 6. D_m/W_s vs. Sc $\cdot$ Re.

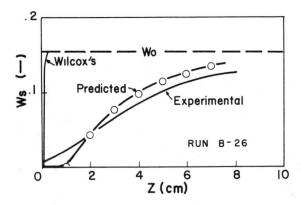

Figure 7. Predictability of the filtration model, W_s vs. Z.

4. Using the proposed model along with the two generalized correlations for the experimental transmission coefficients in the partially
solidified zone, a fairly close prediction of the solute redistribution
(macro-segregation) of eutectic-forming mixtures after a single zone
pass can be made.

ACKNOWLEDGMENT

Thanks are due to my former students, Y. Fang and T. J. Kulesza for
their assistances in the lab during the initial period of this study.

LEGEND OF SYMBOLS

A   : cross-sectional area, cm^2

a   : slope of a plot, D_m vs. W_s, cm^2/sec

b   : intercept of a plot, D_m vs. W_s, cm^2/sec

C_p : mean heat capacity at constant pressure, cal/gm, °C

d_t : inside diameter of glass tube, cm

d'_t : (Δx + d_t)/2, cm

D   : molecular diffusivity, cm^2/sec

D_e : effective molecular diffusivity, cm^2/sec

D_m : differential transmission coefficient in partially
                solidified zone, cm^2/sec

$D_m(W_s)$: D_m as a function of W_s, cm^2/sec

$\overline{D}_m$: integral transmission coefficient as determined by
                Equation 7, cm^2/sec

g   : acceleration of gravity, 980 cm/sec^2

h   : $(L'$ + $d_t)/2$, cm

j   : solute flux, gm/sec, cm^2

k   : thermal conductivity, cal/sec, cm, °C

k_e : effective thermal conductivity, cal/sec, cm °C

L   : length of melted zone, cm

L'   : distance between the hottest and coldest points in
                melted zone, cm

L_e :   effective boundary layer thickness, cm

T :   temperature, °C

$\overline{T}$:   average temperature, $(T_h + T_c)/2$, °C

ΔT :   greatest temperature difference in melted zone, $T_h - T_c$, °C

t :   time, sec

u :   zone travel speed, cm/sec

W :   weight fraction of solute, (−)

$W_s(z)$: W_s as a function of z

$\overline{W_s}$:   average weight fraction between two limits in refrozen solid, $(W_{s1} + W_{s2})/2$, (−)

ΔW : $W_1 - W_s$, (−)

Δx :   thickness of partially solidified zone, cm

z :   distance along the tube from the point where refreezing starts, cm

Greek Letters

α : D_m/W_s

β :   temperature densification coefficient, $(1/\rho)$ $(\partial \rho / \Delta T)_W$

ρ :   density, gm/cm^3

μ :   viscosity, gm/cm, sec

ϕ_s : W_s/W_o

Dimensionless Groups

Gr   :   Grashof number, $(h^3 \rho^2 g/\mu^2)$ $(\beta \Delta T)$

P   : uL/D

Pe   : uL/D_e

Pr   :   Prandtle number, $\mu Cp/k$

Re   :   Reynolds number, $d_t u \rho / \mu$

Sc   :   Schmidt number, $\mu / \rho D$

Subscripts

a   :   air condition in the cooling ring

c   :   coldest condition

e   :   Eutectic condition, or effective value

h   :   hottest condition

l   :   molten liquid condition

o   :   initial condition

s   :   refrozen solid condition

t   :   glass tube

1   :   point 1, or single zone pass

2   :   point 2

i   :   ith zone pass

LITERATURE CITED

 1. Allred, W. P.; Bate, R. T.   J. Electrochem. Soc. 1961, 108, 258.
 2. Banus, M. D.; Gates, H. C.   J. Electrochem. Soc. 1962, 109, 89.
 3. Dickhoff, J. A. M.   Solid-State Electron 1960, 1, 202.
 4. Hall, R. N.   J. Phys. Chem. 1953, 57, 836.
 5. Herington, E. F. G.   Zone Melting of Organic Compounds, John
     Wiley & Sons, Inc., N. Y. 1962; Chapter 2-8.
 6. Holmes, E. L.; Rutter, J. W.; Winegard, W. C.   Can. J. Phys.,
     1957, 35, 1223.
 7. Kraussold, H.   Forsch. Gebiete Ingenieureu 1934, 15, 186.
 8. Mullen, J. B.; Hulme, K. R.   J. Phys. Chem. Solids 1960, 17, 1.
 9. Morris W.; Tiller, W. A.; Rutter, J. W.; Winegard, W. C.   Trans.
     Am. Soc. Metals 1955, 47, 463.
10. Pfann, W. G.   Trans. Am. Inst. Min. and Met. Eng. 1952, 194,
     747.
11. Pfann, W. G.   Zone Melting, John Wiley and Sons, Inc., N. Y.
     1958, Chapter 2 - 6.
12. Rutter, J. W.; Chalmers, B.   Can. J. Phys. 1953, 31, 15.
13. Thomas, L. J.   Ph.D. Thesis, University of Illinois, Urbana, Ill.
     1962.
14. Thomas, L. J.; Westwater, J. W.   Chem. Eng. Progr. Symp.Series
     1963, 59, 155.
15. Tiller, W. A.; Rutters, J. W.   Can. J. Phys., 1956, 34, 96.
16. Tiller, W. A.   Growth and Perfection of Crystals, John Wiley and
     Sons, Inc., N. Y. 1958; p. 332.
17. Walter, W. G.   Ph.D. Thesis, University of Connecticut, Storrs,
     Conn. 162.

18. Westwater, J. W.; Thomas, L. J. Microscopic Study of Solid-Liquid Interfaces during Melting and Freezing-Motion Picture, University of Illinois, Urbana, Ill. 1962.
19. Wilcox, W. R. Ph.D. Thesis, University of California, Berkeley, Cal 1960.
20. Wilcox, W. R.; Wilke, C. R. A.I.Ch.E. J. 1964, 10, 160.
21. Wilcox, W. R. J. Appl. Phys., 1964, 35, No. 3 (part 1), 636.
22. Wilcox, W. R.; Friendenberg, R. M.; Back, N. Chem. Revs. 1964, 64, 187.
23. Wilcox, W. R. I. F. E. C., 1964, 3, No. 3, 235.
24. Flemings, M. C.; Nereo, G. E. Trans. TMS-AIME 1967, 239,1449-1461.
25. Flemings, M. C.; Mehrabian, R.; Nereo, G. E. Trans. TMS-AIME 1968, 242, 41-49.
26. Flemings, M. C.; Nereo, G. E. Trans. TMS-AIME 1968, 242, 50-55.
27. Mehrabian, R.; Keane, M.; Flemings, M. C. Metall. Trans. 1970, B 1B, 3228-3241.
28. Kou,S.; Poirer, D. R.; Flemings, M. C. Proc. Elect. Furn. Conf. Iron Steel Soc. AIME 1977, 35, 221-228.
29. Maples, A. L.; Poirer, D. R. Metall. Trans. 1984, B 15B, 163-172.
30. Szekely, J.; Jassal, A. S. Metall. Trans. 1978, B 9B, 389-398.
31. Voller, V. R.; Prakash, C. Int. J. Heat Mass Transfer 1987, 30, 1709-1720.
32. Voller, V. R. Numerical Methods in Thermal Problems V (Edited by R. W. Lewis et al) Pineridge, Swansea, 1987, p. 693-704.
33. Bennon, W. D.; Incropera, F. P. Int. J. Heat Mass Transfer 1987, 30, 2161-2170.
34. Bennon, W. D.; Incropera, F. P. Int. J. Heat Mass Transfer 1987, 30, 2171-2187.
35. Bennon, W. D.; Incropera, F. P. Metall Trans. 1987, B 18B, 611-616.
36. Bennon, W. D.; Incropera, F. P. Numer. Heat Transfer 1988, 13, 277-296.
37. Viskanta, R.; Beckermann, C. ASME Annual Meeting, Boston, Mass. 1987, Dec. 14-18.
38. Beckermann, C.; Viskanta, R. PCH 1988, 10, 195-213.
39. Voller, V. R.; Brent, A. D.; Prakash, C. Int. J. Heat Mass Transfer. 1989, 32, No. 9, 1719-1731.

RECEIVED June 1, 1990

Chapter 18

Purity Decrease of L-Threonine Crystals in Optical Resolution by Batch Preferential Crystallization

Masakuni Matsuoka, Hirokazu Hasegawa, and Koichi Ohori

Department of Chemical Engineering, Tokyo University of Agriculture and Technology, Koganei, Tokyo 184, Japan

In the optical resolution of DL-threonine mixtures by batch preferential crystallization, changes of solution concentration and crystal purity were measured. The mechanism of nucleation of the un-seeded enantiomer was discussed to explain the purity decrease of the resolved crystals. From the observation of crystallization behavior of the seed crystals of L-threonine, it was concluded that the existence of the D-enantiomer on the surface of the seed caused the sudden nucleation when they grew to attain sufficient amounts.

Threonine, $CH_3C^*H(OH)C^*H(NH_2)COOH$, is an optically active compound and L-threonine as a useful material for food additives or medicines has been commercially produced by optical resolution of DL-threonine mixtures which are mainly synthesized from copper glycinate and acetaldehyde (1). The optical resolution by batch preferential crystallization is achieved by seeding one of the enantiomers in a supersaturated solution of the racemic mixtures, but the other enantiomer quite often crystallizes in the course of the crystallization. Although studies have been reported on this general subject of the crystal purity decrease (2,3,4), the mechanism has been left unclear. In this study changes in crystal purities and solution concentrations are measured and crystallization behavior of threonines are observed in order to make clear the mechanism of appearance of the other enantiomer as crystals during the preferential crystallization.

Experimental Apparatus

A glass stirred tank crystallizer with 350 ml volume was used, which had a marine type propeller located near the bottom. The rotational speed was 200 rpm throughout the experiments. The crystallizer had a filter at the center of the bottom to quickly separate the solution

0097–6156/90/0438–0251$06.00/0

from the crystalline particles after three hours. It also had a water
jacket to maintain the solution temperature constant during the
crystallization. The solution concentration and the crystal purity
were measured by use of a HPLC (Shodex), with a column of CROWNPAK CR
produced by Daicel Chemical Industry.

Materials and Experimental Procedure

DL-threonine and L-threonine crystals were supplied from Ajinomoto Co.
Inc. and were used without further purification. Excess amounts of DL-
threonine crystalline particles were dissolved in water kept at 55,
57, 58 or 60°C. After decantation and filtration each saturated
solution was placed in the crystallizer maintained at 50°C. The
difference between the saturation temperature and the crystallization
temperature was defined as the initial supersaturation in terms of
supercooling of the solution and was the driving force for the
crystallization.
     After the solution temperature reached at 50°C 1 gram of the seed
crystals were added to the solution, when the crystallization time is
defined to be 0. Each crystallization was conducted for three hours.
Samples of the solution and the crystalline particles were taken at
every 30 minutes and their concentrations or purities of L-threonine
were measured by the HPLC at least for two times.

Washing Effects of Seed Crystals

Since the surface of the seeds was covered by a number of tiny
crystalline particles as shown in **Figure 1-a**, the seed crystals
were all washed by water and dried by ethanol just before usage. The
surface appearance after washing was smooth as shown in **Figure 1-b**,
although there were small holes in the surface where tiny crystalline
particles were found to be trapped. The effect of washing the seed
crystals was examined as preliminarily tests by comparing the changes
in crystal purity during the crystallization between the resolution
experiments at $\Delta t=8°C$ using washed and unwashed seed crystals. As
shown in **Figure 2**, the time when the purity decrease started was
definitely delayed when the seed crystals were washed. This suggests
that the tiny particles on the surface of the seed crystals are the
mixture of the L- and D-threonine enantiomers and that the most of
them were removed by the washing process and that the purity decrease
is the results of the breeding of the D-enantiomer which may occur
when sufficient amounts of the D-enantiomer come to exist on the
surface of the seed crystals during the resolution.

Changes of Solution Concentration and Crystal Purity during Crystallization

After the addition of the seed crystals the relative solution
concentration was periodically measured. In **Figure 3** the changes of
L-threonine concentration in the solution were compared for four
experimental runs with different initial supersaturations. Here the
relative concentration is defined as the ratio of L-threonine to the
total enantiomers. For the case of lower initial supersaturations
such as $\Delta t=5°C$, the concentration decreased monotonically. On the
other hand, the concentration - time curves have minima for the

(a) unwashed seed   (b) washed seed

Figure 1. SEM photomicrographs of the crystal surfaces of unwashed and washed seed crystals.

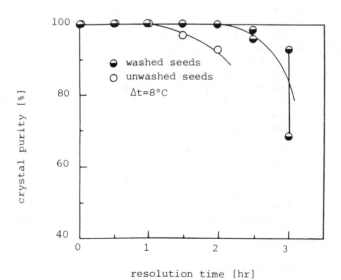

resolution time [hr]

Figure 2. Changes of crystal purity with time for crystallization with washed and unwashed seed crystals.

solutions with higher initial supersaturations. Due to indefinite reasons the starting concentrations in Figure 3 were not exactly 50% but were scattered within a range of 1.5%, however each curve was accurate in that the reproducibility of the concentration measurement under the same conditions was confirmed.

In **Figure 4** the corresponding changes in the crystal purity are shown. For the solutions with the initial supersaturation $\Delta t = 5°C$, the purity was always 100% for three hours, while purity decrease started very soon for the solution having higher initial supersaturation such as $\Delta t = 10°C$. Since with different initial supersaturations, the crystallization rate are very different, the purity decrease should be compared on the basis of either the supersaturation change or the crystal mass grown in the solution rather than the crystallization time.

When the crystal purity is plotted against the total crystal mass in the slurry calculated from the mass balances, the purity decrease seems to start at some constant value of the crystal mass as shown in **Figure 5**. As mentioned earlier in the text, there are possibilities of existence of the D-enantiomer as small particles on the surface of the seed crystals. If we assume that the breeding of the D-enantiomer starts only when that enantiomer has grown to a certain size, the amount of the L-enantiomer crystals must have also increased to a certain value, the latter being proportional to the former. The crystallization kinetics of the both enantiomers are believed to be the same, the relative amounts of crystals of the both enantiomers must therefore be constant before nucleation of the D-enantiomer starts.

Crystal Size Distribution of Seed and Product Crystals

Both the seed and product crystals are of needle- or rod-like morphology. The seed crystals were taken from the feed stock of L-threonine as supplied. The CSDs are shown in **Figure 6** both for the seed and the product crystals, the latter being taken under the experimental conditions of $\Delta t = 8°C$ and crystallization time $= 5hr$. About one hundred particles were measured from photographs taken under microscope and plotted on the diagrams, the dimensions being measured along the length and width directions which are normal each other.

From the diagrams and also from the photographs the seed and the product crystals are found to be similar each other although the maximum and mean dimensions along the length are slightly larger for the product crystals. This implies that growth of the seed crystals did occur only in the length direction. It is also suggested from the CSD that the crystallization proceeded mostly by the mechanism of macro-attrition since the number of smaller particles increased during the crystallization. In addition in the initial period of crystallization, many thin needle crystals were found to grow on the seed crystals in random directions. Most of these crystals were disappeared because of separation of individual needles from the parent crystals by collision between crystals.

Growth Behavior of Seed Crystals

In order to confirm the idea that the D-enantiomer has already been introduced to the solution as the seed crystals, a piece of seed

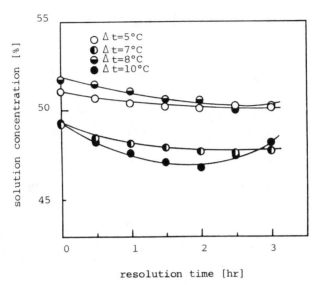

Figure 3. Changes of solution concentration with time for resolutions with different initial supersaturations.

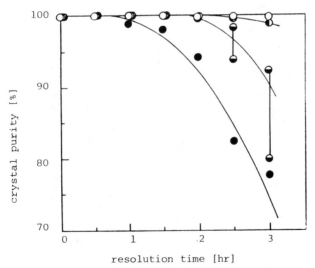

Figure 4. Decrease of crystal purity during resolutions. See Figure 3 for the notation of the keys.

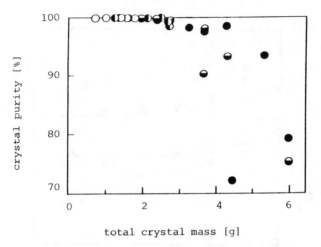

total crystal mass [g]

Figure 5.   Relation between crystal purity and total crystal mass
in the slurry.   See Figure 3 for the notation of the keys.

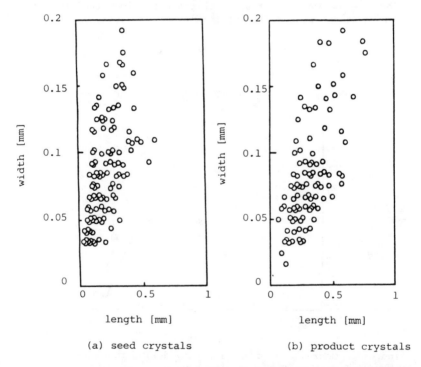

(a) seed crystals   (b) product crystals

Figure 6.   Comparison of size distributions of seed and product
crystals.

crystals was placed in stagnant supersaturated solutions and the crystallization behavior was observed for 114 hours. In all solutions used of which supersaturations were varied from 2 to 8°C, needle-like crystals grew on the seed crystal at first, then secondary nuclei appeared in the solution near the seed and eventually a number of nuclei appeared in the solution in distance from the seed. This is schematically shown in **Figure 7**. Several crystalline samples were taken including isolated particles and agglomerated mass, and their purities were analyzed by the HPLC.

The bundles of needle crystals on the seed were not pure L-threonine but mixtures of the L- and D-enantiomers, meanwhile the isolated single particulate needles in the solution were either L- or D-threonine. Some single pieces of crystals were found to be mixtures of the D- and L-enantiomers. The composition of such mixtures varied in a wide range including almost racemic mixtures (50%) or those concentrated in one of the enantiomers. This was common in all the solutions having different initial supersaturations examined. All the results are summarized in **Table 1** below. Some needle crystals were also mixtures even when they grew on the seed crystal of L-threonine. Particularly agglomerated crystals and those having star-like shapes found in the solution were completely mixtures, in most case they were racemic mixtures.

Table 1   Purity of individual crystals [% L-threonine]

| crystal form | initial supersaturation [°C] | | | |
	2	5	6	8
a) clear needles				
on seeds	-	17.7	100.0(x2) 99.6(x2)	97.4
in solution	98.9, 99.9	99.8(x4) 99.6, 13.2	99.7, 99.6 99.3, 1.2 0.1	98.1, 97.4
b) unclear needles				
on seeds	-	16.4	100.0, 99.8	96.2, 92.3
in solution	73.6(x2) 51.2, 47.8	100.0 99.7(x2) 99.3, 97.4 97.1, 95.4	100.0, 99.8 99.1, 97.8 94.3	42.4, 97.3 34.0
c) needles				
on seeds	-	100.0, 70.3	-	97.3
in solution	96.6, 60.6 34.9	24.7, 20.1 17.7	97.6, 82.0	-
d) agglomerates	55.4	-	50.1	96.7
e) star-likes	48.4	38.0	29.5	-

The crystallization of the D-enantiomer is therefore considered to be induced by crystal growth on the surface of the seed crystal and at the same time initial breeding may play a role that causes small crystals near the seed crystal. The propagation of nucleation in distance from the seed may be caused by convective flow of the solution due to density difference during the crystal growth.

This observation is in accordance with the phenomena of the crystallization in the resolution operation mentioned above in the following points. There are no clear, definite critical supersaturations above which nucleation of D-threonine occurs. Ohtsuki (2), however, reported supersolubility curve for this system, who gave the value of the supersaturation width $\Delta t = 7°C$ at 50°C. Their definition of the metastability was that no nucleation of the enantiomer other than seeded one was observed for two hours of resolution experiments. According to this definition, the super-solubility can be determined to lie somewhere between $\Delta t = 8$ and 5°C from the present experimental data, this being in agreement with his result. If the crystallization proceeds further, however, D-threonine crystals may start to crystallize from the solution even if the initial supersaturation is 5°C. In this sense it is no longer the metastability limit.

Stability of the Supersaturated Solutions

Another experiments were also carried out to see whether spontaneous nucleation would occur or not at the similar supersaturation levels as those in the resolution experiments.

About 100 ml of solutions with known different initial supersaturations were sealed in ampoules and placed in a constant temperature bath. Half of the solutions were mechanically agitated with a magnetic stirrer. The time when the nuclei were observed in each ampoule was recorded. For the solutions sealed in an ampoule where the ampoule was washed in an ultrasonic wave bath with HCl solutions for 40 minutes and both the cap of the ampoule and the stirrer were cleaned by immersing them in fresh boiling water for three hours, i.e. for the solutions kept in the completely cleaned ampoules, the waiting time was more than 5 hours even at $\Delta t = 8°C$, but it was less than 1 hour when they were not cleaned at all as shown in **Figure 8**. Under the partial cleaning conditions where the stirrer alone was cleaned for two hours while the ampoule and the cap were not treated, the waiting time was found to become three times larger on average. In addition the use of a new stirrer lead to the very similar result with that in the complete cleaning.

Furthermore no nucleation was observed in the experimental time in any ampoules at any supersaturation levels if there was no stirrer in the ampoule. This implies that the heterogeneous nucleation was induced by the presence of the stirrer, probably by enhancing the nucleation at the sites where fine scratches exist. This fact suggests that the solution is stable if there are no possible seed crystals in the system. The existence of the D-enantiomer in the seed crystals is therefore very likely to initiate the secondary nucleation after they grow to attain sufficient sizes, which results in the sudden purity decrease of the product crystals.

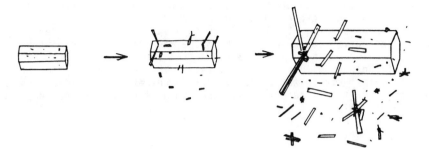

Figure 7. Schematic representation of growth behavior of seed crystals in stagnant solutions.

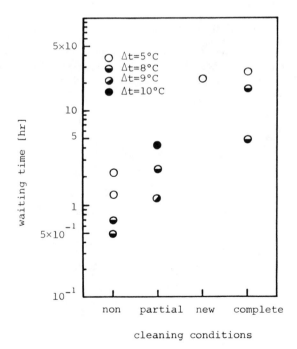

Figure 8. Depression of spontaneous nucleation by cleaning of apparatus.

Conclusion

From the consecutive measurements of solution concentrations and crystal purities during the optical resolution by preferential crystallization, the crystallization of D-threonine other than the seeded component (L-threonine) was observed in the later stage of the resolution. Washing of the seed crystals was found to be effective to delay the purity decrease. D-threonine was believed to be introduced to the solution in the early stage of the resolution probably in the form of inclusions in the seed crystals. The sudden nucleation of the D-enantiomer starts when a sufficient amount of it has grown on the surface of the seed crystals, hence resulting in the decrease of purity of the product crystals.

Acknowledgments

Threonines used in this study were supplied from Ajinomoto Co. Inc. The HPLC and the separation column were supplied from Showa Denko KK and Daicel Chemical Industry respectively.

Literature Cited

1) Sato,M., K.Okawa and S.Akabori: Bull. Chem. Soc. Japan, <u>30</u>,937 (1957).
2) Ohtsuki,O.:Kagaku Kikai Gijutsu (Technology of Chemical Machinery), 1982, No.34 p.31.
3) Denk,E.G. and G.D.Botsaris: J. Crystal Growth, <u>13/14</u>, 493 (1972)
4) Jacques,J., A.Collet and S.H.Wilen:"Enantiomers, Racemates and Resolutions", p.217, John Wiley & Sons, Inc., New York (1981)

RECEIVED May 15, 1990

Chapter 19

Semi-Batch Cooling Crystallization
of Quizalofop-Ethyl with Polymorphism

Akihiro Shiroishi[1], Isao Hashiba[1], Ryo Kokubo[1], Kazuo Miyake[2], and Yuji Kawamura[2]

[1]Nissan Chemical Industries, Ltd., 7–1, 3-chome, Kanda-Nishiki-cho, Chiyoda-ku, Tokyo, Japan
[2]Department of Chemical Engineering, Hiroshima University, Shitami, Saijo-machi, Higashihiroshima, Japan

It was found that two different crystal forms, namely a platelike α-form and a hairlike β-form, were obtained by cooling crystallization of quizalofop-ethyl dissolved in ethyl alcohol. The solubilities, supersolubility and stabilities of the two crystals were investigated. Two solubility curves for the α-form and the β-form respectively were found to intercept at 284K. A metastable supersaturation zone up to 13–15K at 314K was found. It was confirmed that a little amount of α-form crystals suspended in a saturated solution at temperature above 316K gradually changed into β-form crystals. Considering those results, authors developed a semi-batch crystallization process to obtain α-form crystal selectively in which crystallization temperature was lowered enough below 293K.

Quizalofop-ethyl (1), (4) which has been developed as an effective herbicide by Nissan Chemical Industries, Ltd. is needed as its crystalline state in commercial production. But no physical properties, solubilities and/or any other information useful for the crystallization of the compound have not appeared to be available.

Therefore a few physical properties of quizalofop-ethyl, such as its solubility and its supersolubility in ethyl alcohol, were studied. Two types of crystals with different shapes were obtained by cooling crystallization and the stabilities of the crystals were tested in saturated solutions at various temperatures.

Platelike α-form crystals easily separate from the solvent in commercial operation but hairlike β-form crystals are troublesome because their suspension is extremely viscous. So the main purpose of this work is to develop a modified crystallization process to obtain α-form crystal selectively.

0097–6156/90/0438–0261$06.00/0
© 1990 American Chemical Society

Physical Properties of Quizalofop-ethyl

The chemical structure for quizalofop-ethyl is shown below. The

molecular weight is 372.5 and the chemical equation is given by $C_{19}H_{17}O_4N_2Cl$. The compound is optically active and only the D-formed will be discussed in this investigation.

Two different types of crystals were obtained by cooling crystallization of quizalofop-ethyl dissolved in ethyl alcohol at room temperature. As shown in Photograph 1, one type consists of plate-like crystal of 20-50 μm in size and another one is hairlike crystal of 1-2 μm in short axis size. As shown in Figure 1, different X-ray patterns were obtained for the two crystals. Melting points and heats of fusion were also measured for the two crystal types by DTA and DSC, which are listed in Table I. These results indicate that the different shapes of the two crystal types are related to differences in crystal lattice.

Table I. Physical Properties of Quizalofop-ethyl Crystals

Crystal Forms	Melting Point (K)	Heat of Fusion (J/mol)
α	348.15	2.573×10^4
β	352.65	2.413×10^4

Solubility

Apparatus and Procedure for Solubility Measurement. Chemicals used were ethyl alcohol of special grade reagent and quizalofop-ethyl crystal of purity 99.5%, specific rotatory power 99.5% e.e. α-form crystal of quizalofop-ethyl was prepared by cooling crystallization at 283K from the quizalofop-ethyl solution (5 gr. of quizalofop-ethyl was dissolved in 100 gr. of ethyl alcohol.) β-form crystal was also prepared in the same way at 313K from the solution in which 25 gr. of quizalofop-ethyl was dissolved in 100 gr. of ethyl alcohol.

Experimental apparatus is shown in Figure 2. 100 gr. of ethyl alcohol was fed into ajacketted glassware vessel (vessel volume: 220 ml) with agitator. The vessel was heated by recycling heat transfer medium through the jacket. When the temperature reached 2-3°K lower than the settled temperature, a large amount of sample crystal (α-form or β-form) was poured into the vessel which was enough to make 3-7% suspension. After the suspension thus obtained was heated up to the settled temperature, a small amount of the solution was pippeted off through a glass filter (pore size: 100-150 μm) at every 30 minutes which was equipped in the suspension. The concentration of the solution became constant after two or three hours, the numerical value was assumed to be solubility. The concentration was determined by gravimetric analysis.

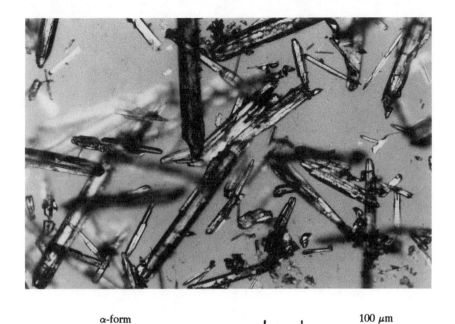

α-form   100 μm

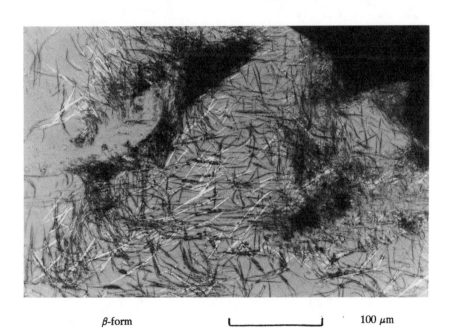

β-form   100 μm

Photograph 1.   Shape and form of quizalofop-ethyl crystals.

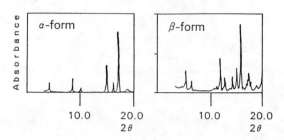

Figure 1. X-ray diffraction pattern of quizalofop-ethyl
crystals.

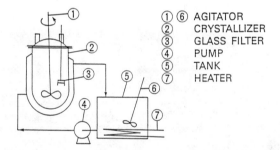

Figure 2. Experimental apparatus for solubility measurement.

Results. Solubilities of α-form crystal and β-form crystal are shown in Figure 3 with full line and dashed line. It can be seen that α-form crystal is more soluble than β-form crystal at above 284K and that two curves intercept at 284K.

Supersolubility

Apparatus and Procedure for Supersolubility Measurement. Same apparatus used in solubility measurement was also used in this experiment. Various amounts of quizalofop-ethyl was dissolved into 100 gr. of ethyl alcohol. In order to dissolve the compound completely, the solution was heated up to 10K higher than its saturation temperature. After keeping the solution as such for about 20 minutes, the solution was cooled at the rate of 0.5K/min. As well known, if the solution was supercooled beyond its saturation temperature, crystallization did not occur immediately. But as soon as a few crystals began to deposit, many crystals precipitated succeedingly almost at the same time. The temperature at which crystals began to deposit was observed. Thus produced crystals were withdrawn and analyzed for α-form content by using DSC and X-ray.

Results. Primary nucleation temperature is plotted against the prepared concentration shown in Figure 3. The curve means supersolubility itself. Even if α-form crystal or β-form crystal was used, obtained supersolubility data were quite the same. Temperature of supersaturation for α-form crystal or β-form crystal was about 13-15K. It was found that higher crystallization temperature than 309K gave mixture of α-form and β-form crystal, while lower temperature than 304K gave α-form crystal in the majority.

Stability of the Crystals Suspended in the Saturated Solution

It is known that many carboxyl acid crystals have polymorphs (3) and that the crystals suspended in saturated solution sometimes display a solid-transformation (2). If α-form crystal or β-form crystal in saturated solution changes into other crystal form within experimental time scale, solubilities for α-form crystal or β-form crystal do not have substantial meaning. So stability of α-form crystal or β-form crystal suspended in the saturated solution was investigated.

Apparatus and Procedure. Same apparatus was used as used previous section. By reference to the solubility data, several solutions were prepared using ethyl alcohol and quizalofop-ethyl which gave different saturation temperatures. A kind of 120 ml of thus prepared solution was poured into the vessel and was heated up to 5K higher than its saturation temperature, then the temperature was lowered to saturation temperature while 5 gr. of α-form crystal was warmed to the same temperature as said temperature elsewhere. It was added to the solution. Keeping the temperature of the suspension at the saturation temperature, about 10 ml of the suspension was pipetted off at hourly intervals. Withdrawn sample was quickly filtrated and analyzed for α-form content by means of DSC. It was checked that the concentration before adding crystals was almost equal to the concentration at the final stage of each experiment.

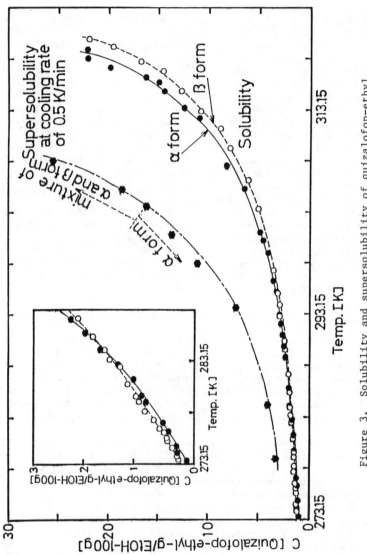

Figure 3. Solubility and supersolubility of quizalofop-ethyl crystals.

They both gave saturated concentration for α-form crystal. Changing the saturation temperature, similar experiments were performed. Instead of adding 5 gr. of α-form crystal, 1 gr. of α-form crystal and 1 gr. of β-form crystal were used and another series of data were collected.

Results and Discussion. Experimental results are shown in Figure 4. Circular marks in this figure show the data when α-form crystal is added. It seems that a part of α-form crystal changes into β-form crystal in the saturated solution at 316.2K or 312.3K with elapse of time. Triangular marks show the data when mixed crystals (α-form/ β-form = 1/1) are added. It also seems that a part of α-form crystal changes into β-form crystal at 303K and that a part of β-form crystal changes into α-form crystal at 278.4K.

It is unknown whether suspended α-form crystal or β-form crystal in the saturated solution displays a solid-solid transformation or not. The reason for above experimental results may be assumed as follows in the standpoint of industrial crystallization. Since β-form crystal is less soluble than α-form crystal at high temperature above 284K. The state that α-form crystal is suspended in the saturated solution is considered to be supersaturated for β-form crystal. So the state has the potential to take place primary nucleation and crystal growth for the β-form. In this way, β-form crystal may be produced in the suspension of α-form crystal. Rewarding to the formation of β-form crystal, a part of α-form crystal suspended may be dissolved. Opposite phenomena may take place at low temperature below 284K.

Semi-Batch Cooling Crystallization

In order to obtain α-form crystal preferentially semi-batch cooling crystallization at low temperature below 284K is considered more preferable as investigated above sections. Authors studied the process in the beaker scale test and in the commercial operation.

Apparatus and Procedure for Beaker Test. Agitaotr, thermometer and jacketted titration funnel were mounted to glassware vessel which was dipped into the thermostat. 96 gr. of ethyl alcohol was fed into the vessel, and the temperature was settled at the crystallization temperature. 74 gr. of ethyl alcohol and 37 gr. of quizalofop-ehtyl were fed to titration funnel and dissolved by heating with circulating hot water at 330K through the jacket. Thus obtained highly concentrated solution was fed into the vessel at the constant rate, keeping the temperature of the vessel constant. Semi-batch crystallizations were carried out under three titration rate of 3.70, 0.925 and 0.370 gr./min. After titration finished, the suspension was aged for about one hour. Obtained crystals were analyzed by DSC and crystal size of them was measured by using Coulter Counter.

Results of Beaker Test. α-form content and average crystal size of product crystal are shown in Figure 5 and Figure 6 against the crystallization temperatures. It is shown that α-form content increases according to decrease of crystallization temperature and

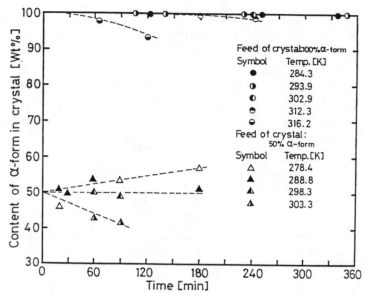

Figure 4.   Composition change of polymorphs in saturated
solution.

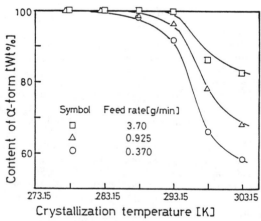

Figure 5.   Relationship between content of α-form crystal and
crystallization temperature in beaker test.

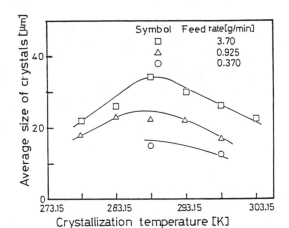

Figure 6. Relationship between average size of crystals and crystallization temperature in beaker test.

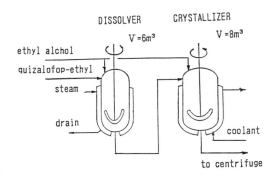

Figure 7. Flowchart of commercial plant.

increase of titration rate. It is also shown that crystallization temperature 288K gives maximum crystal size and that high titration rate gives large crystal size.

Crystallization in Commercial Plant. Flowchart of commercial plant is shown in Figure 7. Small amount of ethyl alcohol was fed into the crystallizer and cooled the solvent below 293K. Highly concentrated quizalofop-ethyl solution was prepared in the dissolver by heating the solution through the jacket. Thus obtained hot and concentrated feed solution was supplied into the crystallizer at constnat rate. Keeping the temperature of crystallizer below 293K by circulating coolant through the jacket of the crystallizer, semibatch crystallization was carried out. The suspension was aged for one hour after feeding was finished. Obtained slurry was centrifuged and the cake was dried to be a final product.

Results for Commercial Operations. The content of α-form was up to
99% and average size of the crystal was about 24-35 μm. The formation of β-form crystal in commercial operation induced considerable
increase of the viscosity of the suspension. The features of the
semi-batch cooling crystallization process are as follows. Even if
crystallization temperature is considerably lowered in order to
avoid the formation of β-form crystal, and also even if the feed
solution is highly concentrated at high temperature above 333K,
obtained crystal size is large enough to separate the solvent by
centrifuge.

Conclusion

Quizalofop-ethyl was found to have two types of crystals with
different shapes which were crystallized from the ethyl alcohol
solution at room temperature. Solubilities and supersolubility for
the two crystals were investigated. Two solubility curves for α-
form crystal and β-form crystal were found to intercept at 284K.
A little amount of α-form crystal suspended in saturated solution
above 312K was found to be changed into β-form crystal within experimental time scale. A unique semi-batch crystallization process
was proposed to obtain α-form crystal selectively. Crystal properties from commercial plant gave good agreement with that of
beaker test. The process may give a new hint in industrial crystallization of other organic compounds having polymorphism.

Literature Cited

1. Sakata, G; Makino, K.; Morimoto, K.; Hasebe, S. J. Pesticide
    Sci. 1985, 10, 69
2. Sato, K.; Suzuki, K.; Okada, M. J. Crystal Growth 1985, 72,
    699-704
3. Sato, K.; Yoshimoto, N.; Arishima, T. J. Japanese Association
    Crystal Growth 1989, 16, No.1, 67-76
4. Ura, Y.; Sakata, G.; Makino, K.; Kawamura, Y.; Kawamura, Y.;
    Ikai, I.; Oguti, T. U.S. Patent 4629493, 1986

RECEIVED June 19, 1990

Chapter 20

Growth Rate of *S*-Carboxymethyl-D-cysteine and Its Optical Resolution

M. Yokota[1], T. Oguchi[1], K. Arai[1], Ken Toyokura[1], C. Inoue[2], and H. Naijyo[2]

[1]Department of Applied Chemistry, Waseda University, 3–4–1 Ohkubo, Shinjuku-ku, Tokyo 169, Japan
[2]Life Science and Research Laboratory, Showa Denko K. K., 2–24–25 Tamagawa, Ohta-ku, Tokyo 146, Japan

Growth experiment of S-carboxymethyl-D-cysteine(D-SCMC) in a DL-SCMC supersaturated solution were carried out by a batchwise agitated crystallizer and the growth rates in longitudinal and lateral directions and optical purity of D-SCMC crystal were measured. The growth rates of crystals with partially broken surfaces or with fines adhered to it were larger in both directions than those of smooth surface. Further, the optical purities of crystals with broken surfaces and with fines adhered on the surfaces were lower than those with smooth surfaces. From these results, the kinetics of growth of D-SCMC crystals and the mechanism of inclusion of impurity during their growth process was considered.

Amino acid having the configuration of the optical isomers, sometimes interact as a physiological active material in a living matter and becomes an important product in a pharmaceutical industries. SCMC are generally produced through synthetic method. When bio-engineering technique is applied for production of SCMC, racemic substance of DL-SCMC is expected to be produced on lower cost. But separation of D- and L-SCMC should be followed for these processes. Crystallization is one of the useful method for separation of optical isomers. But SCMC is a racemic compound and difficult to crystallize by conventional method. When sodium chloride or potassium chloride was added into a supersaturated solution of racemic SCMC, nucleation easily occured. When seed of D- or L-SCMC crystal grew in this solution, purity of grown parts of D- or L-SCMC seed crystals in these tests was from 95.0 to 99.9 %(1). When these process are applied for optical resolution of D- or L-SCMC crystal, purer product should be crystallized for industrial purposes and mechanism of inclusion of impurity in crystal is expected to be turned out.

0097–6156/90/0438–0271$06.00/0

In this study, D-SCMC seed crystals were put in a racemic SCMC
supersaturated solution in a batchwise agitated vessel and growth
rates in longitudinal and lateral directions and the optical purity
of D-SCMC crystals were measured. The growth rates and optical purity
were discussed considering surface states of grown crystal observed
by a microscope. The kinetics of crystal growth were measured and a
model of inclusion of impurity was proposed.

EXPERIMENTAL

0.735 grams DL-SCMC crystals were dissolved in a 10 wt% aqueous solution of sodium chloride and the solution, saturated at 313 K, was
prepared. This solution was put in a 600 ml of glass vessel with a
two-blade impeller and cooled down in a water bath at a constant rate
of 0.1 K/min. When the solution reached the desired level of supercooling, about one gram D-SCMC seed crystals were introduced into the
solution. The crystals were suspended by moderate agitation of the
impeller and grew for thirty minutes. After thirty minutes, all suspended crystals were taken out and weighed. Size distribution and
optical purity of these crystals were measured. Then these crystals
were returned to the crystallizer where they continued to grow for
another thirty minutes. These operations were repeated until the seed
crystal grew by 10 %. The crystal size was measured in longitudinal
(La) and lateral(Lb) directions as shown in Figure 1-(b), which is a
simplified model of the actual D-SCMC seed crystal photographed in
Figure 1-(a). In these tests, some seed crystals identified by fixed
colored thread on the surface were also suspended in a magma crystal
bed and the change of crystal size was measured.
    When a series of tests carried for the particular seed was over,
some crystals obtained by the series of the growth tests, were put in
the saturated solution of D-SCMC at 303 K together with the seed
crystal of D- and L-SCMC as dipping test and the surface states of
the crystal in the solution were observed through a microscope.
    Optical purity(O.P.) of grown parts of the crystals were ana-
lyzed by high performance liquid chromatography. Experimental conditions for growth tests are summarized in Table I.

RESULTS AND DISCUSSIONS

Growth rates of suspended D-SCMC crystals. The size distribution in
longitudinal and lateral directions of D-SCMC crystals obtained in
growth tests are shown in Figure 2-(a), (b) respectively. The size
distributions shifted to larger direction with time for initial five
hours having similar shape. When the operation was continued for
seven hours, some fines appeared and the size distribution curves
were deformed. On longitudinal length, a population of smaller and
larger parts increased. On the other hand, a population of larger
parts showed a trend of increase on lateral directions. These trends
were clear and population of smaller length also became to be found,
when operation became longer to twelve hours. From observation of
sampled crystal by microscope, larger crystal grew for more than five
hours, were seen to be stuck by fines especially on the end of the
longitudinal length. These fines were supposed to affect on crystal
growth rate. In lateral directions, fines were not found to stick on
the end of crystal clearly, but population of larger crystal increased when fines appeared.

Table I   Experimental conditions and the ratios of concentration
of L-SCMC and sodium chloride in the mother liquor
or grown parts of the crystal measured from broken crystal

run no.	supercooling [K]	r.p.m.	mother liquor			grown crystal		
			L-SCMC	NaCl	L/NaCl	L-SCMCx10^{5}	NaClx10^{4}	L/NaCl
1	5.01	200	0.3654	10.001	0.0365	2.817	7.718	0.0365
2	4.98	300	0.3654	10.001	0.0365	2.823	7.718	0.0366
3	5.00	400	0.3654	10.001	0.0365	2.821	7.719	0.0365
4	9.99	200	0.3655	10.001	0.0365	2.814	7.718	0.0365
5	10.01	300	0.3656	10.001	0.0366	2.820	7.718	0.0365
6	10.01	400	0.3656	10.001	0.0366	2.825	7.718	0.0366
7	14.98	200	0.3655	10.001	0.0365	2.820	7.718	0.0365
8	14.97	300	0.3656	10.001	0.0366	2.819	7.719	0.0365
9	14.99	400	0.3655	10.001	0.0365	2.820	7.718	0.0365
10	19.98	200	0.3655	10.001	0.0365	2.816	7.719	0.0365
11	19.97	300	0.3655	10.001	0.0365	2.814	7.719	0.0365
12	19.99	400	0.3654	10.001	0.0366	2.818	7.718	0.0364

(a):photographic picture

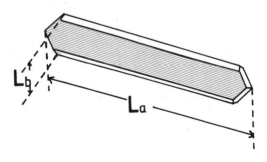

(b):modeling shape

Figure 1. D-SCMC seed crystal.

The average crystal size calculated from size distribution was plotted against the growth time in Figure 3. The crystals grew linearly with time for initial periods, but non-linear growth with time was observed after these periods. These deviations were supposedly due to the generation of fines. Therefore, growth rates in each direction were calculated from the slope of the straight lines in the initial operating period and plotted against supercooling of the solution in Figure 4-(a), (b). On tests operated under 400 r.p.m., fine crystals appeared after two hours' operation, but average crystal sizes in each directions were plotted on their particular straight lines, respectively, before fines became to appear. The growth rates were also calculated from these straight lines, and plotted in Figure 4. Crystal growth rates obtained under 200, 300 r.p.m. were almost same, but those under 400 r.p.m. were 60 and 40 % larger than those under 200 to 300 r.p.m. in longitudinal or lateral directions, respectively. From authors work for potassium alum.(2), the appearance of fines affect the growth rate. In this study, growth rates shown in Figure 4 were observed before fines appeared. But some embryos are supposed to be born before fines were confirmed. Therefore even when these growth rates were observed before fines appeared generation of embryos is assumed to affect on growth rates and these differences is supposed to be come from generation of embryo correlated to generation of fines. Then growth rates of longitudinal and lateral directions, Ga and Gb, were correlated with supercooling as Equations (1), (2) for 200, 300r.p.m. impeller speed.

$$Ga = 1.1 \times 10^{-2} \times \Delta T^{0.9} \tag{1}$$

$$Gb = 5.1 \times 10^{-5} \times \Delta T^{1.9} \tag{2}$$

Correlative equations of Ga and Gb for 400 r.p.m. also obtained as Equations (3), (4).

$$Ga = 1.8 \times 10^{-2} \times \Delta T^{1.2} \tag{3}$$

$$Gb = 6.9 \times 10^{-5} \times \Delta T^{2.3} \tag{4}$$

From these results, good transparent crystals grew when generation of fines were almost negligible under moderate impeller speed; growth rates appear to be controlled dominantly by surface reaction. When some embryos were assumed to appear, they affect the crystal growth mechanism. These embryo are supposed to affect the power number of supercooling and make growth rate faster.

Growth behavior of identified D-SCMC crystal marked by colored thread in a multi-suspended bed. On case 1, the crystal had smooth surface, grew as plot in Figure 5-(a) and the slope of the straight line on initial period was almost same to those of the average growth rate in Figure 3. When operation became longer, the slope of these lines became decreased and these trend was come from the decrease of supercooling. These decrease was supposed to be come from the growth of suspended crystals in a batchwise operations. When these growth rates obtained under the supercooling less than those of initial solution are plotted in Figure 4, these plots were a little bigger

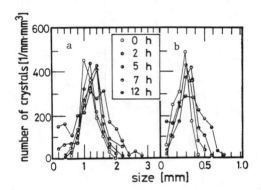

Figure 2. Size distribution of crystals grown under operational conditions of 15 K initial supercooling and 300 r.p.m. (a) : longitudinal direction, (b) : lateral direction

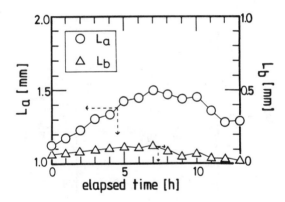

Figure 3. Examples of change of crystal size against operating period. Average crystal size was calculated from CSD in Figure 2.

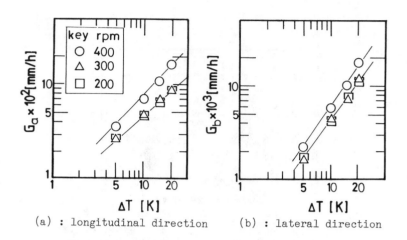

(a) : longitudinal direction   (b) : lateral direction

Figure 4. Correlation between crystal growth rate and initial supercooling.

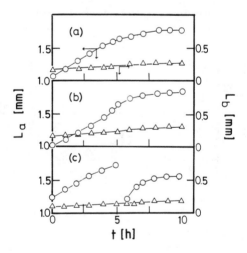

Figure 5. Change of size of an identified crystal in suspended crystals against elapsed time: a, smooth surface; b, some fines stuck on them; and c, broken crystal.

than those of extrapolated value of the straight lines in Figure 4.
These trends are supposed to come from effect of fines on crystal
growth.

On case 2, when fines stuck on growing crystal, slopes of the
lines in each directions suddenly increased after 8 hours' operation
as shown in Figure 5-(b) and then these slope gradually decreased to
the original ones same to the those of the crystal had smooth surface
within about two hours.

On case 3, growing crystal was broken suddenly on ten hours
operation and longitudinal length of the remained parts of crystal
identified by a thread became from 1.75 to 1.2 mm. Growth rates shown
by the slope of the lines in Figure 5-(c) suddenly increased almost
1.5 times. These fast growth rate continued for about two hours and
then gradually decreased to the original one. But growth rates in
lateral directions were not affected by crystal breakage.

From these results, if fines generate and breakage happen,
growth rate changed much and the whole growth rate of a suspended
multi crystals is supposed to be affected by them and become complicated. But growth rates of the crystal affected by stuck fines or
damage gradually return back to original ones and when data will be
accumulated, reasonable growth rate might become estimated.

Mechanism of inclusion of impurity. Optical purity of D-SCMC crystal
were observed for grown parts on the whole tests that seed were repeatedly used and plotted against the supercooling of the initial
operation as a parameter of impeller speed from 200 to 400 r.p.m. in
Figure 6. As shown in Figure 6, the optical purity of D-SCMC crystal
lowered, when the impeller speed and supercooling were higher. The
optical purity of about ten crystals of D-SCMC obtained after the
whole tests of the particular seed were plotted against initial super
cooling in Figure 7. Plots in Figure 7 shows classification of seed crystals by their surface states as follows.
Optical purity of the grown parts of crystals with smooth
surface was 99.0-99.9 %. But those of D-SCMC crystals
stuck by some fines and broken pieces of D-SCMC were from
90.8 to 96.7 %, From the results of dipping tests, the
optical purities of grown crystals with smooth surface
were almost same before and after dipping test and they
were from 99.0 to 99.9 %. In Figure 8-(a), photographic
picture of surfaces of D-SCMC crystals, on which two
fines were recognized to stick, was shown. From Figure 8-
(a) it would recognized that two hours later of dipping
tests, one of them began to dissolve and finally disappeared, another one remained itself before and after
tests. The optical purity of before dipping tests were
from 90.2 to 96.3 %, but after tests, these values were
almost 99.9 % as shown in Figure 8-(b). Sodium chloride
was not recognized in the grown parts before and after
tests. In Figure 9-(a), the optical purity of grown parts
of the broken pieces was shown and they were from 91.7 to
96.3 % before dipping tests. Further sodium chloride was
recognized in the crystals. The ratios of concentration
of L-SCMC and sodium chloride in the grown parts of the
crystals and in the mother liquor were summarized in
Table I. From Table I, it was found that the ratios of L-

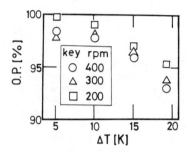

Figure 6. Optical purity of grown parts of the D–SCMC crystal against initial supercooling.

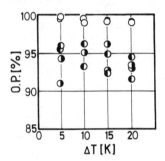

Figure 7. Optical purity of individual crystals against initial supercooling obtained at 400 rpm. ○, smooth surface; ◑, fine stuck on the surface of grown crystal; and ◑, broken piece.

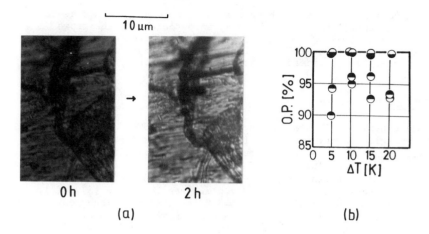

Figure 8. Optical purity of grown parts of the crystal on which some fines stuck before and after dipping test. ●, before dipping tests and ◐, after dipping test. a, photographic picture of surface of grown crystal before and after 2 h of dipping test and b, data.

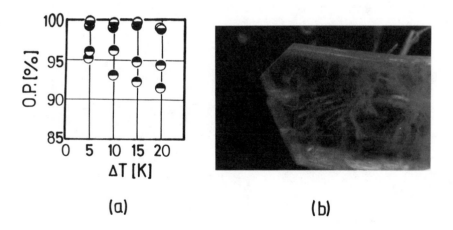

Figure 9. Optical purity of grown parts of the broken pieces. ●, before dipping test and ◐, after dipping test. a, data and b, photographic picture of crystals after tests.

SCMC and sodium chloride in the grown parts were almost same as those of mother liquor. After dipping tests, some pockets in the grown parts was recognized as shown in Figure 9-(b) and the optical purity were recovered till 98.9-99.7 %. Inclusion of mother liquor in the broken pieces was considered affect the optical purity during growth process.

CONCLUSIONS

From the experimental data obtained by growing D-SCMC crystals in a batchwise agitated crystallizer, following results were obtained:
1. Crystal growth rates in longitudinal and lateral directions at 200 300 r.p.m. were formulated as a function of supercooling (see Equations (1), (2)). Crystal growth rates in both directions at 400 r.p.m were higher than those at 200, 300 r.p.m. (see Equations (3), (4)). The mechanism of the growth of the D-SCMC crystal was discussed considering the behavior of embryo in a supersaturated solution.
2. The sticking phenomena of L-SCMC fines and crystal breakage phenomena affects on optical purity of D-SCMC crystals during growth process. Optical resolution of D-SCMC can be done successfully by careful crystallization in a racemic solution with sodium chloride.

Literature Cited
1. Yokota, M.; Toyokura, K. et al. J. crystal growth, in press.
2. Toyokura, K. et al. Proc. World Congress III of Chem. Eng., 1986, P 1020.
RECEIVED May 15, 1990

Chapter 21

Observation of High-Pressure Crystallization of Benzene from Benzene—Cyclohexane Mixture

Hideo Narahara, Kenji Yamada, Ken Toyokura, Masato Moritoki[1], and Nobuhiko Nishiguchi[1]

Department of Applied Chemistry, Waseda University, 3–4–1, Ohkubo, Shinjuku-ku, Tokyo 169, Japan

High pressure crystallization characterized by merits of purer product and high yield was studied by the laboratory equipment of the high pressure crystallizer on whose wall two optical glasses were set and crystallization of benzene from benzene-cyclohexane mixture was observed under high supersaturation. Between 10 and 50 MPa of supersaturation, benzene crystals appeared on the glass surface and the fringe of the crystals moved on it. Moving rate of the fringe of crystals was correlated as power function of relative supersaturated pressure, and the power number was between 1.67 and 1.85. When high supersaturation between 20 and 60 MPa was kept, fine crystals were observed to be floated in inlet flow of the melt solution; and the purity of the obtained crystal was over 99.9 mol%.

High pressure crystallization, where supersaturation is made by compression to several hundreds megapascals, has been studied for industrial purposes recently. Less energy consumption for product, purer crystal and higher yield of product, have been reported as several merits of this crystallization in comparison with conventional crystallization (1). But fundamental phenomena of this crystallization are not clear yet. In this study, the phenomena of high pressure crystallization were observed; and the moving rate of the fringe of crystal grown on glass wall of the crystallizer was measured. The benzene concentration of the crystals obtained by high pressure crystallization was determined after the uncrystallized melt was removed from the crystal particles by replacement of water.

[1]Current address: Kobe Steel, Ltd., 1–3–18, Wakinohama-cho, Chuo-ku, Kobe 651, Japan

0097–6156/90/0438–0281$06.00/0
© 1990 American Chemical Society

EXPERIMENTAL EQUIPMENT

A laboratory equipment used for this study consisted of pressing,
optical and separating cells as shown in Figure 1. The pressing and
separating cells were cylindrical high pressure vessels of twenty
milliliters, and the liquid in them could be compressed to 500 megapascals by piston in the vessel. The optical cell consisted of a
cylinder and two sapphire glasses, which were set on the walls of it.
The pressing cell and the optical cell were put in a thermostat bath
in order to keep the melt at a desired temperature. These three
cells were connected in series by high pressure pipes. Individual
thermocouples and pressure gauges were equipped on each cell.

EXPERIMENTAL METHODS

1) Observation of high pressure crystallization
A benzene-cyclohexane mixture was fed into the optical cell through
the pressing cell and then compressed to the equilibrium pressure.
When the temperature of melt in the cell became constant, the melt
mixture was quickly compressed to the desired pressure. The crystallization took place was observed through a video camera system.
Crystals grew along the wall surface, and the moving rate of crystal
fringe on glass surface was measured. The benzene concentration of
feed melt was 70.0, 80.0 and 90.0 mole percent for tests at 283°K.
Experimental data for mixtures, whose initial concentration was 80.0
mole percent of benzene, were also obtained at 278, 283 and 293°K.
These tests were carried out with and without seeding. Seed crystals
used in these tests were prepared through the following method: the
nucleation was occured under low supersaturation in the optical cell
at first and nuclei grew up to fine crysyals. Then, the pressure was
released to make Some nuclei melted in unsaturated conditions. After
some fines had disappeared, the melt mixture was compressed again;
and remained fines continued to grow. These melting and growing
operations were carried out several times; and relatively large
crystals were obtained and used as seed crystals.
2) Tests of the purity of grown crystals
In these tests, seed crystals made by the same way as described
above, were placed in a supersaturated solution in the optical cell.
The seed crystals grew mainly on the wall surface of the optical cell
under relatively low supersaturation. Under high supersaturated
conditions, crystal growth was recognized both in the moving melt and
on the surface of the wall. While crystallization took place in the
cell, the piston of the pressing cell moved, and the melt mixture
flew into the optical cell. When the movement of the piston stopped,
it was considered that the crystallization was over, and the crystals
in the optical cell were washed by the water almost saturated by
mixing with benzene on a preliminary treatment, and uncrystallized
melt among crystal particles was also replaced by the water. Then
benzene crystals and the water were cooled down to about 263°K, and
the pressure in the cell were released to atmospheric pressure. The
benzene crystals were taken out from optical cell and the benzene
crystals were dried on a filter paper completely. The wiped benzene
crystals were melted and their composition was analyzed by gas
chromatography.

S.G. Sapphire Glass
P.G. Pressure Gauge
S   Stroke Gauge
T   Thermocouple
V1 ⎤
V2 ⎟ Valve
V3 ⎟
V4 ⎦

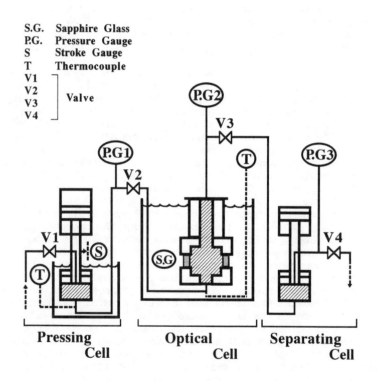

Figure 1. Schematic diagram of experimental apparatus

RESULTS AND DISCUSSION

1) Observation of high pressure crystallization
Various phenomena of crystallization were observed in the optical
cell and the shapes of crystals were modeled as shown in Figure 2.
In these tests, 170 runs were carried out and the experimental
results are plotted in Figure 3.   The marks of □, +, ◇, △, × and ▽
in Figure 3 indicate the phenomena shown in Figure 2.

In some test runs that the melt mixture in the optical cell was
compressed to a desired pressure and kept under the same pressure for
fifteen minutes, no crystal appeared.   As shown in Figure 3, these
test results were often obtained for the system with 80.0 mole percent benzene at 283°K, under the supersaturation of 20 megapascales.
But in some other tests using different melt compositions, some
nucleation and crystal growth were observed under almost same operational conditions.

In relatively low supersaturation, dendritic crystals were grown
on the glass of optical cell.   The shape of crystals is expressed by
the modeling sketch which was like skeleton of leaf as shown in Figure 2-b).   This crystal appeared under operating conditions shown as
the mark + in Figure 3.   This phenomenon was often observed under
relatively low supersaturated conditions, and also at high temperatures when melts with high concentration of benzene were used.

When the operational supersaturation is slightly increased,
triangle crystals with a skeleton in the central part as shown in
Figure 2-c), appeared.   These operational conditions are shown by the
mark ◇ in Figure 3.

Triangle shaped crystals as shown in Figure 2-d), also appeared
in some tests.   These crystals appeared under the operational conditions of the mark △ in Figure 3.   When a relatively low supersaturation was imposed, most crystals grew on the wall.   In these tests,
the heat of compression is supposed to be released through the wall
of the optical cell and supersaturation of the melt on the wall is
considered to become higher.   Isothermal high pressure crystallization operation had the trend that benzene crystal appeared on the
wall surface, due to the temperature gradient across the wall.

When a higher supersaturation was imposed, two different phenomena as shown in Figure 2-e) and 2-f) were found.   Phenomenon in
Figure 2-e) showed that triangle crystals grew on the glass surface
and some fines were suspended in the melt that flowed into the
optical cell.   This phenomenon was found on tests of mark × in Figure
3.   Another phenomenon of Figure 2-f) was that vigorous nucleation
occured and much fines were found to be suspended in the melt.   After
the nucleation, another shape of crystal in Figure 2-f) appeared on
the optical glass and grew quickly.   This phenomenon was found for
operational conditions marked by ▽ in Figure 3.   These phenomena did
not occure repeatedly, but the trend that nucleation is apt to occure
under relatively higher supersaturation, was confirmed.

The test results with seed crystals were similar to those
without seed.   The crystals grew mainly on the wall with or without
seeding; and this phenomenon is considered to be caused by the high
supersaturation of the melt on the wall.

The supersaturation caused by compression of the melt was
supposed to spread uniformly throughout the melt, and homogeneous
crystallization was expected.   But in these crystallization tests,

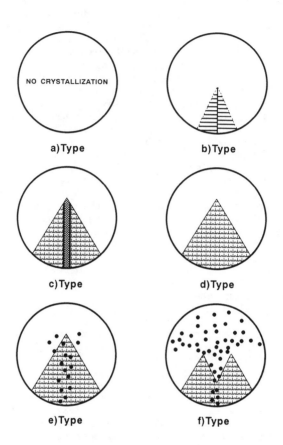

Figure 2. Classification of crystal type. a) No crystallization. b) Dendritic crystal grown on the glass of optical unit that was observed under relatively low supersaturation. c) Triangle crystal with a skeleton in the central part of it on the glass surface. d) Triangle crystal grown on the glass surface. e) Triangle crystal grown on the glass surface with fine crystals in the melt flowed into the optical unit. f) Vigorous nucleation in the melt under relatively high supersaturation.

nucleation and growing phenomena were different under relatively low supersaturation; and heat transfer was supposed to affect the crystallization. When high supersaturation was suddenly imposed to the melt, homogeneous crystallization was assumed to occur since the effect of heat transfer was considered to be small. When high pressure crystallization is considered in a industrial scale, large crystallizer might be used and the effect of heat transfer is assumed to be less on crystallization than on a laboratory scale crystallization. In general high pressure industrial crystallization, the operation is near adiabatic; and since supersaturation is generally very large, homogeneous nucleation in the melt may be favored.

The shape of moving boundary of crystals was different if the composition of melt mixture was changed. When pure benzene was compressed, the shape of moving boundary of crystal was roundish. The fringe shape made by 90.0 mole percent benzene was slightly sharp, and the crystals formed from the melt 80.0 and 70.0 mole percent benzene had sharper moving boundary. But the details of these difference are not clear and future studies are expected.

The distance of movement of the fringe of a crystal formed on the sapphire glass was recorded by video film, and the moving rate of the fringe was calculated from the time required for the fringe of the crystal to pass the distance between two particular points on the optical glass, $(\Delta l/\Delta \theta)$. This moving rate of the fringe was defined as the crystallization rate in this study.

Data of crystallization rate were a little bit scattered; and this was supposed to be caused by complicated phenomena of both transfer of crystallization heat and change of composition of the melt due to crystallization of benzene. At the initial stage of crystallization, the effects of crystallization heat and the changing of composition are small, and therefore the crystallization rate would be fast. As the crystallization continues, the temperature of the slurry of suspended crystals increases and the benzene concentration decreases, and so crystallization rate decreases. Therefore, the maximum crystallization rate under the same operational conditions is obtained at the initial stage of operation. In Figure 4, the maximum crystallization rates are plotted against α, relative supersaturated pressure for tests where nucleation took place without seed crystals. The relative supersaturated pressure, α is defined by Equation 1,

$$\alpha = \frac{P}{P_e} - 1 \qquad (1)$$

where P is the operational pressure, and P_e is the equilibrium pressure. The crystallization rates obtained from the growth of seed crystals at 283°K were plotted against the relative supersaturated pressure in Figure 5. The same correlation was made in Figure 6 for the temperatures 278,283 and 293°K with 80.0 mole percent melt.

As may be seen Figures 4, 5 and 6, the crystallization rate increased with relative supersaturated pressure, and the power number of the correlative lines between these two variables were almost same, as shown by Equation 2.

$$\frac{\Delta l}{\Delta \theta} = 1.06 \times \alpha^{1.77} \qquad (2)$$

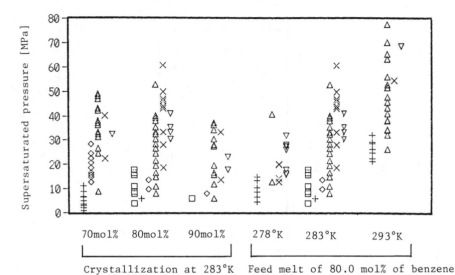

Crystallization at 283°K   Feed melt of 80.0 mol% of benzene

☐   a :No crystallization

+   b :Dendritic crystal

◇   c :Triangle shaped crystal with skeleton

△   d :Triangle shaped crystal

✕   e :Triangle shaped crystal with fine crystals in
        the melt mixture

▽   f :Vigorous nucleation in the melt mixture

Figure 3. Dependence of crystal types on operational condition

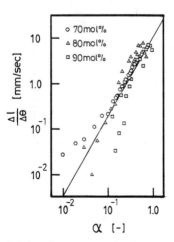

Figure 4. Correlation between relative supersaturated pressure
and crystallization rate (unseeded, 283 K)

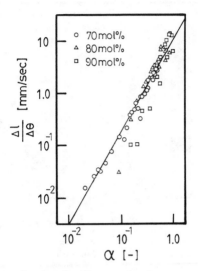

Figure 5. Correlation between relative supersaturated pressure
and crystallization rate (seeded, 283 K)

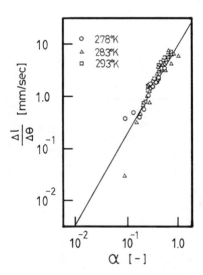

Figure 6. Correlation between relative supersaturated pressure
and crystallization rate (unseeded, 80.0 mol% of benzene)

It was found that crystallization rates do not depend on the presence of seed crystals, the composition of feed melt and the operational temperature. The maximum crystallization rate observed in this study was 13.73 [mm/sec]. However, the growth rate of cooled type crystallization in the same system was about 3.0×10^{-4} [mm/sec].(2) The maximum supersaturation corresponding to 1.0 of relative supersaturation in Figure 4, is 60 megapascals, and 60 megapascals is estimated to correspond to 15°K from equilibrium data in the paper (3) reported by the author. The supersaturation corresponding to the maximum growth rate in a cooled type crystallizer, was about 3°K. This phenomenon that the supersaturation made by compression to high pressure becomes higher value quickly, is supposed to come from incompressibility of the liquid mixture. The incompressibility of the mixture makes large uniform supersaturation before nucleation and higher crystal growth rate possible.

2) Tests of the purity of grown crystals

The experimental conditions and results of the analysis of the purity of separated benzene crystals are shown in Table 1. In tests of No. 1-1 to 1-4, and 3-1 and 3-2, the melt was compressed to the pressure shown in Table 1 and kept on the same value, without seed crystals. Nucleation occured on the wall and crystals grew there. In tests of No.2-1, 2-2 and 3-3, seed crystals were made as described above; they grew both inside the optical cell and on the wall. In these tests, since the melt around benzene crystals was replaced by the water, the crystals were taken out without serious destruction. The shapes of benzene crystals were dendritic, and purity of it was over 99.9 mole percent, independent from the operational conditions and the feed compositions as shown in Table 1. Therefore, crystals obtained by high pressure crystallization is considered to be very pure due to the complete removement of mother liquid from crystal surface.

Table 1. Experimental conditions and tests results
of purity of benzene crystal

Run No.	Feed Conc. [mol%]	Super Press. [MPa]	Purity [mol%]
1-1	90.0	29.4	99.94
1-2	90.0	39.2	99.95
1-3	90.0	49.0	99.99
1-4	90.0	58.8	99.99
2-1	90.0	49.0	99.99
2-2	90.0	53.9	99.98
3-1	80.0	31.4	99.99
3-2	80.0	50.9	99.99
3-3	80.0	90.2	99.98

CONCLUSION

The experiments under high supersaturation effected by compression
at high pressure were carried out. Crystallization rates
defined as the moving rate of the fringe of a crystal on the sapphire
glass were very fast. The power number in the correlation of crystallization rate against relative supersaturated pressure did not
depend on feed composition, supersaturation and operational temperature. When relatively higher supersaturation was given, nucleation
in inlet flow of melt was observed. The purity of crystals formed by
high pressure crystallization was over 99.9 mole percent even when
crystals were obtained under high supersaturation. From these
results, the high pressure crystallization is concluded to be a quick
process for separation of pure products from their mixture.

LEGEND OF SYMBOLS

P   operational pressure   [MPa]
P   equilibrium pressure   [MPa]
Δl distance which fringe of crystal moved between two particular
     points on optical glass   [mm]
$\Delta \theta$ time required for the fringe of the crystal to pass distance
     between two particular points on the optical glass   [sec]

LITERATURE CITED

(1) Toyokura K. : "Kagaku Kogaku Ronbunshu" 12, (5), 622 (1986)
(2) Toyokura K. : "Industrial Crystallization '87" edited by J.Nyvlt
                and Zacek, Elsevier, 561 (1989)
(3) Toyokura K. : "Industrial Crystallization '87" edited by J.Nyvlt
                and Zacek, Elsevier, 485 (1989)
RECEIVED June 27, 1990

CRYSTALLIZATION AND PRECIPITATION OF INORGANIC COMPOUNDS

Chapter 22

Crystallization of Gypsum from Phosphoric Acid Solutions

E. T. White and S. Mukhopadhyay[1]

Department of Chemical Engineering, University of Queensland, Brisbane 4067, Australia

A computer model has been generated which predicts the behaviour of a continuous well mixed gypsum crystallizer fed with a slurry of hemihydrate crystals. In the crystallizer, the hemihydrate dissolves as the gypsum grows. The solution operating calcium concentration must lie in the solubility gap. Growth and dissolution rates are therefore limited.

Measurements were undertaken of the solubility of each phase in acid solutions, of the growth rate of gypsum crystals and the dissolution rate of hemihydrate. The growth rate depends on the square of the supersaturation and on temperature with an activation energy of 64 kJ/mol. The nucleation rate appears to vary linearly with supersaturation.

As the shape of the needle-like hemihydrate crystals changes as they dissolve, it is necessary to convert to the crystal width as a measure of size. In terms of this measure, the dissolution rate is first order with undersaturation and shows only a small temperature effect (activation energy of 10 kJ/mol).

With this data, the computer program is capable of predicting the operation of a hemihydrate - dihydrate crystallizer installation.

The world's phosphorus consumption is in the order of 40 m.t.p.a. (as P_2O_5). About 90% of this involves the fertilizer industry (1). The primary natural source of phosphorus is rock phosphate; the major chemical produced from it is phosphoric acid.

[1]Current address: Chemistry Department, State University of New York at Buffalo, Buffalo, NY 14214

"Wet" processes using sulphuric acid are the most common means of producing phosphoric acid. The simplified overall reaction ($\underline{2}$) is,

$$Ca_3(PO_4)_2 + 3\ H_2SO_4 + 3x\ H_2O \rightarrow 2\ H_3PO_4 + 3\ CaSO_4 \cdot xH_2O \downarrow$$

where x maybe 0, 1/2 or 2. Depending on conditions, the calcium sulphate can precipitate out as the anhydrite, the hemihydrate or the dihydrate (gypsum). All have very low solubilities. Commercial processes have been developed producing any one of these as the byproduct ($\underline{1}$).

One of the most commercially viable processes is the Nissan hemihydrate-dihydrate process. The conditions chosen for the digestion of the rock phosphate with sulphuric acid produces the hemihydrate phase. The slurry of fine hemihydrate crystals then passes to a crystallization stage where conditions are chosen (mainly temperature) so that gypsum is the stable species and it crystallizes out as substantially larger crystals, which aids subsequent filtration and washing.

It is the purpose of this study to model gypsum crystallization under Nissan process conditions.

The Process

Figure 1 shows a process diagram for a typical Nissan H installation ($\underline{3}$). The slurry produced by the digester contains about 30% solids (mainly hemihydrate with a little undissolved rock) suspended in an acid liquor, typically about 30% by wt P_2O_5 with a few percent of sulphuric acid. The crystallizers (three in this case), each with a residence time of about 6 hours, operate at 60 to 70°C and may be considered to be well mixed. The calcium sulphate recrystallizes in these vessels as gypsum. Some of the product from the last crystallizer (typically half) is recycled.

The liquor carries a wide variety of impurities (eg. Al, Fe, Mg, F, Si) originating from the original rock. A number of these may affect the crystal morphology ($\underline{4}$, $\underline{5}$). For laboratory studies a synthetic liquor was used containing about 30% P_2O_5, 2 to 6% H_2SO_4 and added impurities as 0.5% F, 0.05% Fe and 0.02% Al. The additions of these impurities resulted in reasonably shaped product gypsum crystals, comparable to those produced commercially.

The Crystal Products

Gypsum crystals have an SG of 2.32, are colourless and belong to the monoclinic system. Figure 2 shows photographs of industrial and laboratory gypsum crystals. The crystals are elongated and can range up to 40 μm in size.

Hemihydrate crystals have an SG of 2.71, are white in colour and belong to the rhombohedral system. Figure 3 shows photographs of industrial and laboratory hemihydrate. The crystals are needle-like with sizes up to 20 μm.

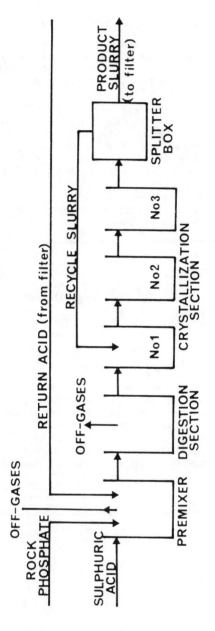

Figure 1. Typical hemihydrate – dihydrate process for phosphoric acid.

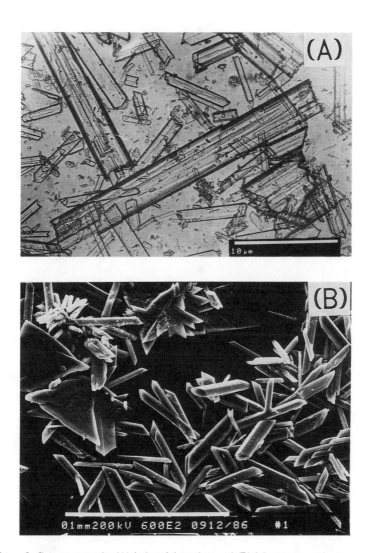

Figure 2. Gypsum crystals; (A) industrial product and (B) laboratory product.

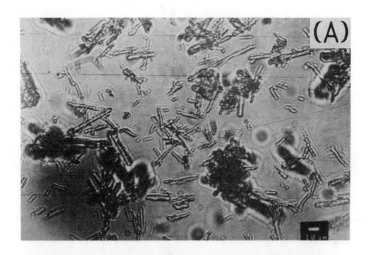

Figure 3. Hemihydrate crystals; (A) industrial product and (B) laboratory product.

The Solubility Gap

Two calcium sulphate phases coexist in the crystallizer, the hemihydrate fed in and the gypsum being formed. Figure 4 shows the solubility of these two phases as a function of temperature for one particular acid liquor. The two solubility lines cross at the transition temperature (75°C in this case). Above this temperature, the hemihydrate is the stable phase (has the lower solubility) and below, gypsum.

Thus digestion of the rock phosphate would be undertaken above the transition temperature to form hemihydrate crystals. The crystallizers however would be operated at a temperature below the transition value and so gypsum crystals are produced.

If hemihydrate is to dissolve in a crystallizer, the calcium concentration must lie below the solubility line for the hemihydrate. On the other hand, if gypsum is to grow, the calcium concentration must lie above the gypsum solubility line, to provide the operating supersaturation. For example, operating at 62°C (Figure 4), point A could represent the operating calcium concentration. A" - A then represents the driving force for dissolution and A - A', the driving force for growth. For hemihydrate dissolution and gypsum growth to occur simultaneously, A must lie in the range A" - A', the solubility gap.

The position of A will depend on process conditions. For a well mixed crystallizer at steady state, point A will adjust until the rate of hemihydrate dissolution equals the rate of gypsum growth (in mole units). In simplified terms,

Rate hemihydrate dissolution = Rate gypsum growth

$$K_H \ S_H \ \phi_H (A"-A) \ = \ K_G \ S_G \ \phi_G (A-A') \tag{1}$$

where K is the rate coefficient, S the crystal surface area and ϕ the dependence on driving force. The subscript H refers to hemihydrate and G to gypsum. It is assumed that the change in the solution calcium concentration between inlet and outlet is small, which is true in practice.

Equation 1 is the key relation in understanding the operation of such a crystallizer and follows a similar development by Garside (6).

The solubility gap is very small (eg. 0.01% Ca^{2+}). This is the total driving force available for both mechanisms and thus places limits on the maximum rates. If the total driving force (at 62°C, Figure 4) were available for growth, the maximum growth rate (using the correlations presented later), would be less than 0.3 $\mu m/hr$. Likewise the maximum dissolution rate (based on the full driving force) would be about 10 $\mu m/hr$. Thus large residence times are a necessity to achieve gypsum product crystals of a suitable size. Typical residence times in practice are about 6 hr per crystallizer. Altering the residence time will only shift the position of A in the solubility gap and does not have the large effect on growth rate experienced in single species MSMPR operation.

There is a further important consequence of equation 1. There must, of necessity, be an amount of undissolved hemihydrate remaining in a continuous crystallizer to provide the S_H value in equation 1. For the same production rate, if A"- A were reduced,

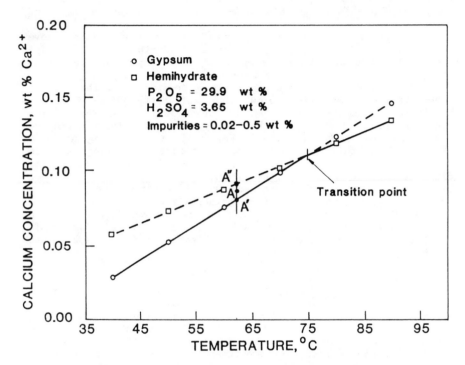

Figure 4. Solubilities of gypsum and hemihydrate for a particular solution as a function of temperature. A" - A is the solubility gap and A represents the operating conditions.

then S_H must increase to satisfy equation 1 and thus there would be more undissolved hemihydrate leaving the crystallizer. In practice, somewhere of the order of 10% of the entering hemihydrate to each crystallizer remains undissolved on discharge.

The main processing options open to the crystallizer designer are the solubility gap (transition temperature, acid content), the operating temperature and the values of the rate coefficients (affected by impurities) and crystal surface areas (eg. altering crystal content). The computer model generated in this study allows these effects to be evaluated.

Information Required For Modelling

A model for a gypsum crystallizer will require the following information,

A. Solubilities: to evaluate the solubility gap.
B. Gypsum growth rates.
C. Hemihydrate dissolution rates.

Each was investigated and will be discussed in turn.

Solubility

The solubility of hemihydrate and gypsum in phosphoric-sulphuric acid solutions is reviewed by Mukhopadhyay and White (in press). While there is a lot of data on the solubility of these species in water, data on the solubility in acid solutions is limited. Data is given by Taperova (7), Taperova and Shulgina (8), Ikeno et al. (9), Linke (10), Dahlgren (11), Kurteva and Brutskus (12), Bevemzhanov and Kruchenko (13), Zdanovskii and Vlasov (14) and Glazyrina et al. (15).

Further measurements were made by Khan (16) and Mukhopadhyay (17). A large excess of either gypsum or hemihydrate crystals was added to an acid solution of measured composition. This was agitated at constant temperature until equilibrium was reached. Duplicate runs were carried out both from undersaturated and supersaturated solutions. After settling, the solution was analysed for calcium content by atomic absorption.

Since the need for solubility data in this work is to evaluate the solubility gap, the available data were correlated in terms of,

(a) the transition temperature, T_t, (where the solubilities of hemihydrate and gypsum are equal).

(b) the calcium concentration at this transition temperature, Ca_t.

(c) the change of solubility with temperature for gypsum, (assuming a straight line relation), $m_G = (Ca-Ca_t) / (T-T_t)$

(d) the change of solubility with temperature for hemihydrate, $m_H = (Ca-Ca_t) / (T-T_t)$

Figure 5 shows all data from a number of sources for the transition temperature. The values are well correlated by plotting against the total acid content (as wt% P_2O_5 + H_2SO_4). The other quantities above have been correlated by,

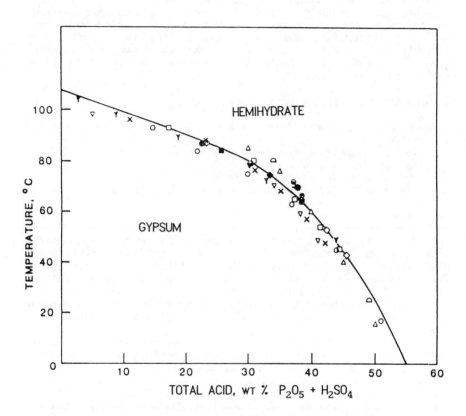

Figure 5. Transition temperature for gypsum - hemihydrate, as a function of total acid content (wt % P_2O_5 + H_2SO_4).

$$Ca_t(S+2.2) = (T_t/90)^2$$
$$m_G = 0.0035 \exp [-S/5] \tag{2}$$
$$m_H = m_G/2$$

where $S = \%$ by wt H_2SO_4.

The use of this correlation fits 95% of the measured data within $\pm$ 0.04% Ca for gypsum and $\pm$ 0.07% Ca for hemihydrate. The solubility gap, ΔCa, can be evaluated by

$$\Delta Ca = (m_G-m_H)(T-T_t) \tag{3}$$

Gypsum Growth

Previous work with acid solutions had been undertaken by Amin and Larson ([18]), Hoa ([19]) and White and Hoa ([20]). Quite a number of studies (eg. [21]-[25]).have dealt with crystallization of gypsum from aqueous (non-acid) solutions.

In the present study, gypsum growth kinetics were studied in a 1 litre, agitated, isothermal, batch crystallizer. It was constructed of plastic as the fluorine tends to etch glass vessels, taking silica into solution. Initially a calcium-rich solution, containing calcium hydrogen orthophosphate dissolved in 30% P_2O_5, was mixed in the crystallizer with a sulphate-rich solution, containing 7% H_2SO_4 in 30% P_2O_5 and the added impurities to give 700 ml of supersaturated solution.[5] The initial calcium supersaturation was over 1%, considerably in excess of the solubility gap values, and thus industrial levels.

2 g of sieved gypsum seed, preheated to the operating temperature, was added. Samples were taken every few minutes, using weighed plastic syringes, for crystal sizing, solution analysis and crystal content determinations. Crystals were separated from solution using a membrane filter. For sizing, crystals were suspended in a methanol electrolyte containing 2% lithium chloride (and saturated with gypsum) and were sized with an electronic sensing zone sizer (Particle Data) using a 300μm orifice. Solutions were analysed for calcium by atomic absorption, sulphate by barium addition and atomic absorption for residual barium, while phosphate was determined by the vanado-molybdate method. Greater detail is given by Mukhopadhyay ([17]).

Figure 6 shows the size distributions for the samples taken from one of the runs, presented as the cumulative number oversize per ml of slurry. From the lateral shift of the size distributions, the growth rate can be determined. Figure 7 shows values of growth rate, G, supersaturation, s, and crystal content determined during the run. As a material balance check, the crystal contents were evaluated from direct measurements, from solution analyses and from the moments of the size distribution. The agreement was satisfactory. No evidence of size dependent growth or size dispersion was observed.

In Figure 8, the growth rates have been plotted against supersaturation. For each temperature, a square law dependence was found. This is shown directly in Figure 9. Accepting a square law, the corresponding growth rate constants, k_G, were evaluated, where,

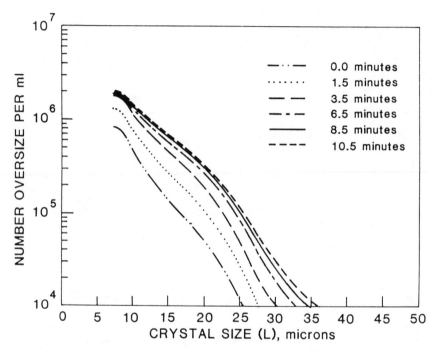

Figure 6. Size analyses of samples for gypsum growth run at 50°C. Size as volume equivalent size.

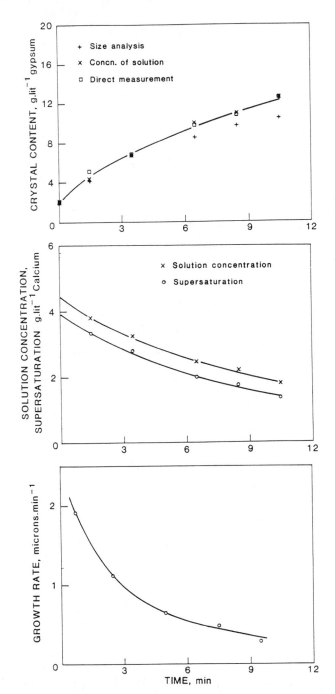

Figure 7. Crystal content, supersaturation and growth rate vs time for gypsum growth run at 50^{q}C.

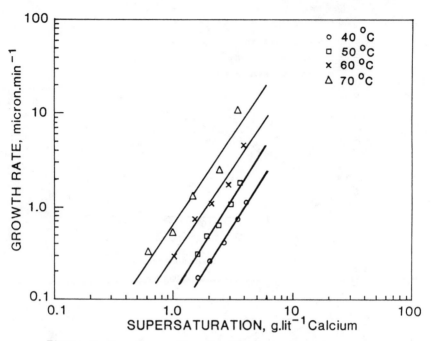

Figure 8. Gypsum growth rate vs supersaturation.

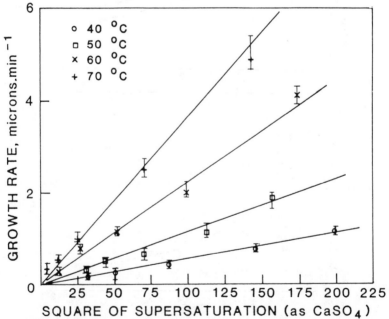

Figure 9. Gypsum growth rate vs square of supersaturation.

$$k_G = G / s^2 \qquad (4)$$

These are shown on an Arrhenius plot (Figure 10) and exhibit an activation energy of 64 kJ/mol ($\pm$ 8%). The high activation energy and the second order dependence, implies that surface kinetics is controlling in the growth mechanism.

Also shown on Figure 10 are the previous results of White and Hoa (20). They differ by a factor of five from the present results. Despite considerable effort, no reason could be found for such a large difference. This discrepancy needs further investigation.

The growth rate of gypsum crystals from this work is given by

$$G = 0.28 \ s^2 \ exp \ [- 7750 \ (1/T - 1/333.16)] \qquad (5)$$

This fits 95% of the present data within $\pm$ 35%.

Nucleation rates could be estimated from the increase in the number of particles, obtained from sizings with a 95μm orifice tube, of the low size end of the size distributions. The rate of increase of numbers in the batch (Figure 11) is independent of temperature, and shows an approximate first order dependence on supersaturation. Assuming a first order dependence on suspension density, M_T, gives the correlating relation as,

$$Bo = 3.5 \ x \ 10^8 \ M_T \ s \qquad (6)$$

where Bo is the number of nuclei formed per minute per litre of suspension and M_T is expressed as g of gypsum crystal per litre of suspension. This agrees with 95% of the limited present data within $\pm$ 35%.

The relation found by White and Hoa (20) is also shown on Figure 11. Again there is an unexplained five fold discrepancy.

Hemihydrate Dissolution Rate

There have been a number of studies on the rate of dissolution of crystals in agitated tanks (eg. 26–31). Predominantly the results indicate dissolution is mass transfer controlled. For mass transfer in stirred tanks, among the many studies, that of Levins and Glastonbury (32) is widely considered.

To determine the rate of dissolution of hemihydrate crystals, the same vessel was used as for the crystallization study. The vessel was filled with the sulphate-rich solution (zero initial calcium concentration). An amount of sieved hemihydrate seed crystals, about 10% in excess of that required to saturate the solution, was added. At very short time intervals, samples were taken using a similar procedure to that for the gypsum growth investigation. Samples were separated into crystals for size analysis (with a 190μm orifice) and crystal content and solutions for analysis. Further details are given by Mukhopadhyay (17).

Figure 12 shows the size analyses from one run. Crystal content and solution analysis measurements were in good agreement. The size shown in Figure 12 is the volume equivalent size, which is acceptable for crystal growth where the crystals retain the same elongated shape as they grow. However when needle-like crystals dissolve, there is no longer preferable transfer on particular faces, so the shape changes. If allowance is not made for this, incorrect dissolution rates will result. Thus for purposes of

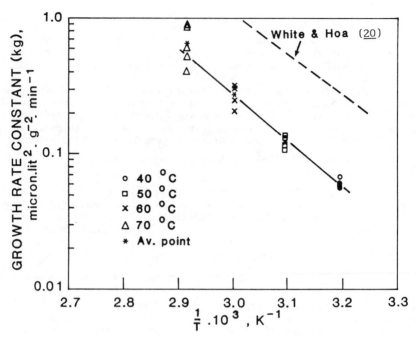

Figure 10. Arrhenius plot for gypsum growth rate constants.

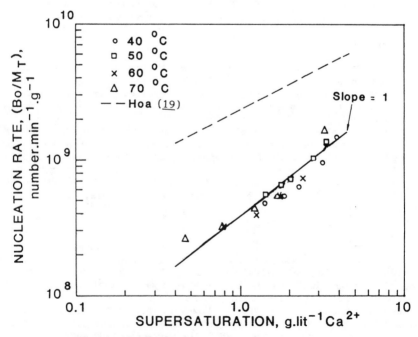

Figure 11. Nucleation rates for gypsum.

calculating the dissolution rate, the crystals width was used as the measure of size. Crystals were assumed to be elongated square prisms, with an initial aspect ratio of 15. The size distributions, replotted in terms of the crystal width size were used to evaluate the dissolution rate. Greater detail is given by Mukhopadhyay and White (33).

A plot of the dissolution rate against driving force, ΔC, is shown in Figure 13. This shows a linear dependence on undersaturation and a slight dependence on temperature (activation energy of 10 kJ/mole). This indicates that the dissolution is mass transfer controlled. The results can be correlated by

$$D = 1.8 \ \Delta C \ \exp \ [-1260(1/T - 1/333.16)] \qquad (7)$$

where D is the dissolution rate in μm/min, evaluated in terms of the crystal width, ΔC is the undercooling as g Ca^{2+}/litre of solution and T is absolute temperature. This fits 95% of the experimental results within $\pm$ 30%.

A comparison of the measured growth and dissolution rates is shown in Figure 14. Note that the dissolution rate is in terms of the hemihydrate crystal width, while the growth rate is in terms of gypsum volume equivalent size. Also note that the abscissa is undersaturation in one case and supersaturation in the other. While values are of the same order at high driving forces, industrial operation would have driving forces about 0.1 g/litre and below, where the rates differ by two orders of magnitude.

Separation of Hemihydrate and Gypsum Crystals

Both hemihydrate and gypsum crystals coexist in a crystallizer in the Nissan process. Samples can be readily taken. The fraction of hemihydrate is determined on a bulk sample by evaluating the amount of hydrated water in the sample (2 moles for gypsum, 0.5 for hemihydrate).

To size the two species however, it is necessary to separate them. A sink-float technique has been developed to separate the two. A carbon tetrachloride – tetrabromoethane mixture having an SG about 2.5 was used. The dry crystal mixture is added to the dense organic liquid, then put under vacuum to remove entrapped air bubbles. The suspension is then centrifuged for several hours to separate out the gypsum which floats and the hemihydrate which sinks. Tests on synthetic mixtures showed that a very clean separation is possible.

Modelling

A computer model was written for a continuous, well mixed, isothermal crystallizer fed with a hemihydrate crystal slurry.

As far as the gypsum crystals are concerned, the analysis is identical to that for a seeded MSMPR (34). The information required is the growth rate and the mean residence time. For the hemihydrate, the analysis is that for a continuous seeded MSMPR dissolver (35), which parallels that for the crystallizer. The information needed is the dissolution rate and the mean residence time.

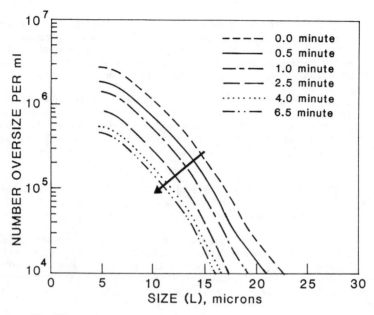

Figure 12. Dissolution of hemihydrate crystals. Size analyses for run at 60 °C.

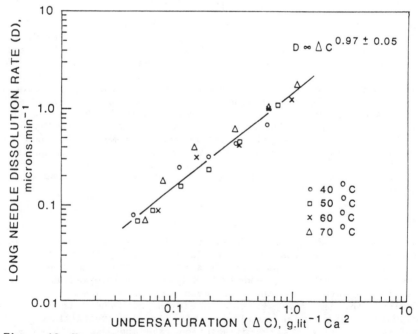

Figure 13. Hemihydrate dissolution rate vs undersaturation. Rate expressed as change in crystal width.

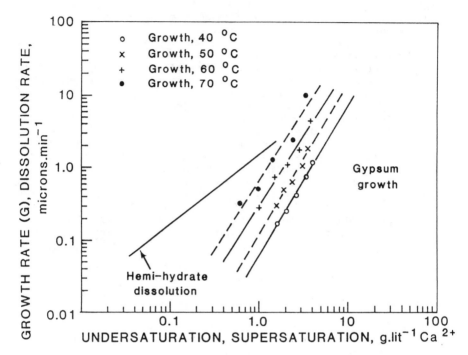

Figure 14. Comparison of rates of hemihydrate dissolution and gypsum growth. Note that dissolution is based on the crystal width and growth on the volume equivalent size.

The growth and dissolution rates depend on the driving forces within the solubility gap and are interrelated by the material balance of relation 1 (in suitable form).

The population balances were solved in moments form to establish the steady state conditions. Then the full size distributions were evaluated for the particular seed size distributions of the feed. Due allowance was made for the different measures of size for the two species.

A listing of the program is given by Mukhopadhyay (17). A flowsheet is shown as Figure 15.

A prior model was described by Marshall et. al. (<u>36</u>), but this did not include the hemihydrate population balance. A version of the present model was given by Steemson et. al. (<u>37</u>), centred around a simulation package developed for alumina plant modelling. This version included hemihydrate dissolution but used an assumed correlation for the dissolution rate.

The present model was checked against the plant data shown by Marshall et. al. (<u>36</u>) and gave a reasonable fit. It was then used to show the effects of changing either operating variables or the process equipment. Figure 16 shows the effect of changing the recycle ratio, the fraction of the exit slurry recycled from the splitter box (Figure 1). The effect is small. Figure 17 shows the effect of replacing the splitter box with a classifier. The abscissa, D_{50}, is the centre cut size of the classifier.

The results given are for a specific application. The model is generally useful and can be used to explore the effect of any possibility the designer or operator may consider.

Conclusions

1. The operating calcium concentration for a gypsum crystallizer fed with hemihydrate crystals must lie in the solubility gap between the solubilities of the two species. This places severe limits on the range of growth and dissolution rates possible.

2. A mechanistic model of the crystallizer requires information on solubility and the rates of gypsum growth and hemihydrate dissolution.

3. A correlation is given which allows the solubility of either species in acid solutions to be estimated, and thus the magnitude of the solubility gap.

4. Gypsum growth rates show a square law dependence on supersaturation with an activation energy of 64 kJ/mol. Growth appears to be surface kinetics controlled.

5. The nucleation rate for gypsum shows a first order dependence on supersaturation.

6. Hemihydrate crystals change shape as they dissolve, and it is necessary to convert from a volume equivalent size to a crystal width. The dissolution rate, in terms of crystal width, shows a first order dependence on undersaturation and a low activation

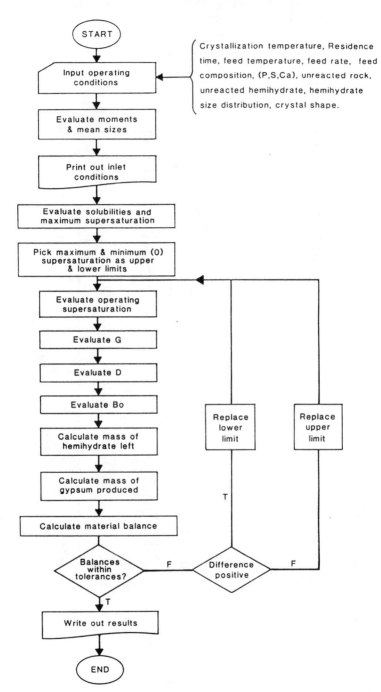

Figure 15. Flowsheet for computer model of hemihydrate - dihydrate crystallizer.

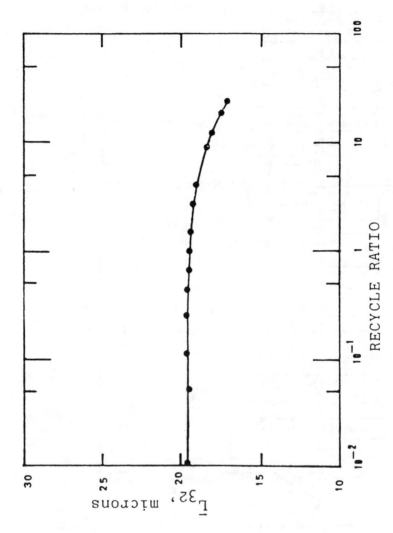

Figure 16. Model prediction of effect of splitter recycle ratio on the mean product size.

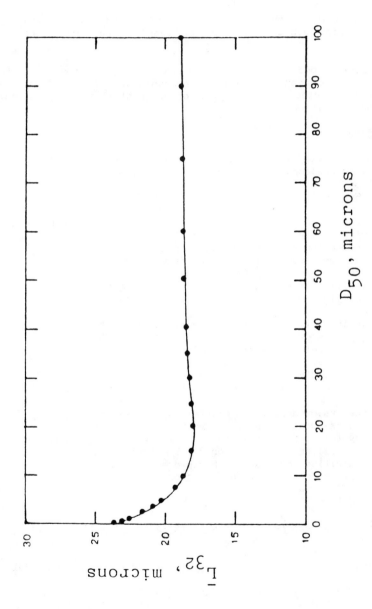

Figure 17. Model prediction of effect of classifier cut size on the mean size.

energy of 10 kJ/mole. Dissolution seems to be mass transfer controlled.

7. A sink-float technique has been developed for separating hemihydrate from gypsum crystals for samples from mixed species crystallizers.

8. A computer model has been produced that incorporates the solubility, growth and dissolution mechanisms. It predicts the behaviour of a single crystallizer or a crystallization section, given input conditions.

Acknowledgment

The financial support of the Australian Research Council is gratefully acknowledged.

Legend of Symbols

A, A', A'': Calcium concentrations, operating, satd. with gypsum, satd. with hemihydrate (Figure 4).
B_0 : nucleation rate, #/min . 1 of slurry.
C : calcium concentration, g Ca^{2+}/l soln.
ΔC : undersaturation, g Ca^{2+}/l soln.
Ca : equilibrium calcium solubility, g Ca^{2+}/g soln.
Ca_t : equilibrium calcium concentration at transition temperature g Ca^{2+}/g soln.
D : hemihydrate dissolution rate (crystal width), μm/hr.
G : gypsum growth rate (vol. equiv. size), μm/hr.
k_G : growth rate constant (equation 4).
K : rate coefficient (equation 1).
L : crystal size (vol. equiv. size), μm.
m : rate of change of solubility with temperature, g Ca^{2+}/g soln. $^{\circ}$C
M_T : suspension density, g crystal / ml of slurry
s : supersaturation, g Ca^{2+}/l soln
S : crystal surface area (equation 1); wt % H_2SO_4 (equation 2)
T : temperature, C, (equation 2), K (equation 4, 5)
T_t : transition temperature, hemihydrate-gypsum, $^{\circ}$C.
ϕ : driving force function (equation 1)

Literature Cited

1. Becker, P. Phosphate and Phosphoric Acid, Raw Materials and Economics of the Wet Process; Fert. Sci. and Technol. Ser.; Dekker: New York, 1983; Vol. 3.
2. Beveridge, G.S.G.; Hill, R.G. Chem. and Proc. Eng. 1968, July, 61.
3. Pratt, C.J. British Chemical Engineering 1964, 9 (8), 535.
4. Adami, A.; Ridge, M.J. J. Appl. Chem. 1968 18 (12), 361.
5. Edinger, S.E. J. Crystal Growth 1973, 18, 217.
6. Garside, J. "Hydrate Transformation in Continuous Crystallization Processes", Nth. Westn. Branch Papers, Instn. Chem. Enginrs. 1982, 2.
7. Taperova, A.A. J. App. Chem. USSR, 1940, 13 (5), 43.

8. Taperova, A.A. ; Shulgina, M.N. J. Appd. Chem. USSR, 1945, 18 (9), 521.
9. Ikeno, S.; Keiji, K.; Tsuda, T. Sekko to Sekkai 1958, 33, 1607.
10. Linke, W.F. Solubilities of Inorganic and Metal Organic Compounds; Amer. Chem. Soc.; Washington, 4th ed. 1965, Vol. 1.
11. Dahlgren, S.E. J. Agr. Food Chem. 1960, 8, 411.
12. Kurteva, O.I.; Brutskus, E.B. J. Appd Chem., USSR 1961,34 (8), 1636.
13. Bevemzhanov, B.A.; Kruchenko, V.P. Russian J. Inorg. Chem. 1970, 15, 10.
14. Zdanovskii, A.B.; Vlasov, G.A. J. Appd. Chem. USSR 1971 44 (1), 15.
15. Glazyrina, L.N.; Savinkova, E.I.; Grinevich, A.V.; Akhmedova, L.E. Zh. Prikl. Khim. 1980, 53 (11), 2524.
16. Khan, S.; Solubility of Calcium Sulphate in Acid Solutions, M.Eng. St. Thesis, Uni. of Queensland, 1986.
17. Mukhopadhyay, S.; Crystallization Kinetics of Gypsum from Phosphoric Acid Mixtures, PhD Thesis, Uni. of Queensland, 1988.
18. Amin, A.B.; Larson, M.A. Ind. & Eng. Chem. Proc. Des. and Dev. 1968, 7 (1), 133.
19. Hoa, L.T. Crystallization Kinetics of Calcium Sulphate Dihydrate from Phosphoric Acid Solutions, M.Eng.Sc. Thesis, Uni. of Queensland, 1974.
20. White, E.T.; Hoa, L.T. Proc. Second Australian Conf. on Heat and Mass Transfer, Sydney, 1977, 401-408.
21. Schierholtz, O.J. Canad. J. Chem. 1958 36, 1057.
22. Liu, S.T.; Nancollas, G.H. J. Colloid and Interface Science 1975 52 (3), 593.
23. Zielinski, S. In Industrial Crystallization 81; Jancic, S.J. and De Jong (Eds.), North-Holland Publishing Company. 1982
24. Toyokura, K.; Shibauchi, T.; Uchiyama, M.; Tokumoto, S.;Fujiwara, S. AIChE Intl. Chem. Engng., 1983 23, 1, 65.
25. Randolph, A.D.; Vaden, D.E.; Stewart, D. CEP Symposium Ser., No. 240, 1984 80 110.
26. Garside, J.; Mullin, J.W. Trans. Instn. Chem. Engrs., 1968 46, T11.
27. Mullin, J.W.; ʻGaska, C. Canad. J. Chem. Engng. 1969 47, 483.
28. Nienow, A.W.; Unahabhokha, R.; Mullin, J.W. Chem. Eng. Sci., 1969 24, 1655.
29. Mullin, J.W.; Garside, J.; Das, S.N. Ind. & Eng. Chem. Fund., 1974 13, 299.
30. Christoffersen, J.; Christoffersen, M.R. J. Cryst. Growth, 1976 35 (1), 79.
31. Hamzat, S.M.; Nancollas, G.H. Langmuir, 1985 1 (5), 573.
32. Levins, D.M.; Glastonbury, J.R. Chem. Eng. Sci. 1972 27, 537.
33. Mukhopadhyay, S.; White, E.T. Proc. Chemeca 87, Fifteenth Aust. Chem. Engin. Conf. Melbourne, 1987, pp. 60.1-60.9.
34. Randolph, A.D.; Larson, M.A. Theory of Particulate Processes, Academic Press, New York 1988 2nd. ed..
35. Steemson, M.L.; White, E.T. J. Cryst. Growth, 1987 82, 311.
36. Marshall, R.J.; Stone, P.C.; White, E.T. Proc. Chemeca 77, 5th Aust. Chem. Engin. Conf., Canberra, 1977, 116.
37. Steemson, M.L.; Mukhopadhyay, S.; White, E.T. Proc. Chemeca 85, 13th. Aust. Chem. Engin. Conf., Perth, (1985) pp. 225-230.

RECEIVED May 17, 1990

Chapter 23

Crystallization of Ice from Aqueous Solutions in Suspension Crystallizers

Yuping Shi[1], Baomin Liang[1], and Richard W. Hartel

Department of Food Science, University of Wisconsin, Madison, WI 53706

The formation and growth of ice nuclei from aqueous solutions in suspension crystallizers is important for the development of freeze concentration processes. The maximum rate of growth of ice crystals can be accomplished by maintaining the optimum growth conditions in the crystallizer. Since the growth of ice crystals is limited by the rate of heat transfer as well as counterdiffusion of solute molecules from the crystal surface, the optimum growth conditions are determined by the heat and mass transfer conditions existing in the crystallizer. In particular, the optimal growth conditions occur when a balance exists between the rate of heat generation due to phase change and the rate of heat transfer removal from the system. The growth of ice crystals in 6% lactose solutions in a batch or semi-batch crystallizer was determined by following the crystal size distribution using image analysis of periodic samples. In this study, the rate of heat generation was altered during the growth experiments by removal of a portion of the crystal distribution to alter the total crystal surface area. The morphology, rate of crystal growth and change in size distribution will be discussed in terms of maintaining a heat balance in the suspension crystallizer.

The freeze concentration process is based on the partial solidification of water into ice in a fluid food product followed by the removal of the solid ice phase from the concentrated liquid phase. This process has some inherent advantages over evaporation and reverse osmosis for concentrating fluid foods as well as other process streams (1). One advantage is that essentially none

[1]Current address: Laval University, Quebec City, Quebec G1K 7P4, Canada

0097–6156/90/0438–0316$06.00/0
© 1990 American Chemical Society

of the volatile flavors and aromas are lost in the process. Additionally, since a low temperature process is employed, minimal product quality losses occur due to thermal degradation (2) resulting in a high quality product. However, commercial freeze concentration processes generally suffer from the drawback of being capital intensive (3). That is, the process is quite expensive initially and the relatively high operating costs result in unfavorable economics. Further advances in freeze concentration technology will be required to reduce these economic concerns and to make freeze concentration more competitive with evaporation and reverse osmosis (4-5). A simple process for rapidly growing large, uniform and spherical crystals would help to make freeze concentration a more competitive technology.

Several parameters have been seen to influence the crystallization of ice crystals in subcooled aqueous solutions. The primary factor is the extent of subcooling of the solution. Other factors include the agitation rate, the types and levels of solutes in solution. Huige (6) has summarized past work on conditions under which dusk-shaped and spherical crystals can be found in suspension crystallizers. The effects of heat and mass transfer phenomena on the morphology of an ice crystal growing in a suspension have not been fully understood.

For ice crystal growth that is solely determined by heat transfer rates, the growth rate would be linearly related to the bulk subcooling (interface temperature minus the bulk temperature). However, when the surface incorporation kinetics or mass transfer considerations come into play, there is no such simple relation between bulk subcooling and growth rate. Many researchers have attempted to find a simple correlation between the crystallization conditions and the ice crystal growth rate (6-7) under various conditions. These attempts generally result in relationships in which the growth rate is related to the bulk subcooling to some power, usually between one and two. The actual relationship between bulk subcooling and ice crystal growth rate is probably not so easily defined due to the confounding effects of heat and mass transfer rates.

Heat Balance

A heat balance during suspension ice crystal growth involves equating the energy term due to heat generation from phase change during growth to the rate of heat transfer removal from the crystallizer to the refrigerant. At any instant in time during a batch crystallization or for steady state continuous crystallization, the heat generated by phase change, q_g, will be equal to the product of the latent heat of formation, ΔH_f, and the mass rate of ice deposition, m_G.

$$q_g = \Delta H_f \, m_G = r_i \, \rho H_f \, A_s \, G \qquad (1)$$

In this case, the linear growth rate is usually a function of the mass and heat transfer conditions at the surface of the crystal. The linear growth rate may be related to the heat and mass transfer coefficients at the crystal surface neglecting any surface incorporation effects as ($\underline{8}$)

$$G = \left\{ \rho_i \left(\frac{C_i}{\rho k_c} \right) \cdot \left(\frac{dC}{dT} \right)_{eq} \left(\frac{\Delta H_f}{h_i} \right) \right\}^{-1} [C_i(T_b) - C_b] \tag{2}$$

The transfer coefficients can be found from correlations relating the flow properties, the geometry of the system and the physical properties of the fluid. Typical correlation equations for crystals growing in a suspension may be found in the literature ($\underline{9\text{-}10}$).

Combination of Equations 1 and 2 allows calculation of the rate of heat transfer from the growing crystal surface to the bulk solution. Under heat balance conditions, this rate of heat generation must be balanced by the amount of heat removed from the crystallizer by convection and conduction. This will be determined by the overall heat transfer coefficient, U, between the bulk solution and the refrigerant including convective resistances between the fluid and both sides of the crystallizer wall (refrigerant side and product side) as well as the conductive resistance across the crystallizer wall.

$$q_r = U A_c (T_b - T_c) \tag{3}$$

U is effected by the degree of agitation in the crystallizer (product side convective coefficient) and the flow velocity of refrigerant across the outside of the crystallizer (refrigerant side convective coefficient). Increasing either of these parameters will increase the rate of heat transfer removal from the crystallizer. A complete energy balance requires incorporation of any heat inputs or losses to the system. In this case, the agitation of the stirrer will constitute a heat input which can be calculated by the power number. In addition, heat input from the environment through the top of the crystallizer must also be accounted for. These terms can be compiled into a q_{loss} term. The desired heat balance then becomes

$$q_g = q_r - q_{loss} \tag{4}$$

In a batch crystallization process, a steady state heat balance can not be maintained due to the changing operating conditions. In particular, the surface area of the growing crystals will change during the crystallization process, resulting in an ever increasing heat generation term. However, the bulk subcooling (freezing temperature minus bulk solution temperature) will also change, resulting in a change in the linear growth rate. The magnitude of these two changes depends on the rate of heat transfer removal. In order to maintain the desired heat balance, conditions within the crystallization system must be varied accordingly. The alternatives for varying process parameters include the agitation rate (affects the convective coefficients and thus, the growth rate as well as the rate of heat transfer removal), the refrigerant temperature (affects the heat transfer removal rate) and the crystal surface area (affects the rate of heat generation). Decreasing the refrigerant temperature progressively throughout the run will allow a heat balance to be maintained as will continuous removal of a portion of the crystal suspension. In addition, increasing the agitation rate may be necessary to maintain the mass transfer rate to the crystal surface as the solution concentration increases. Maintaining a heat balance through the variation of these parameters will be discussed in greater detail in a later section.

In this study, the heat balance outlined above is applied to correlate the conditions of ice crystal growth with the morphology, size distribution and growth kinetics of ice crystals growing under various operating conditions. Conditions of growth resulting in disc-shaped to spherical morphology are investigated to determine the effects of heat balance operations on overall crystal growth.

Experimental

A laboratory scale, semi-batch crystallizer was used for the studies on 10% sucrose solutions to determine the general regions of heat balance. The crystallizer was constructed from a plastic beaker with a working volume of 200-220 mL. Agitation was provided by a helical impeller. The vessel was immersed in a refrigerated glycerine/water bath to maintain constant temperature below the freezing point and the entire experimental apparatus was kept in a cold room (3-5°C) to minimize thermal losses to the environment. Temperature in the crystallizer was controlled manually by monitoring both the refrigerant bath temperature and the bulk solution temperature. Temperature control was limited to ± 0.1°C in this way. Due to the lack of accurate temperature control, bulk subcooling values must be considered as only approximate values. Nuclei were introduced into the crystallizer by subcooling a portion of the liquid in a separate vessel to well below the spontaneous nucleation point. This was generally lower than -2.2°C. A portion of these nuclei were then introduced into the crystallizer and the refrigerant temperature set to the desired point. Subcooled sucrose solution was then continuously fed to the crystallizer while crystal free solution was continuously removed. This allowed continual growth of the seed crystals in 10% sucrose solutions. The growth of these nuclei was followed by removing a slurry sample from the crystallizer and placing the sample on a cooling stage on the microscope. The temperature of the cooling stage was maintained at the point where no melting or freezing of the sample occurred at the center of the sample. This could generally be accomplished with little difficulty as verified by visual observation. It was found that the center portion of the sample was unchanged for approximately 3 min of observation under most conditions. Therefore, with careful technique and some haste, representative samples from the crystallizer could be studied in this way. The average size and distribution of the ice crystals in the slurry sample were obtained by image analysis of photomicrographs made from each sample.

This same crystallizer was operated in a batch or semi-batch mode for study of ice crystal growth in 6% lactose solution. In this study, nuclei were generated in the method described above and then input to the batch crystallizer maintained at constant temperature. These seeds were allowed to grow under controlled conditions where a balance was approximately maintained between the rate of heat generation and the rate of heat transfer removal. As the ice crystals grew, samples were periodically removed and observed under the temperature controlled microscope stage. Photomicrographs of each sample were taken for image analysis to yield average size and distribution.

In order to maintain approximate heat balance conditions, several parameters may be adjusted according to the growth conditions. These parameters are the agitation rate of the stirrer, the temperature of the refrigerant water bath and the total amount of seed crystals growing in the crystallizer. In these studies on ice crystallization from lactose, a portion of the growing seeds were removed as needed in an attempt to maintain a balance of heat generation conditions within the crystallizer. Since the surface area was not controlled directly in terms of the actual heat balance, the changes in this parameter were effected based solely on the experience of the operator and thus, a true heat balance was probably not accomplished throughout the duration of the experiment. Calculations made after the experiments verify that the conditions oscillated around an approximate heat balance. Nevertheless, these experimental conditions allowed for a reasonably quantitative study of the crystallization behavior under heat balance conditions. Further development of the experimental technique will be required to quantify these results in a more exact sense.

Results and Discussion

Crystal Morphology. A summary of the results for ice crystal growth in the semi-continuous crystallizer for 10% sucrose solutions is given in Table I. This table demonstrates the variety of heat balance conditions that could be obtained in the crystallizer as well as the results for the typical growth rates and crystal habit generated under these conditions. In addition, information relative to the occurrence of nucleation (due either to a heterogeneous or a secondary nucleation mechanism) is given. Under conditions of nonheat equilibrium, the crystal morphology varied from dendritic or needle shaped at large imbalances to thin slices and discs at a low level of heat imbalance. The growth rates decreased accordingly with this level of imbalance although quantifying these growth rates was impossible due to the uneven nature of the crystal shape and the varying conditions in the crystallizer. Nucleation accompanied the growth of these seeds under the nonequilibrium growth conditions. This was most likely due to the relatively high degree of subcooling (heterogeneous nucleation) as well as to the uneven nature of the crystal surface and the relative ease of breaking or shearing these protuberances off of the surface and into the solution (secondary nucleation). Once in the

Table I. Summary of Experimental Results for Ice Crystal Growth From 10%
Sucrose Solutions Under Various Heat Equilibrium Conditions

Heat Equilibrium State	Growth Rate	Nucleation	Morphology
Nonheat Equilibrium			
$q_r >> q_g$	high	nucleates	needles, dendrites
$q_r > q_g$	medium	nucleates	leaf-shaped
$q_r > q_g$ (slightly)	low	nucleates	thin slice, disc-shaped
Heat Equilibrium ($q_r = q_g$)			
Low heat transfer:	very low	secondary nucleation	disc-shaped, nuclei form
Medium heat transfer:			
Case A	low (0.04 mm/h)	no nucleation	small discs, cobbles, spheres
Case B	medium (0.12 mm/h)	no nucleation	discs, cobbles, spheres
Case C	high (0.24 mm/h)	no nucleation	discs, drums spheres
High heat transfer:	very high (1.2 mm/h)	no nucleation	irregular

solution, these broken protuberances were able to grow into stable crystals under the high subcoolings at these conditions.

Conditions of heat balance can be obtained at a variety of subcoolings depending on the extent of the total mass deposition rate. This is dependent on the inherent growth rate as well as on the number (or surface area) of the seed crystals. Conditions which yield a high rate of heat transfer removal require a high mass deposition rate (either large subcooling and/or large crystal surface area) to maintain the heat balance. At low rates of heat transfer removal, the rate of mass deposition must also be lower (lower inherent growth rate and/or lower crystal surface area). During operation under heat balance conditions at the lower growth rates (less than ca. 0.04 mm/h), it was found that disc-shaped and irregular crystals grew into small discs. However, secondary nucleation of nondisc-shaped crystals was also observed to occur under these conditions. How these conditions are related to the instability of the growth of the crystal surface is still under question. At higher growth rate conditions (higher heat transfer removal and larger surface area), these irregularities in the surface during growth were smoothed out and more uniform crystals were grown. For Case A (ca. 0.04 mm/hr) in this medium heat transfer range, small disc-shaped crystals or irregular grains grew into cobbles and spheres. No secondary nucleation was observed under these conditions. At slightly higher growth rates in Case B (ca. 0.12 mm/hr), the crystals were seen to grow into smooth disc or cobble-shaped grains with the presence of some spherical crystals. At even higher growth rates of Case C (ca. 0.24 mm/hr), many spheres were formed with some disc and drum shaped crystals. The habit of ice crystals growing in this medium heat transfer range was quite similar over a range of growth rates. Secondary nucleation was found not observed at these high growth rates. At much higher growth rates (ca. 1.2 mm/hr), where the rate of heat transfer was exceedingly high, irregular shaped grains were once again formed indicating that some stability point had been exceeded. In this range, the crystal growth mechanism resulted in nonspherical growth of the surface and irregular crystals resulted. These results are quite qualitative since the rates of heat transfer and generation were not monitored but rather, were based solely on the relative conditions of refrigerant temperature, crystal surface area and inherent growth rate. However, these results give a feel for the types of morphological changes that were found to occur during ice crystallization under different conditions of heat balance.

Ice Crystal Growth. In order to quantify these results for the production of large disc and spherical crystals, several batch experiments on 6% lactose solutions were undertaken. The experimental conditions and results are shown in Table II. In these experiments, nuclei were generated at -2.5°C (except for Run 5a at -4.0°C) and input to the batch crystallizer controlled at various refrigerant temperatures. As these crystals grew, the total crystal surface area was controlled manually in order to maintain a heat balance for a constant value of the refrigerant temperature. Slurry removal rate for these experiments was based on operator experience rather than any quantitative measure of the heat balance. Samples of the ice crystal slurry were taken periodically and analyzed for crystal size distribution. A sharp change in slope of the average size vs. time plot could be seen after the start of slurry removal due to the change in growth conditions. As seen in Table II, conditions were investigated which allowed the formation of large disc-shaped or spherical crystals resulting in significant concentrations in relatively short periods of time. These conditions generally fell into the heat balance conditions of Case C given in Table I. For example, in Run 4, large uniform crystals of 1.25 mm diameter were formed, resulting in a concentration up to 15.2% lactose in a little over 6 hours. This was accomplished again by controlling the rate of heat generation through adjustment of the crystal surface area.

Table II.   Batch Ice Crystal Growth Conditions and Results Under Heat Equilibrium in 6% Lactose

Run	C_i (%)	T_N (°C)	T_c (°C)	Time (h)	C_f (%)	L_f (mm)
R0	6.0	-2.5	-0.8	5.6	6.2	0.27
R1	6.3	-2.5	-1.0	6.3	7.7	0.53
R2	6.4	-2.5	-1.8	6.9	9.9	0.65
R3	6.6	-2.5	-2.8	6.0	12.1	0.92
R4	6.4	-2.5	-3.9	6.3	15.2	1.25
R5a	6.4	-4.0	-4.8	6.25	15.0	1.67
R5b	6.0	-2.5	-4.8	4.75	13.9	1.50

In Figure 1, the change in average crystal size with time is shown during the batch period of operation. That is, over this time period, no crystal slurry was removed and the crystallizer operated under purely batch conditions. Typical batch behavior was observed, with the average size increasing with time. After a period of faster growth, the rate of change of crystal size was seen to gradually decrease, particularly at the lower levels of heat transfer removal rate (Runs R0, R1 and R2). At these conditions, the bulk temperature remained essentially constant. The bulk driving force, however, was continually decreasing due to the lowering of the freezing temperature caused by the increasing solution concentration. Therefore, the change in average crystal size was seen to slowly decrease. There are two competing terms in the heat generation equation, Equation 1, that influence this change. The first is the crystal surface area, A_s. This increases as the crystals grow. The bulk subcooling, however, controls the linear rate of change in crystal size

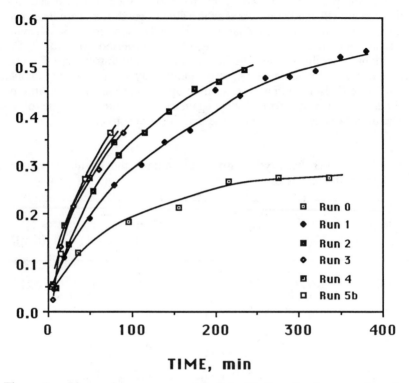

Figure 1.   Change in average crystal size with time for ice crystal seeds grown in 6% lactose under batch crystallization conditions.

and this decreases due to the increasing solution concentration. The overall effect is that the rate of linear growth of ice crystals in a pure batch system decreases since the bulk subcooling term becomes dominant. In order to maximize the rate of growth, conditions must be changed to maintain a balance between heat generation and heat transfer removal. One technique for accomplishing this is to maintain a constant crystal surface area while maintaining a uniform heat removal rate.

The dependence of average crystal size with time over the initial batch operation period was determined for Runs R0-R3. The data was fit to a power-law model of the form

$$L = L_o + at^b \tag{5}$$

Linear regression results show that b is approximately 0.5 for each experimental condition and that a increases with the rate of crystal growth due to environmental conditions. The refrigerant subcooling, ΔT_c, was seen to directly effect the value of a, whereas, since the bulk subcooling changed with time, there was no clear correlation between a and ΔT_b. Further work will be necessary to delineate these relationships for both batch and semi-batch conditions.

When some crystal surface area was removed, by removal of some of the crystal slurry, the conditions of heat balance were changed. Since the total surface area was reduced, the relative ratio of heat generation to heat transfer removal was also reduced. This resulted in an increased driving force for growth (see Equation 4) and an increased rate of mass deposition onto the existing surface area. Thus, the average crystal size was seen to increase. Under these conditions, the bulk temperature was seen to decrease slightly during the course of semi-batch operation. This decrease in bulk temperature resulted in a slightly decreasing rate of heat transfer removal. Under optimal heat balance conditions, sufficient surface area would be removed to maintain a constant rate of heat transfer removal and thus, maintain maximum rate of crystal growth. Figure 2 shows how the average crystal size changed after slurry removal was initiated. Slurry removal for Runs R3, R4 and R5b began after approximately 90 minutes of batch operation. At this time, the average size was seen to increase significantly due to the altered heat balance conditions. Continued removal of slurry from the crystallizer resulted in continued increases in crystal size. If slurry removal were stopped, the system would gradually approach a new heat equilibrium and the growth of crystals would slow down as seen previously in the batch portion of the experiments. These results demonstrate that maintaining a high driving force for growth by maintaining heat balance conditions at a high level allows more rapid growth of crystals than in purely batch operation. These results also demonstrate that maintaining a heat balance via withdrawal of seed crystals may not result in the optimal growth conditions. Maintaining a constant or ever increasing rate of heat transfer removal at the same time that the heat balance is maintained by slurry withdrawal should result in the maximum growth of ice crystals from suspension.

Another parameter of importance to freeze concentration processes is the wideness of the crystal size distribution as measured using the variance of the distribution. Figure 3 shows an example of how the variance of the size distribution changes with time during the course of growth. For Run R1, a slight widening of the distribution was observed during the course of the batch operation. A similar widening was observed for Run 5b, in which semi-batch operation was employed. Each of the experimental runs exhibited this behavior to some extent. However, there is insufficient data on these changes to make any clear distinctions between the distributions observed in batch vs. semi-batch operation. Potential mechanisms for these changes include ripening effects (large crystals grow at the expense of the small ones), secondary nucleation processes (formation of nuclei by contact mechanisms), growth rate dispersion or agglomeration processes (fusion of several small crystals together to make large ones). The relative contribution of these mechanisms and how to control them are important points for further development of freeze

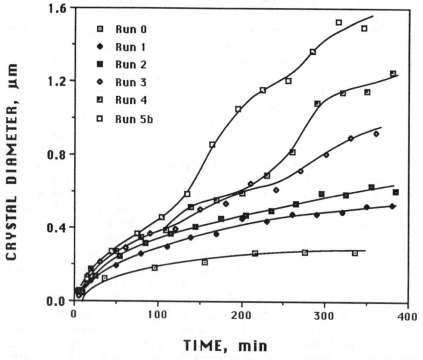

TIME, min

Figure 2. Change in average crystal size with time for ice crystal seeds grown in 6% lactose under semi-batch conditions with removal of crystal slurry.

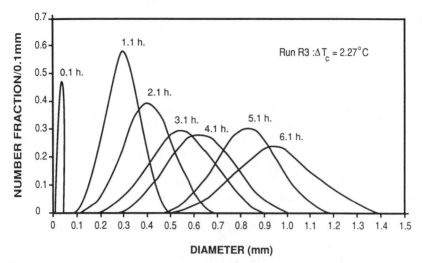

Figure 3. Changes in size distribution with time during growth of ice crystal seeds in 6% lactose under batch or semi-batch growth conditions. concentration processes based on suspension growth. In addition, the relation of this observed widening to the heat balance conditions needs to be further explored.

Summary

These results provide an interesting basis for future work on the study of the effects of heat balances on the crystal growth processes during suspension ice crystallization. The results of this study suggest that if a balance between the rate of heat generation by phase change and the rate of heat transfer removal is maintained, control over the ice crystal morphology will result. In addition, the growth rates of ice crystal seeds can be maximized by controlling the heat balance conditions within the suspension. In particular, if a high rate of heat transfer removal is maintained and appropriate growth conditions also maintained, a crop of large, uniform, spherical or disc-shaped ice crystals can be produced in a relatively short period of time. This would be advantageous in the development of simple freeze concentration units due to the relative simplicity of the equipment, the facilitation of the separation step and enhanced economics that would be realized. Further studies to quantify this phenomenon and to better understand the relationship between the surface growth conditions and the morphological stability criteria are needed.

Acknowledgments

This work was supported, in part, by the Department of Food Science at the University of Wisconsin-Madison and the Center for Dairy Research through funding from WMMB.

Legend of Symbols

a	Parameter in Equation 5, mm/h^b.
A_c	Crystal surface area, m^2.
A_s	Heat transfer surface area, m^2.
b	Parameter in Equation 5.
C_b	Bulk solution concentration, kg solute/kg solution.
C_f	Final bulk concentration kg solute/kg solution.
C_i	Concentration at crystal surface in equilibrium with T_b, kg/kg.
$(dC/dT)_{eq}$	Slope of solution freezing curve at the freezing point, kg/(kg-h).
G	Linear growth rate, mm/h.
h_i	Heat transfer coefficient at crystal surface, W/m^2-K.
ΔH_f	Latent heat of formation, J/kg.
k_c	Mass transfer coefficient, m/s.
L_f	Final average crystal size after completion of run, mm.
L_o	Initial average crystal size of seeds, mm.
m_G	Mass deposition rate, kg/h.
q_g	Rate of heat generation by phase change, J/s.
q_{loss}	Rate of heat loss to environment, J/s.
q_r	Rate of heat transfer removal by refrigerant, J/s.
T_b	Bulk solution temperature, °C.
T_c	Refrigerant temperature, °C.
T_f	Freezing point of solution, °C.
T_N	Temperature of nucleation of seeds, °C.
ΔT_b	Bulk subcooling temperature $(T_b - T_f)$, °C.
ΔT_c	Refrigerant subcooling $(T_c - T_b)$, °C.
U	Overall heat transfer coefficient, W/m^2-K.
ρ	Solution density, kg/m^3.
ρ_i	Density of ice, kg/m^3.

Literature Cited

1. Chowdhury, J. Chem. Eng. 1988, 95(6), 24.
2. Muller, J.G. Food Technol. 1967, 21, 49.
3. Van Pelt, W.H.J.M.; Swinkels W.J. In Food Engineering and Process Applications, Vol. 2, Unit Operations; LeMaguer M.; Jelen P., Eds., p. 275, Elsevier Appl. Sci., London, 1986; p275.
4. Schwartzberg, H.G., paper 73d, AIChE National Mtg., Denver, CO, 1988.
5. Swinkels, W.J., paper 73e, AIChE National Mtg., Denver, CO, 1988.
6. Huige, N.J.J. Ph.D. Thesis, Eindhoven, The Netherlands, 1972.
7. Fernandez, R.; Barduhn A.J. Desalination 1967, 3, 330.
8. Wey, J.S.; Estrin J. Ind. Eng. Chem. Proc. Des. Dev. 1973, 12(3), 236.
9. Brian, P.L.T.; Hales H.B. Amer. Inst. Chem. Eng. J. 1969, 15(3), 419.
10. Brian, P.L.T.; Hales H.B.; Sherwood T.K. Amer. Inst. Chem. Eng. J. 1969, 15(5), 727.

RECEIVED May 17, 1990

Chapter 24

Kinetics of Secondary Nucleation of Alumina Trihydrate in a Batch Crystallizer

H. M. Ang and P. I. W. Loh[1]

Department of Chemical Engineering, Curtin University of Technology,
Bentley 6102, Western Australia, Australia

Literature has revealed limited kinetic data on secondary nucleation of alumina trihydrate in the precipitator of the Bayer Process for alumina production. A batch agitated, isothermal, three litre crystallizer was used in the study. A Coulter-Counter was utilized as the particle sizing equipment. The effects of seed density, supersaturation and temperature on secondary nucleation were investigated. Maximum nucleation rates were found to occur at about 70°C and for any crystallization temperature, the nucleation rate passed through a maximum. The correlated equation for the effective secondary nucleation rate of alumina trihydrate is

$$B = 9.8 \times 10^{16} \, e^{-20700/RT} \, A^{1.7} \, \Delta C^{0.8} \quad \text{for nuclei} \geq 1.2 \, \mu m)$$

The low dependence of the nucleation rate on the concentration driving force (ΔC) seems to indicate that secondary nucleation of alumina trihydrate is removal-limited.

The Bayer Process, patented by Karl Josef Bayer in 1888 (Pearson ([1])) is still widely adopted in refining alumina trihydrate from bauxite. Despite its long history, the unit operations involved in the process have remained basically unchanged with the reversible reaction below as its backbone.

$$Al_2O_3 \cdot 3H_2O + 2NaOH \underset{\text{precipitation}}{\overset{\text{digestion}}{\rightleftharpoons}} 2NaAlO_2 + 4H_2O$$

Bauxite is first digested in a strong caustic medium to produce a green liquor containing the alumina.

This paper is concerned with the study of the reverse reaction, the precipitation of alumina. The effects of varying supersaturation, seed surface area and temperature on secondary nucleation rates were studied in a batch precipitator.

---

[1]Current address: Dampier Salts, Dampier, Western Australia, Australia

The industrial importance of the nucleation lies in its effects on the yield and product size distribution of alumina crystals. The yield is significant for plant economic reasons while off-target product size distribution adversely affects the downstream operations (e.g. washing and separation) as well as not meeting the customers' product specifications.

This study is intended to fill in some of the existing gaps in our understanding of the secondary nucleation of alumina trihydrate.

Theory

Five methods are available to determine the nucleation rates using a batch crystallizer. They are:
(i)   super-cooling (Kenji, Katsumi and Noriake (2), Mullin and Ang (3))
(ii)   moment analysis (Tavare and Garside (4))
(iii)   s-plane analysis (Tavare and Garside (4))
(iv)   induction period method (Mullin and Ang (3))
(v)   first principle method (Misra and White (5))
        The first four methods are deemed unsuitable for this study because
(a)   the batch crystallizer in this study will be operated isothermally and thus ruling out the supercooling method.
(b)   the moment and s-plane methods require samplings to be done over a small time interval (a few minutes) for the assumptions made to be valid, especially constant growth rate and constant population density at size = 0 over the time interval. Using the single channel Coulter Counter Industrial Model D, one complete particle size analysis, where number counts are made at some 10 particle sizes, and including sample preparation will take 35-45 minutes. This time interval is considered too long for the assumptions to be valid. The solution to this problem is probably to collect as many samples as possible during the run and analyse them subsequently but was not adopted owing to the amount of time required to analyse all the samples. However, a multi-channel Coulter Counter Model T may render these two methods feasible.
(c)   The induction period method which relies on visual observation of the onset of nucleation could not be adopted because of the difficulty of pinpointing nucleation in a high seed density system used in the present study.
        The first-principle method used in the present study will be outlined below. Population balance analysis on a perfectly mixed batch crystallizer with negligible crystal breakage and agglomeration yields the familiar nucleation rate equation used by Misra and White (5)

$$B^\circ = \frac{dN}{dt} \tag{1}$$

where B° = nucleation rate
       N = total number of crystals of size > 0 μm at any time t

Thus, numerous nucleation rates can be determined from the slopes of a plot of N vs t, one rate at one time selected, as shown in Figure 1.

Current particle sizing technology limits the smallest particle measurable to around 1 μm. A number of extrapolation methods have been suggested by Misra and White (5). Workers like Jones, Budz and Mullin (6) cautioned extrapolating to zero size and recommended the use of effective nucleation rate instead. It was thus decided not to extrapolate back to zero size but to find the effective nucleation rate basing on the smallest particle measurable. In contrast, the nucleation rate calculated using the total number of crystals > 0 μm is known as the true nucleation rate.

The secondary nucleation kinetics could be empirically correlated by equation (2).

$$B° = k_n \, A^a \, \Delta C^b \tag{2}$$

where k_n = nucleation rate constant
A = crystal surface area per litre of the crystallizer content (m²/l)
ΔC = supersaturation difference (C - C*)
C = alumina concentration (g/l)
C* = equilibrium alumina concentration at a given temperature (g/l)
a = exponent for A
b = exponent for ΔC

The crystal surface area, A, could be calculated from the specific crystal surface area, A_s (m²/g of crystals) by equation (3).

$$A = A_s \times \text{crystal concentration} \tag{3}$$

where crystal concentration = (weight of seed added + weight of crystals precipitated out)/litre

A_s, the specific crystal surface area (m²/g of solids) can be obtained by the method recommended in the Coulter Counter Model D Operating Manual:

$$A_s = \frac{\Sigma(\Delta n \, \bar{d}^2) \, \phi}{\Sigma(\Delta n \, \bar{d}^3) \, \rho} \quad \text{m}^2/\text{g of solids} \tag{4}$$

where Δn = number of particles in the size range examined
$\bar{d}$ = arithmetic mean diameter of the size range examined (in μm)
ϕ = the Heywood Shape factor or specific surface coefficient
= 16 for alumina trihydrate (British Standards 4359:3 1970 Table 13)
ρ = relative density of solids (g/cm³)

Experimental

The laboratory batch crystallizer used in this study is shown in Figure 2. It consisted of a cylindrical vessel (ID = 155 mm, height = 250 mm) of three litre working capacity agitated by a variable speed 4-bladed (variable-pitch) impeller. For temperature control, the crystallizer was immersed in a constant temperature water-bath.

Preliminary work involved finding the minimum impeller speed to attain a uniform suspension in the crystallizer and the extent of attrition at this minimum impeller speed. It has already been reported by Loh, Ang and Kirke (7) that an impeller speed of 350 rpm was sufficient to maintain a uniform suspension in the crystallizer and the bulk of attrition occurred in the first 4 hours. The extent of attrition was not very significant in any typical nucleation run.

Supersaturated sodium aluminate solutions were prepared using an autoclave which was operated at 150°C and about 600 kPA. The hot caustic solution was filtered through three Whatman filter papers of 4 μm pore size under positive pressure. Three litres of the filtered solution was transferred to the crystallizer, maintained at a constant temperature, which was set for a particular run.

Once temperature equilibrium was established, 600 g of alumina seeds were added to the crystallizer to make up a slurry of solids density 200 g/l. Samples for size distribution analysis and solution concentration determinations were taken periodically for some 24 hours.

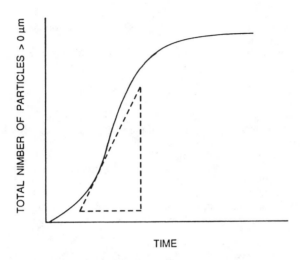

Figure 1. Total number of crystals vs time.

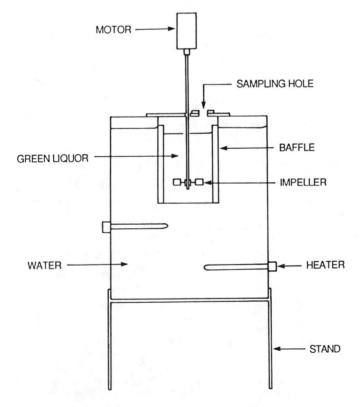

Figure 2. Laboratory batch crystallizer set-up.

The sample collected from the crystallizer was diluted with filtered electrolyte (2.0 wt% aqueous NaCl) and decanted to separate the fines from the coarse crystals. The dilution and decantation were repeated 5-7 times till 200 ml of the decanted sample was obtained. Preliminary tests by Loh (8) have shown that these repeated decantantions were sufficient to recover at least 95% of the fines present in the original sample. Both the fine and coarse fractions were then analysed with the Coulter Counter using a 50 µm orifice tube for the former and a 280 µm orifice tube for the latter. By adjusting the Coulter Counter's settings, a set of particle number counts at different sizes (successive sizes differing by a factor of $2^{1/3}$) was made for each of the 2 orifice tubes used.

After appropriate dilutions, samples collected for concentration determinations were analysed using the sodium gluconate - potassium fluoride method on a dosimate-titrator (Kirke, E.A., Alcoa of Australia, Kwinana, personal communication, 1986). Total alkaline, total caustic and alumina concentrations were determined using this method.

Results and Discussion
Nucleation rates were obtained for varying seed densities, seed sizes, temperature and supersaturation in the batch crystallizer.

Effect of Seed Density. Figure 3 shows the results obtained for batch nucleation runs at 70°C for various seed densities. The seeds used were taken directly from alumina trihydrate bags supplied by Alcoa of Australia Ltd.

Observations made from the results obtained were:

(i)   Figure 3 (a) shows that the total number of particles (≥ 1.2 µm) generated increases with seed density. This increase, however, is not directly proportional to the initial seed density used as illustrated in Table I.

Table I. Effect of Seed Density on Total Number of Nuclei

Seed Density (g/l)	Mass of Seeds Used (g)	Total No of Particles (≥ 1.2 mm)	No. of Particles (≥ 1.2 µm) Generated per gram of Seed
50	150	75×10^6	0.5×10^6
150	450	450×10^6	1.0×10^6
200	600	800×10^6	1.3×10^6
250	750	925×10^6	1.2×10^6

(ii)   The seed density does not only have an effect on the number of nuclei generated but also on the size of the final product. The same degree of desupersaturation was observed at the end of the 4 runs (i.e. the same amount of alumina had precipitated out for each run). Simple mass balance will indicate overall larger crystals when less crystals are formed and vice versa. The crystal size distribution obtained confirm that finer crystals were produced when a higher seed density was used.

(iii)   The higher the seed density, the steeper is the slope of the graphs shown in Figure 3(a) and thus the higher the nucleation rates.

(iv)   For the run at 50 g/l, it was observed that an induction period of about 50 minutes resulted. This was obtained from the desupersaturation vs time

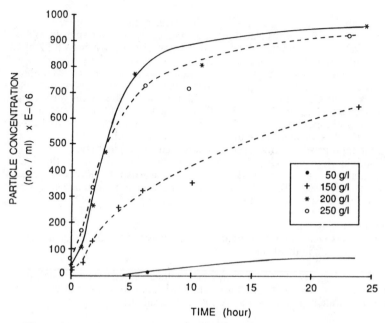

Figure 3(a). Effect of seed density on nucleation at 70°C. Number of crystals (≥ 1.2 μm) vs crystallization time.

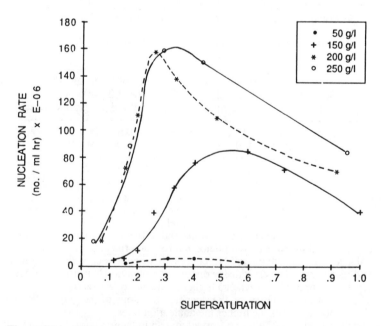

Figure 3(b). Effect of seed density on nucleation at 70°C. Nucleation rate vs supersaturation.

curve (see Figure 6). However, it has been reported by (Brown (9)) that the nucleation process can be suppressed by increasing the seed area to 20 m²/l. Brown explained that at high crystal surface area, the aluminate ion clusters were preferentially used in the orderly crystal growth process to the extent that no nucleation resulted. This phenomenon, unfortunately, was not observed in this study though the 200 g/l seed density has a surface area of ~ 25 m²/l.

(v)   It is worth noting in Figure 3(b) that nucleation rate passes through a maximum for any batch. Further discussion of this same phenomenon is documented in the section of Effect of Temperature.

Effect of Seed Size. The effect of seed size on the nucleation rate was investigated by using screened seeds of sizes in the range of 53 - 63, 63 - 75 and 75 - 90 μm. The seed density and temperature used in this study were 200 g/l and 70°C respectively. The results are illustrated in Figures 4(a) and 4(b).

(i)   Figure 4(a) shows that the yield was not significantly affected by the seed size for the range tested.

(ii)   For the same seed density, Figure 4(b) shows that the nucleation rates (accelerating phase) increased with decrease in the seed size. This may be due to the larger crystal surface area associated with the smaller seeds, providing more sites for the nucleation to occur. This observation is in agreement with the higher nucleation rate observed at larger seed densities noted in the section on effect of seed density.

Effect of Temperature. To study the effect of temperature, the nucleation runs were conducted by keeping all the other variables constant i.e. seed density = 200 g/l, impeller speed at 350 rpm and solution concentration adjusted such that the initial absolute supersaturation ($\Delta C/C^*$) ~ 0.9 at the temperatures studied. The results are shown in Figures 5(a) and 5(b).

The nucleation rate increased from 65°C to 70°C and dropped from 70°C to 80°C. Thus 70°C seems to be the optimum temperature for maximum nucleation. Published work on alumina trihydrate by Misra and White (5) and Brown (9, 10) revealed that the nucleation rate decreases with increasing temperature, at greater than 70°C by the former but from 50 to 75°C by the latter. This nucleation rate dependence on temperature differs with normal chemical reaction where the reaction rate increases with increase in temperature. It is not clear whether their studies at different temperatures in the published work were conducted at constant initial absolute supersaturation ($\Delta C/C^*$) for all the temperatures studied or at constant initial concentration. The latter would account for the higher nucleation rates obtained at lower temperatures as the $\Delta C/C^*$ is higher at lower temperatures since C^* decreases with temperature.

Figure 5(a) shows that a much slower nucleation rate was found at 80°C. Negligible or no nucleation at 75°C and above were reported by Misra & White (5) and Brown (9, 10). While not arguing the validity of our result or theirs, it is agreed that there is a marked decrease in nucleation rate at 75°C and above. The lower seed densities (10 - 100 g/l) used in their studies may explain why no nucleation was observed as the induction period would have been very large. In our study, an extra run at 80°C was conducted but at a higher seed density of 400 g/l. The results (Loh (8)) showed a much higher nucleation rate, comparable to the one at 65°C.

Breakdown of hydrogen bondings at 75°C and above has been proposed to explain the absence of nucleation observed by Misra and White, and Brown. This study has found that the conditions were unfavourable for nucleation at these high

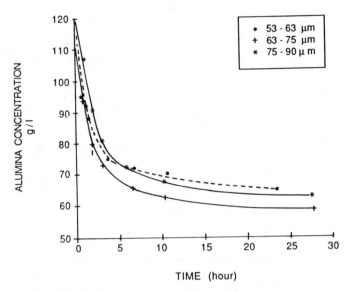

Figure 4(a). Effect of seed size on nucleation at 70°C, 200 g/l seed density. Desupersaturation curve.

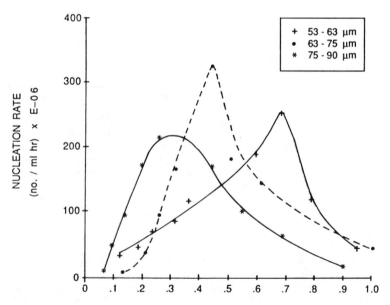

Figure 4(b). Effect of seed size on nucleation at 70°C, 200 g/l seed density. Nucleation rate vs supersaturation.

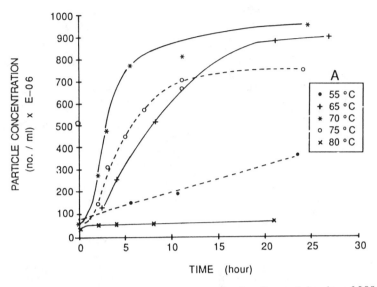

Figure 5(a). Effect of temperature on nucleation for seed density of 200 g/l. Number of crystals (≥ 1.2 µm) vs crystallization time.

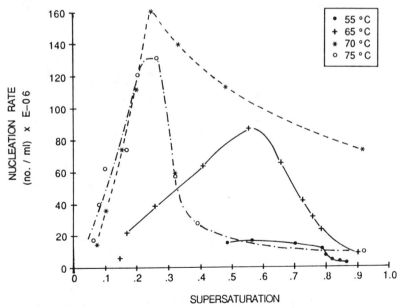

Figure 5(b). Effect of temperature on nucleation for seed density of 200 g/l. Nucleation rate vs supersaturation.

temperatures but nucleation can and does occur depending on the seed density used. Different nucleation mechanisms may operate at these higher temperatures.

Apart from the significant effect of temperatures on nucleation rate, it has also an effect on the yield as shown in Figure 6. After time t, less alumina has precipitated out at 65°C than at 70°C or 75°C. Thus a lower yield was obtained at 65°C but significantly lower yields resulted at 55°C.

Figure 5(b) illustrates that the nucleation rate goes through a maximum in any batch run. These results seem to agree with the findings of Brown (9, 10). It is not expected that a higher nucleation rate results at a lower supersaturation. This may be due to the fact that the total number of particles is based on 1.2 µm and not 0 µm. The start of nucleation resulted in a large number of nuclei being formed but they were smaller than 1.2 µm. These nuclei were not detected by the current measurement technique. As time proceeded, these nuclei grew in size and more and more of them became detectable by the Coulter Counter used. However, Brown (9) has extrapolated the total number of nuclei back to zero size by using a line of best fit. His plot of N vs t obtained still shows the same trend noted in Figure 5(b).

It has already been stated in the section on effect of seed density, that an appreciable induction period existed especially at low seed densities such as 50 g/l. Thus nucleation is definitely not instantaneous. Even in the runs considered here where higher seed densities of 200 g/l were used, the seeds may likewise need to be activated at their active sites before causing significant nucleation to occur. Recent experimental work conducted by the author has confirmed that even at seed densities of 200 g/l an induction period of about 30 minutes was observed. Ang (11) has reported observing nucleation to occur slowly at the beginning and reaching more significant rates soon after. Just like seeds which need to be activated, nuclei that are formed may also need time to be activated before they grow to measurable sizes. The current experimental results tend to support the view that maximum nucleation rates do not necessarily occur at the start of the precipitation process in spite of the fact that its supersaturation is a maximum at that point.

It is also possible that in the initial stages of the batch run, where supersaturation is at its highest, agglomeration may be occurring thus reducing the actual particle concentration and resulting in a lower nucleation rate as observed. One of the authors, Ang, is currently looking into the phenomenon of agglomeration in the precipitation of alumina trihydrate.

Effect of Supersaturation. In the study of the effect of supersaturation, the experiments were conducted at 70°C, impeller speed of 350 rpm and seed density of 200 g/l. The supersaturations studied were 0.45, 1.0 and 1.4. The results are shown in Figures 7(a), (b) and (c).

(i)   The number of nuclei generated (Figure 7(a)) and the nucleation rate (Figure 7(b)) increased with higher initial supersaturation. This is apparent from higher concentration driving forces at higher supersaturations. Finer crystals resulted at higher initial supersaturations.

(ii)   Figure 7(c) shows that higher initial supersaturations resulted in a faster desupersaturation rate as well as a higher crystal yield.

(iii)   The greater nucleation rates resulting from higher initial supersaturations may be explained as follows (Nyvlt et al., (12)):

    (a)   The micro-roughness of the seed crystal increases with increasing supersaturation, resulting in higher probability of more nuclei breaking loose on collisions.

(b)   At higher supersaturations the critical nucleus size becomes

exponentially smaller. Thus more molecular clusters will survive
to form nuclei.

(c)   The adsorbed layer on the crystal surface is thicker at higher
supersaturations. The thicker adsorbed layer will contain more
potential nuclei.

Correlation of Secondary Nucleation Rate. The nucleation rate equation (2) was
correlated by using multiple regression analysis at 70°C. Only data corresponding
to the accelerating phase of nucleation rate was used in the correlation. The rate
equation obtained at 70°C is

$$B° = 5800 \ A^{1.7} \ \Delta C^{0.8} \tag{5}$$

Assuming that the exponents of A and ΔC remain constant with temperature
(Larson and Garside (13)), these exponents were used to obtain the rate constant,
k_n, at 55 and 65°C. Table II shows the rate constants calculated. r^2 is the
correlation coefficient at each of the temperatures analysed.

Table II. Rate Constants at Different Temperatures

Temperature (°C)	k_n	r^2
70	5800	0.83
65	4300	0.79
55	1500	0.71

From the data in Table II, an Arrhenius plot was made resulting in a straight line
whose intercept k = 9.8 x 10^{16}, and slope = E/R = -10428. The activation energy
E = -20,700 cal/gmol.

The final rate equation is given by equation (6)

$$B° = 9.8 \ x \ 10^{16} \ e^{-20,700/RT} \ A^{1.7} \ \Delta C^{0.8} \tag{6}$$

A comparison between the experimental and calculated nucleation rates is
illustrated in Figure 8 which shows that the nucleation rates calculated using
equation (6) compare favourably with the experimental results. The 95%
confidence limits (i.e. ± 1.96 x standard deviation) are indicated by the two
broken lines.

Equation (6) indicates that secondary nucleation is significantly affected by
the crystal surface area A (m^2/l). The nucleation process is usually viewed as a
process where copious amounts of fines are produced, resulting in a
corresponding increase in the specific crystal surface area (m^2/g) and crystal
surface area/volume of the crystallizer content (m^2/l). Results obtained in this
study concur with this nucleation view. However, Brown (9) reported that
nucleation can be suppressed at high crystal surface areas (~ 20 m^2/l). Brown's
observation was not evident in this study though the seed density of 200 g/litre
used in this study had a surface area of ~ 25 m^2/l. The lower dependence of
secondary nucleation on supersaturation or concentration driving force, ΔC, in
comparison to crystal surface area, A, is obvious from equation (6).

The low exponent of ΔC (i.e. 0.8) is characteristic of a secondary nucleation
process (Randolph and Sikdar (14), Garside and Davey (15)). Homogeneous and
heterogeneous nucleations have a higher dependence on ΔC. This was made
evident from the work of Mullin and Ang (3) who correlated B = K ΔC^b using
nickel ammonium sulfate, and found b = 3.2, 2.0 and 1.1 for homogeneous,

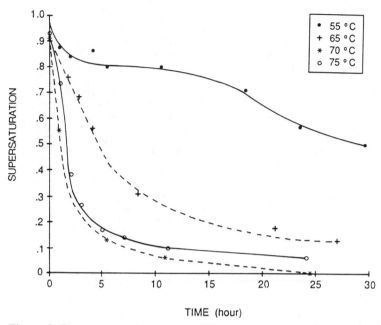

Figure 6. Desupersaturation curve at different temperatures for seed density of 200 g/l.

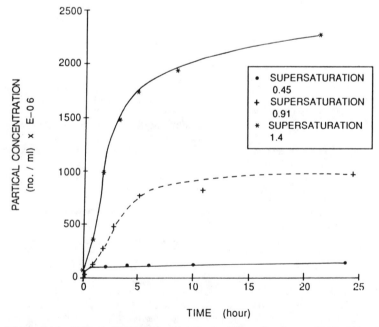

Figure 7(a). Effect of initial supersaturation on nucleation at 70°C, 200 g/l seed density. Number of crystals (≥ 1.2 μm) vs crystallization time.

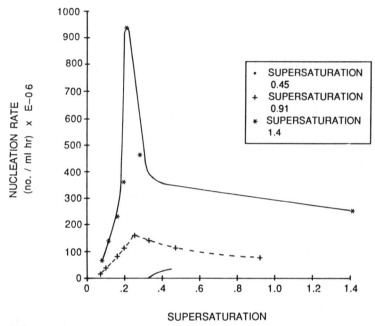

Figure 7(b). Effect of initial supersaturation on nucleation rate at 70°C, 200 g/l seed density.

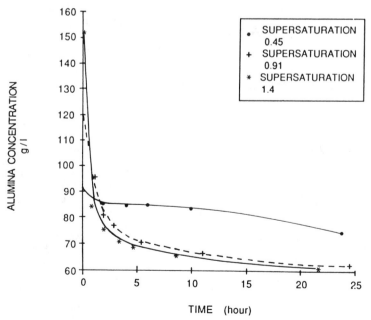

Figure 7(c). Effect of initial supersaturation on desupersaturation at 70°C, 200 g/l seed density.

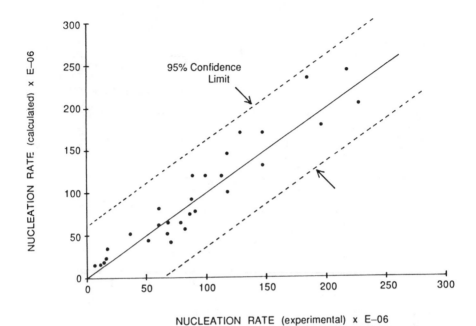

Figure 8. Comparison of experimental and calculated nucleation rates.

heterogeneous and secondary nucleations respectively. Typical values of b for secondary nucleation lie between 0.5 and 2.5 (Garside and Davey (15)). With b < 1, the secondary nucleation of alumina trihydrate is probably removal-limited (Randolph and Sikdar (14)). However, Halfon and Kaliaguine (16) found $B° \propto \Delta C^2$. The difference may be due to different liquors used. Halfon and Kaliaguine used re-digested plant liquors while synthetic sodium aluminate liquors were used in this study. Furthermore, the nucleation rates obtained in this study are effective rates (based on 1.2 μm size). It is not clear whether the nucleation rates obtained by Halfon and Kaliaguine were true or effective rates. Using potassium sulfate system, Jones et al., (6) have found b to have values of 2.25 and 1.0 for nucleation rates based on 0 μm size and 50 μm size respectively. In comparing nucleation orders, b, the nucleus reference size should be stated in the nucleation rate correlation as the nucleation order is dependent on the nucleus reference size.

Misra and White (5) tried b = 2 in their work and found poor correlation with the experimental data (the value of b = 2 was arrived at on the assumption that the nucleation rate was similar to the growth rate of alumina trihydrate which was found to be proportional to ΔC^2. They suggested more experimental work to better correlate the nucleation rate of alumina trihydrate.

The activation energy for secondary nucleation of alumina trihydrate is 20,700 cal/gmol as given by equation (6). No published data is available for comparing the value obtained but the activation energy for growth quoted by Misra and White (5) is 14,300 cal/gmol. Thus the value of 20,700 cal/gmol for nucleation seems reasonable as it is well accepted that activation energy for nucleation is larger than that for growth.

Conclusions
1.   Higher seed densities, finer seed sizes, higher initial supersaturations and

higher temperatures (up to 70°C) all tend to result in higher nucleation rates and finer product sizes.

2. A significant induction period of about 50 min was observed in a low seed density run of 50 g/l.

3. Nucleation rates increase with temperature up to a maximum at 70°C after which nucleation rates decrease with temperature.

4. For any fixed batch crystallization temperature, the effective nucleation rate passes through a maximum even at high seed densities. It is suggested that the induction period required to activate the seed surfaces may be responsible for the lower initial nucleation rate observed when the supersaturation was higher. It is also suggested that agglomeration may have caused the observed phenomenon.

5. The effective secondary nucleation rate of alumina trihydrate in a batch crystallizer for temperature 55 to 70°C is given by:
$B^{\circ} = 9.8 \times 10^{16}$ e$^{-20,700/RT}$ A$^{1.7}$ $\Delta C^{0.8}$ (nuclei ≥ 1.2 μm)

6. At a given temperature, the crystal surface area has a bigger contribution to the nucleation rate than ΔC. The low dependence of the nucleation rate on the concentration driving force implies that the secondary nucleation of alumina trihydrate is removal-limited.

7. The activation energy for secondary nucleation of alumina trihydrate was found to be 20,700 cal/gmol.

Acknowledgments
The authors wish to thank Alcoa of Australia for granting Philip Loh a scholarship to undertake this study. The assistance of Mr E A Kirke during the early stages of the study is acknowledged.

Literature Cited
1.   Pearson, T.G., Royal Institute of Chemistry Lectures Monographs and Reports, 1955 No. 3.
2.   Kenji, S., Katsuma, A., and Noriaki, K., Industrial Crystallisation 1981, Edited by S.J. Jancic and E.J. De Jong, North-Holland Publishing Company, 1982.
3.   Mullin, J.W., and Ang, H.M., Faraday Discussions of the Chemical Society No. 61 Precipitation, 1976, 141.
4.   Tavare, N.S., Garside, J., Chemical Engineering Research and Design, 64, 1986, 109.
5.   Misra, C., White, E.T., Chemical Engineering Progress Symposium Series, 1971, 67, No. 110, 53.
6.   Jones, A.G., Budz, J., Mullin, J.W., AIChE Journal, 1986, 32, 2002.
7.   Loh, P.I.W., Ang, H.M., Kirke, E.A., Chemeca 88 Australia's Bicentennial International Conference for the Process Industries, 1988, 304.
8.   Loh, P.I.W., Master of Engineering Thesis, Curtin University of Technology, Western Australia, 1988.
9.   Brown, N., Journal of Crystal Growth, 1975, 29, 309.
10. Brown, N., Light Metals, 1977, 2, 1.
11. Ang, H.M., PhD Thesis, University of London, London, 1973.
12. Nyvlt, J., Sohnel, O., Matuchova, M., Broul, M., Kinetics of Industrial Crystallization, Elsevier Science Publishing Co., 1985.
13. Larson, M.A., Garside, J., The Chemical Engineer, 1973, 318.
14. Randolph, A.D., Sikdar, S.K., Industrial and Engineering Chemistry Fundamentals, 1976, 15, 64.
15. Garside, J., Davey, R.J., Chemical Engineering Communication, 1980, 4, 393.
16. Haflon, A., Kaliaguine, S., Canadian Journal of Chemical Engineering, 1976, 54, 160.

RECEIVED May 17, 1990

Chapter 25

Reactive Crystallization of Magnesium Hydroxide

Hideki Tsuge and Hitoshi Matsuo

Department of Applied Chemistry, Keio University, 3–14–1, Hiyoshi, Kohoku-ku, Yokohama 223, Japan

Crystallization of magnesium hydroxide by a
continuous mixed suspension mixed product removal
crystallizer was conducted to make clear the
characteristics of reactive crystallization kinetics
of magnesium hydroxide, which was produced by the
precipitation from magnesium chloride with calcium
hydroxide. The following operating factors were
investigated affecting the crystallization kinetics;
the initial concentration of feeds, residence time
of reactants, feed ratio of reactants, and
concentrations of hydroxide and chloride ions.
It was clarified that the nucleation rate and the
growth rate are correlated by the power law model
and that the kinetic order in the power law model
is correlated with concentrations of OH⁻ and Cl⁻.

The production of sparingly soluble materials by simultaneous
reaction and crystallization has been used widely in the
chemical industries. To make clear the characteristics of
crystallization of sparingly soluble materials by chemical
reactions is important for better design and more efficient
operation of reactive crystallizers. Many works on
crystallization kinetics have been made in continuous mixed
suspension mixed product removal (CMSMPR) crystallizers[3]. The
nucleation rate B^0 and the crystal growth rate G have been
correlated by the following power law model:
$$B^0 = kG^i \qquad (1)$$
The present work deals with the reactive crystallization of
magnesium hydroxide, a well-known sparing soluble material, from
magnesium chloride with calcium hydroxide. Magnesium hydroxide
is produced industrially by the precipitation from brine with

0097–6156/90/0438–0344$06.00/0
© 1990 American Chemical Society

calcium hydroxide and is used industrially as desulfurization agents and materials of steel plant refractory, while few studies have been made on the crystallization kinetics of magnesium hydroxide. Dabir, Peters and Stevens[2] studied the kinetics of magnesium hydroxide by a CMSMPR crystallizer mainly for the lime-soda ash water softening process, but their experimental ranges were rather narrow. Packter[4] discussed the crystallization kinetics of magnesium hydroxide in batch crystallizer. Therefore, it is important to clarify the crystallization kinetics of magnesium hydroxide for better understanding of its crystallization process.

The objectives of the present paper are as follows;
1) to discuss the effect of the operating factors, that is, the initial concentration of feeds, residence time of reactants, feed ratio of reactants, and concentrations of hydroxide and chloride ions on the crystallization kinetics of magnesium hydroxide,
2) to correlate the kinetics with the power law model,
3) to make clear the effect of anion concentrations on the kinetic order in the power law model.

Experimental

Experimental apparatus and procedure. Figure 1 is a schematic diagram of the experimental apparatus. The crystallizer was a 1 liter stirred tank reactor made of acrylic resin and is considered to be a continuous MSMPR reactor. The reactor was 0.1m in diameter and the liquid height 0.14m. The impeller used was of the 6-blade turbine type and operated at 450 rpm to ensure complete mixing. Feed solutions were pumped into the crystallizer continuously to produce magnesium hydroxide. The product was continuously withdrawn from the crystallizer. The reaction temperature was maintained at 25 °C by constant temperature bath.

Sampling was begun after 10 residence times, when the steady-state had been reached. Crystals obtained were photographed by the scanning electron microscope (SEM) and their sizes were analyzed by a digitizer. Irrespective of crystal form, the maximum length of crystal was used to describe the size of individual crystal.

Aqueous solutions of magnesium chloride (reagent grade) and calcium hydroxide (reagent grade) react as follows:
$$MgCl_2 + Ca(OH)_2 = Mg(OH)_2 + CaCl_2$$

Experimental conditions. Table 1 lists the experimental conditions. C_0 shows the apparent initial concentration of reactants in the crystallizer. Series I, II and III were conducted with stoichiometric feed ratio, with changing the feed ratio of reactants and with sodium chloride addition under the constant C_0 of magnesium chloride and calcium hydroxide, respectively.

Table 1 Experimental conditions

Series	Run No.	C_0 [mol/m^3]		
		MgCl$_2$	Ca(OH)$_2$	NaCl
I	1	2.02	2.01	0
	2	4.90	4.90	0
	3	10.2	10.2	0
	4	14.7	14.7	0
II	5	20.7	9.92	0
	6	10.2	4.89	0
	7	10.2	10.2	0
	8	10.2	12.2	0
	9	10.2	14.5	0
III	10	10.2	10.2	0
	11	9.88	9.88	19.7
	12	9.90	9.90	48
	13	9.79	9.79	78.8
	14	9.79	9.79	98.5
	15	9.80	9.80	186

The residence times of reactants in the crystallizer were 5, 10, 20 and 30 minutes.

Analysis of the CSD Data

From the population balance for a MSMPR crystallizer operated under the steady-state condition, the population density n for size-independent crystal growth is given as
$$n = n^0 \exp(-1/G\theta) \qquad (2)$$
where G and θ are growth rate and residence time.

Figure 2 shows the CSD plotted on semilogarithmic coordinates, whose ordinate shows the number of crystals N. In the present CSD analysis, population density n could not be obtained directly, but N was obtained by counting individual crystals in a fixed size range from SEM microphotographs. As the volume and size range, in which N crystals are involved, are constant for each run, N is considered to be proportional to n and the following equation can be assumed:
$$N \propto \exp(-1/G\theta) \qquad (3)$$

While the CSD in Figure 2 seems to show the maximum, the data points of larger particles were measured more accurately in this CSD analysis so that the linear relation of larger particles in Figure 2 was adopted to analyse the CSD. The increase of accuracy of measurements of CSD or the discussion of the CSD analysis by Bransom model[1] will be made further.

The linear correlation indicates that Equation 3 is satisfied and crystal growth obeys the ΔL law. From the slope of CSD, the growth rate G is obtained. The dominant particle size l_m and the nucleation rate B^0 are related as follows[5] ;
$$l_m = 3G\theta \qquad (4)$$
$$B^0 = 9P_r / (2\rho_c f_v l_m^3 V) \qquad (5)$$
where ρ_c, f_v and V are, respectively, crystal density, volume shape factor and the volume of crystallizer. The production rate P_r in Equation 5 is calculated from Equation 6 by the mass balance of magnesium ion;
$$P_r = M(C_0 - C)F \qquad (6)$$
where M, C_0, C and F are molecular weight, initial concentration of Mg^{2+}, concentration of Mg^{2+} in the crystallizer and feed rate, respectively.

Results and Discussion

Volume shape factor. Figure 3 shows SEM microphotograph of the typical crystal of $Mg(OH)_2$ obtained for series I. As the crystal form is composed of disklike units and the crystal structure of $Mg(OH)_2$ is CdI_2 type, the standard unit of $Mg(OH)_2$ crystal is considered to be a disk. The ratio of the length L to thickness D of the disk of crystal unit was measured for each experimental condition, so that it was found that L/D was nearly constant at 6.4. The crystal volumes were calculated for

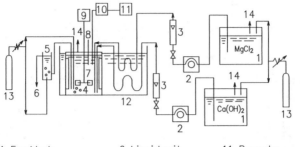

1 Feed tank	6 Liquid exit	11 Recorder
2 Pump	7 Impeller	12 Const. temp. bath
3 Liquid flow meter	8 Thermometer	13 N_2-Cylinder
4 Crystallizer	9 Motor	14 Gas exit
5 Constant head	10 pH meter	

Figure 1: Schematic diagram of experimental apparatus

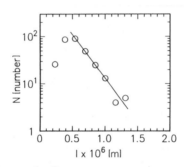

Figure 2: Crystal size distribution

Figure 3: SEM photograph of $Mg(OH)_2$

the unit numbers of disks, n_u, so that the volume shape factors f_v were obtained as a function of n_u as shown in Figure 4. The broken line in the figure shows the volume shape factor of sphere, $\pi/6$.

Average unit number of disk $\bar{n}_u$ increases with increases of the concentrations of OH^- and Cl^-. As $\bar{n}_u$ is a function of both concentrations of OH^- and Cl^-, $\bar{n}_u$ is correlated by two parameter least squares method within an accuracy of $\pm 15\%$ as follows;

$$\bar{n}_u = 7.41[OH^-]^{0.134}[Cl^-]^{0.168} \tag{7}$$

where $[OH^-]$ and $[Cl^-]$ are expressed in mol/l. Equation 7 shows that the crystal surface grows with the increases of OH^- and Cl^- concentrations.

Crystallization kinetics.   Figure 5 shows the relation between the dominant particle size l_m and the residence time θ for each run. l_m decreases linearly with increase of θ in log-log plot. The slope of lines decreases with the addition of NaCl.

Figure 6 shows the relation between the growth rate G and the residence time θ in log-log plot for each run. The parallel straight lines are written as

$$G \propto \theta^{-1.1} \tag{8}$$

The effect of the experimental series on the slope is minute.

Figure 7 shows the relation between the crystal nucleation rate B^0 and θ. B^0 decreases linearly with increase of θ in log-log plot. The slopes of lines increase with the increases of OH^- and Cl^- concentrations, which are caused by the dependency of supersaturation of $Mg(OH)_2$ on the OH^- and Cl^- concentrations.

The phenomena shown in Figures 5~7 are considered as follows: with the increase of the residence time θ, the feed rates of reactants decrease so that both G and B^0 decrease. The decrease of l_m with the increase of θ is caused by the decrease of G rather than the increase of θ.

Figure 8 shows the relation between B^0 and G with θ as a parameter. From the material balance, the relation between B^0 and G can be written by the following equation:

$$B^0 = (1/6f_v\rho_c)M_T\theta^{-4}G^{-3} \tag{9}$$

The broken and dotted chain lines, respectively, show the results of Series I and Series II for constant θ. The solid lines show the relation between B^0 and G with concentrations of OH^- and Cl^- as a parameter and are expressed by Equation 1, so that the kinetic order i is obtained from these slopes.

Kinetic order i increases with increases of OH^- and Cl^- concentrations. As i is a function of concentrations of OH^- and Cl^-, i is correlated by Equation 10 by two parameter least squares method within an accuracy of $\pm 10\%$ as shown in Figure 9, where the ordinate and abscissa show kinetic order i calculated by Equation 10, i_{cal}, and by experiment, i_{exp}, respectively.

$$i = 2.63[OH^-]^{0.124}[Cl^-]^{0.209} \tag{10}$$

where $[OH^-]$ and $[Cl^-]$ are expressed in mol/l.

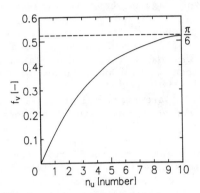

Figure 4: Relation between volume shape factor and unit number

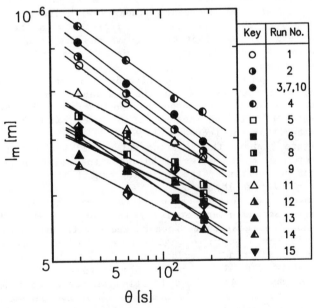

Figure 5: Relation between l_m and θ

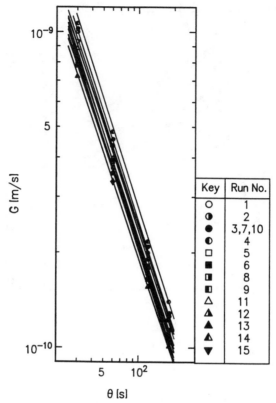

Figure 6: Relation between G and θ

Dabir et al[1] showed that the kinetic order i is written by Equation 11 for the reactive crystallization of $Mg(OH)_2$ produced by magnesium chloride and sodium hydroxide.

$$i = 8100[OH^-] - 0.43 \qquad (11)$$

where $[OH^-]$ is expressed in mol/l. Figure 10 shows the relation between i and $[OH^-]$ of our experimental and Dabir et al's results. The difference of both data is caused by the difference of the reaction system, the initial concentrations of feeds and the measuring method of crystal size, that is, Dabir et al. used a Coulter counter.

Conclusion

Reactive crystallization experiments of magnesium hydroxide were conducted to clarify the characteristics of reactive crystallization kinetics by a continuous MSMPR crystallizer.

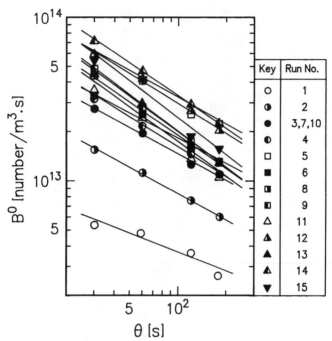

Figure 7: Relation between B^0 and θ

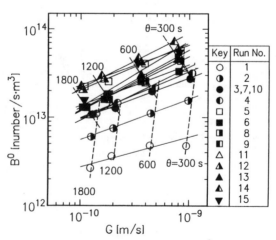

Figure 8: Relation between B^0 and G

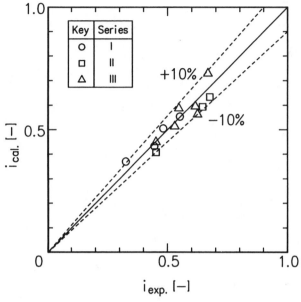

Figure 9: Comparison of kinetic order obtained from experiments and correlations

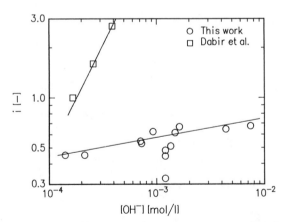

Figure 10: Relation between i and [OH⁻]

Kinetic order i of $Mg(OH)_2$ made by precipitation from magnesium chloride with calcium hydroxide are correlated with concentrations of hydroxide and chloride ions as follows;
$$i = 2.63[OH^-]^{0.124}[Cl^-]^{0.209}$$
where $[OH^-]$ and $[Cl^-]$ are expressed in mol/l.

Legend of symbols

B^0	nucleation rate	$m^{-3}s^{-1}$
C	concentration	mol/m^3
C_0	apparent initial concentration	mol/m^3
D	thickness of disk crystal	m
F	feed rate	m^3/s
f_v	volume shape factor	–
G	growth rate	m/s
i	kinetic order	–
i_{cal}	kinetic order calculated by Equation 10	–
i_{exp}	kinetic order obtained by experiment	–
L	diameter of disk crystal	m
l	particle size	m
l_m	dominant particle size	m
M	molecular weight	kg/mol
M_T	suspension density	kg/m^3
N	number of particle	–
n	population density	$m^{-3}m^{-1}$
n^0	nuclei density	$m^{-3}m^{-1}$
n_u	number of unit	–
$\bar{n}_u$	mean number of unit	–
P_r	production rate	kg/s
V	volume of crystallizer	m^3
θ	residence time	s
ρ_c	crystal density	kg/m^3

Literature Cited
1. Bransom, S.H., D.E. Brown, and G. P. Heeley: Inst. Chem. Eng. Symposium on Ind. Crystallization, 1969, p26.
2. Dabir,B., R.W.Peters and J.D.Stevens: Ind. Eng. Chem. Fundam., 1982, 21, 298–305.
3. Garside, J. and M.B.Shah: Ind. Eng. Chem. Process Des. Dev., 1980, 19, 509–514.
4. Packter, A.: Crystal Res. and Technol., 1985, 20, 329–336.
5. Shirotuka,T. and K.Toyokura: Kagaku Kogaku, 1966, 30, 833–839.

RECEIVED May 17, 1990

Chapter 26

Fluidized-Bed Process for Phosphate Removal by Calcium Phosphate Crystallization

Izumi Hirasawa[1] and Yasunori Toya

Ebara Research Company, Ltd., 4-2-1 Honfujisawa Fujisawa, Kanagawa 251, Japan

We have proposed a fluidized bed type process, which can be applied to phosphate removal from wastewater containing phosphate 2-23 mg/ℓ as P.By the results of experiments using equipment of capacity 1-4m³ /day, factors such as supersaturation, recirculation ratio and space velocity were recognized to affect crystallization rate or phosphate removal efficiency. By mathematical analysis, we could obtain the characteristic equation for fluidized bed process, to agree well with experimental results.

Phosphate removal processes from wastewater have been studied by many workers, in order to protect stagnant water area, such as lakes and coastal region from eutrophication. Among conventional phosphate removal processes, the representative one was flocculation and sedimentation process, which was based on precipitation of insoluble metal phosphate or hydroxide. However, the main problem with this process, is to produce large amounts of sludge, which is difficult to dehydrate.

To cope with these problems, we have developed phosphate removal process using crystallization, which can minimize the amount of sludge and recover phosphate. Mechanism of this process is crystallization of calcium phosphate on the surface of phosphate rocks by contacting supersaturated solution with them. In case of application to wastewater containing 1-3 mg/ℓ phosphate as P, we proposed fixed bed type process, which has demonstrated excellent performance in the sewage treatment.[*1]

Table 1 shows the performance of fixed bed type process, in application to various wastewaters. The merit of this process is stability in ability of phosphate removal and low sludge production. Sludge production of this process is from 1/5 to 1/10 lower than that of the conventional flocculation and sedimentation process.

We have now proposed fluidized bed type process, which can be applied to wastewater, containing from 2 to 23 mg/ℓ phosphate as P. This report reveals fundamental studies on factors affecting phosphate removal and crystallization rate in the fluidized bed process.

[1]Current address: Waseda University, 3-4-1, Ohkubo, Shinjuku-ku, Tokyo 169, Japan

0097-6156/90/0438-0355$06.00/0
© 1990 American Chemical Society

Mechanism

Figure 1 shows the schematic illustration of phosphate removal mechanism. Phosphate in wastewater contacts with seeds made of phosphate rock, after chemical conditioning such as supply of calcium and hydroxide ion. Then calcium phosphate, mainly hydro-xyapatite, crystallizes, according to eq.(1), on the surface of phosphate rocks.

$$10Ca^{2+} + 6HPO_4^{2-} + 8OH^- \rightarrow Ca_{10}(PO_4)_6(OH)_2 + 6H_2O \qquad eq.(1)$$

Table 1   The results applied to various wastewater

	Secondary eff-luent of sewage	Secondary eff-luent of sewage
Capacity (m^3/day)	100	12,000
Concentration of phosphate as P (mg/l) in the influent	$2 \sim 4$	$1 \sim 2$
M-alkalinity in the influent (mg/l)	$28 \sim 123$	$50 \sim 100$
Test periods (year)	5	$1 \sim 6$
Concentration of phosphate as P (mg/l) in treated water	$0.26 \sim 0.35$	$0.30 \sim 0.40$
Sludge production (g/m^3)	6.0	$5.0 \sim 7.0$
Operational conditions	SV 2.5 (l/h) LV 2.5 (l/h) Down flow	SV 2.5 (l/h) LV 2.5 (l/h) Up flow

Characteristics of Fluidized Bed Process

Figure 2 shows the outline of fluidized bed process. Seeds are fluidized by the upflow of influent and recirculated water. Recirculation is necessary to maintain water quality of the influent average, lower phosphate concentration in the influent, and also to obtain flow rate for fluidization.

Mathematical Analysis

By considering material balance of phosphate in the fluidized bed as shown in Figure 3, eq.(2) can be introduced. By integration of eq.(2), eq.(3) and eq.(4) are obtained. And also we can obtain eq.(6) by introducing recirculation factors such as eq.(5).
Therefore we can make eq.(8) as characteristic equation, applied to this fluidized bed process.

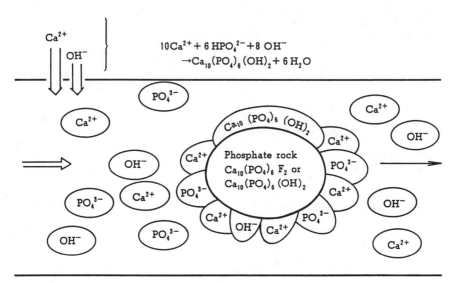

Fig. 1 Schematic illustration of phosphate removal

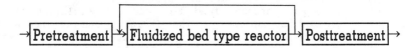

* Seeds are fluidized by up-flow of influent and recirculated water
* Alkali soln. is dosed into the bottom of fluidzed bed reactor, and calcium soln. is dosed into the recirculation pipe.
* Seeds are made of phosphate rock, which has efficient size 0.31mm, and has density 2.4g/cm³.
* Calcium phosphate crystallizes on the surface of the seeds by the following equation.

$$10Ca^{2+} + 6\,HPO_4^{2-} + 8\,OH^- \rightarrow Ca_{10}(PO_4)_6(OH)_2 + 6H_2O$$

Fig. 2 Characteristics of the fluidized system

$$\underbrace{Q.C}_{\text{Input P}} = \underbrace{Q(C+dC)}_{\text{Output P}} + \underbrace{Sdz \cdot K_0 \cdot a \cdot C^n}_{\substack{\text{P fixed on the surface} \\ \text{of the seeds}}} \qquad \text{eq.(2)}$$

$$n = 1 \qquad \frac{C}{C_0{'}} = e^{-K_0 \cdot a \theta'} \qquad \text{eq.(3)}$$

$$n \neq 1 \qquad \theta' = \frac{[(\frac{C_0{'}}{C})^{n-1} - 1]}{K_0 \cdot a \cdot C_0{}^{n-1} \ (n-1)} \qquad \text{eq.(4)}$$

$$C_0{'} = \frac{C_0 + rC}{1 + r} \qquad \text{eq.(5)}$$

$$\frac{C}{C_0} = \frac{1}{(1+r)expK_0 \cdot a \cdot \theta' - r} \qquad \text{eq.(6)}$$

$$\frac{C}{C_0} = \frac{1}{(1+r)exp\dfrac{K_0 \cdot a \cdot \theta'}{1 + r} - r} \qquad \text{eq.(7)}$$

$$Yr = 1 - \frac{1}{(1+r)exp\ [\ \dfrac{K_0 \cdot a}{(1+r)SV}\ ] - r} \qquad \text{eq.(8)}$$

Experimental Apparatus and Procedures
      Figure 4 shows the experimental apparatus having capacity of
1-4 m^3/day, which has influent feed line and recircculation line.
Seeds are fluidized by upflow of the influent and recirculated
water. Alkali solution is dosed into the bottom of the fluidized
bed to maintain pH of the effluent in a adequate range, and calcium
solution is dosed into the recirculation line.
      In this experiment, tap water with added phosphate was used as
influent. Concentration of phosphate was adjusted to an adequate
range from 2 to 23 mg/ℓ. Calcium chloride and sodium hydroxide
solution were added to maintain calcium concentration from 70 to 100
mg/ℓ and pH of the effluent from 9.0 to 9.5. Using this equipment,
we performed experiments to obtain efficiency of phosphate removal,
relationship between phosphate concentration, and crystalization
rate and factors affecting phoshate removal.

Results and Discussion
Crystallization Rate.
      Figure 5 shows the effect of phosphate concentration on crystal-
lization rate R_g. Crystallization rate was obtained by weighing
seeds in the reactor periodically, in which phosphate concentration
drop in the bed was low enough, not to produce concentration distribution, by making the depth of the bed thin. Solubility of hydroxya-
patite is so low, that the influent phosphate concentration is equal
to supersaturation ΔC. As shown in Figure 5, crystallization rate
was proportional to 1.1th order of supersaturation.
      Figure 6 shows the effect of particle size of seeds on
crystallization rate. It became proportional to the 0.3th order of

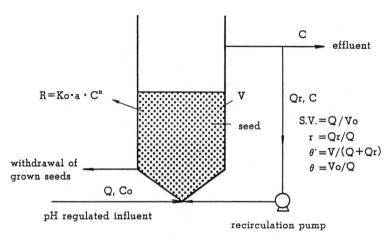

Fig.3 Fluidized bed type reactor

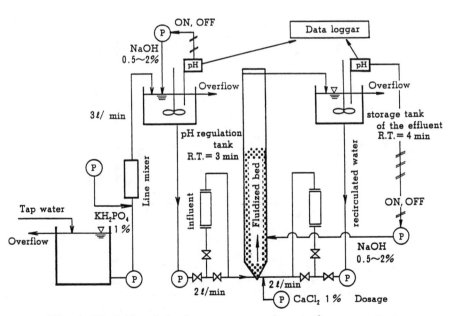

Fig.4 Fluidized bed type experimental apparatus
-Fluidized bed type reactor 100mmϕx3m

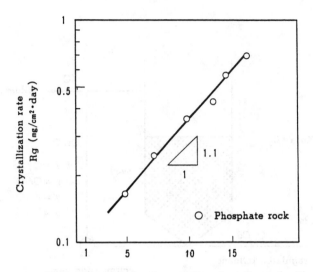

Conentration of phosphate as P in the influent (mg/l)

Fig.5 Effect of phosphate concentration on R_g

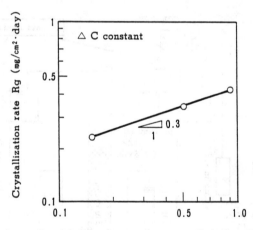

Harmonic mean diameter L (mm)

Fig.6 Effect of seed size on R

particle size. Therefore we could obtain eq.(9) as crystallization rate of calcium phosphate. If we perform experiments in the

$$R_g \propto (\Delta C)^{1.1} L^{0.3}$$ eq.(9)

condition that particle size change is low enough, crystallization rate is assumed to be proportional to supersaturation.

Effect of Phosphate Concentration.
        We performed experiments by changing phosphate concentration in

the influent to obtain Figure 7. Figure 7 shows relationship between PO_4 as P concentration in the influent and phosphate removal efficiency. We could obtain 60-92 % phosphate removal efficiency.

Effect of Recirculation Ratio and Space Velocity.
    Figure 8 shows effect of recirculation ratio and space velocity. Horizontal line shows total PO_4 as P in the influent and vertical line does total phosphate in the effluent. Black circles were data obtained from the experiments in case that recirculation ratio was 0 and space velocity 38 1/h. White circles were obtained, in case that recirculation was 2 and space velocity 13 1/h. Solid line and dotdash line shows theoretical values calculated from eq.(8), to agree well with experimental results. Increasing in recirculation ratio and decreasing in space velocity was recognized to improve phosphate removal efficiency.

Chemical and Physical Analysis of Calcium Phosphate Particles.
    As for properties of calcium phophate particles grown in the reactor, chemical and physical analysis were done to verify availability of this process. Table 2 shows results of chemical analysis, which made clear that our products contained mainly calcium phosphate and some calcium carbonate and could be used as the source of fertilizer.
    Table 3 shows results of physical analysis in which our products had larger specific surface area than one of phosphate rock. It was recognized that they could be some supporting media for substantial substance such as hormone, pesticide and enzyme.
    Porous property of these particles was considered to be based on crystallization mechanism of calcium phosphate, in which fine crystals formed in the bottom of fluidized bed attached to the fluidized seeds by driving force of supersaturation and fluid conditions.

Table 2   Results of chemical analysis of used phosphate rock

	Seed	Waste water from food industry	Tap water $CaCl_2$-NaOH (Case II)	Tap water $Ca(OH)_2$-$CaSO_4$
Water content	0.2	1.0	2.8	2.6
Ignition loss	1.6	4.5	10.3	7.3
Al_2O_3	0.2	0.2	0.1	0.2
SiO_2	1.8	0.4	0.1	<0.1
Fe_2O_3	0.1	0.2	0.1	0.1
Na_2O	2.8	0.5	0.2	0.4
K_2O	<0.01	0.09	0.04	0.04
MgO	0.2	0.3	0.5	0.4
CaO	54.4	54.2	42.0	48.0
P_2O_5	32.9	28.4	40.4	35.2
As_2O_3	<0.0025	<0.0025	—	—
F	1.8	2.6	0.59	2.8
Cl	<0.01	0.01	0.02	<0.01
CO_2	1.4	3.6	5.2	6.8
SO_3	1.3	0.9	0.2	0.6
CODcr	1.4	3.2	0.3	0.3

1) Unit is weight % in the dry solid. dry %
2) Samples are crushed after drying for analysis.

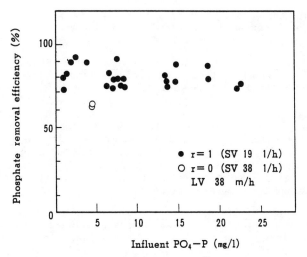

Fig.7 Relationship between PO₄-P concentration in
      the influent and phosphate removal efficiency

Conclusion
     1.We have proposed fluidized bed process and by the results of
experiments, factors such as supersaturation, recirculation ratio
and space velocity affected crystallization rate or phosphate
removal efficiency, and experimental results agreed well with
calculated values from characteristic equation.
     2.From physical and chemical analysis, it was recognized that
our products could be used as source of fertilizer and supporting
media for substantial substance.

Table 3 Properties of used phosphate rock

Samples		Seed	Waste water from food industry	Tap water CaCl₂-NaOH (Case I)	Tap water CaCl₂-NaOH (Case II)	Tap water Ca(OH)₂-CaSO₄
Particle size	e. s. (e. c.)	0.31 (1.7)	0.22 (3.0)	0.56 (1.3)	0.61 (1.3)	0.47 (1.4)
	Mean diameter*	0.48	0.49	0.68	0.72	0.62
	Under 0.3mm(%)	8.8	17.6	0.0	2.7	0.9
Density (g/cm³)		2.4	1.9	2.1	1.9	2.2
Void fraction (−)		0.47	0.52	0.47	0.42	0.51
True specific suface area (m²/g)		14	—	107	149	84
Press Intensity (kg f/cm²·seed)		<3.5	<5.0	<4.5	<1.2	<4.0

* Volumetric mean particle diameter

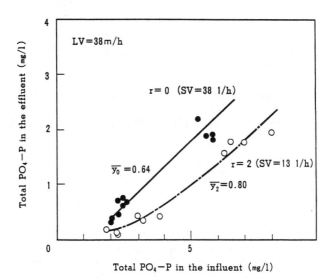

Total PO$_4$−P in the influent (mg/l)

Fig. 8 Effect of recirculation ratio and space velocity

3.We hope our proposed process will contribute to environmental protection of the earth, by recovering phosphate from wastewater.

We have reported only the results of the last investigation of fundamental properties of this process. But there are many problems remaining unsolved such as growing mechanism of calcium phosphate, the reason of difference in phosphate removal efficiency between application to wastewater and tap water, and also how contaminants include into growing calcium phosphate.

Nomenclature
    a : Specific surface area [m^2/m^3]
    C : Phosphate concentration [g-P/m^3]
    ΔC: Supersaturation [g-P/m^3]
    K_0: Overall crystallization rate coefficient [m/h]
    $\bar{L}$: Particle size [mm]
    n : Power number of supersaturation
    Q : Flow rate [m^3/h]
    R_g: Crystallization rate [mg/cm^2·day],[g/m^2·h]
    r : Recirculation ratio [-]
    S.V: Space velocity [1/h]
    V : Volume of seeds [m^3]
    Y : Phosphate removal efficiency [-]
    θ : Space time [h]
            Suffix ; 0 : Influent, Initial r : Recirculation

Literature Cited
    *1 I.Hirasawa etal ; Proceedings of PACHEC'83 Volume IV
                wastewater treatment(II) 259-264(1983)

Received May 17, 1990

Chapter 27

Freeze Concentration of Aqueous Solutions

Tasoula C. Kyprianidou-Leodidou and Gregory D. Botsaris

Chemical Engineering Department, Tufts University, Medford, MA 02155

The development of the freeze concentration process for fruit juices has been hampered by the fact that solute concentrate is entrained by the ice crystals. This incomplete separation of the entrained concentrate from the ice results in a considerable increase of the cost of the process. In this investigation sucrose solutions were concentrated by the formation of an ice layer on the externally cooled walls of the crystallizer. The formation of the layer was initiated by secondary nuclei induced by rotating ice seeds, at subcoolings smaller than the critical subcooling needed for spontaneous nucleation. A minimum in the amount of sucrose entrapped in the ice layer was observed at a subcooling smaller than the critical subcooling for spontaneous nucleation. The effect of soluble pectins on the minimum was also studied.

Freeze concentration involves the concentration of an aqueous solution by partial freezing and subsequent separation of the resulting ice crystals. It is considered to be one of the most advantageous concentration processes because of the many positive characteristics related with its application. Concentration processes such as evaporation or distillation usually result in removal of volatiles responsible for aroma; in addition the heat addition in these processes causes a breakdown in the chemical structure that affects flavor characteristics and nutritive properties. In contrast freeze concentration is capable of concentrating various comestible liquids without appreciable change in flavor, aroma, color or nutritive value (1,2,3) The concentrate contains almost all the original amounts of solutes present in the liquid food.

The application of the freeze concentration process has been successful in various cases such as the preconcentration of wine (4). The application of the process for the concentration of fruit juices, which are mainly consumed for their flavor and taste, seems to be ideal because of the improved volatile retention. Nevertheless, the process is not as widely accepted because of its increased economic requirements. These are related to the difficulty of effectively separating the resulting solid phase from the liquid phase (mother liquor). The formation of a large number of very fine

0097–6156/90/0438–0364$06.00/0

crystals makes their complete separation from the concentrate almost impossible and results in losses of valuable material in the discarded ice. This increases considerably the operating cost of the process. A control of the ice production stage is required in order to minimize the total loss of entrapped material per unit weight of ice.

The freeze concentration process has been mainly studied in suspension crystallizers (5,6) as a crystallization process where secondary nucleation is the prevailing mechanism of ice formation. Since primary and heterogeneous nucleation are of little importance in this process, investigators have focused their research on understanding the mechanisms of the secondary nucleation process and the factors that affect it. It is believed that the processes which contribute to the formation of secondary nuclei are mainly collisions of existing crystals with solid surfaces such as the crystallizer walls, other crystals and the stirrer blades (7,8). In systems which involve relatively high flow velocities of the solution over the crystal surfaces, secondary nuclei are produced because of fluid shear (9,10). Dendrites or other entities are detached from the surfaces of the existing crystals. Some of these entities are larger than the critical size and grow to form new particles. Supersaturation, stirring rate and additives are some of the factors that influence the kinetics of secondary nucleation (5,6.)

In some applications of the freeze concentration process, ice formation is taking place on the cooling surfaces of scraped surface heat exchangers (1). Scraping paddles mounted on the rotating shaft subsequently scrape the solid ice layer and push it into the liquid where crystal growth is taking place. A combination of an external heat exchanger and a stirred tank is adopted in some industrial applications such as the Grenco process (1). In this process the goal of minimizing the loss of concentrate in the ice is promoted by feeding fine crystals from the scraped surface heat exchanger to a recrystallizer that contains a suspension of larger crystals. The difference in the melting temperatures of the various sizes of crystals results in growth of the larger crystals and dissolution of the smaller ones (11). Thus Ostwald ripening is taking place (12).

Our study involved the combination of an ice layer formation on the cooled walls of a crystallizer and of secondary nucleation. The latter was initiated by rotating ice seeds. The crystallizing liquids were aqueous solutions of sucrose and of pectins and sucrose. Fruit juices are composed of various organic species the most important of which are sugars in the form of sucrose, glucose and fructose. It is clear that the study of fruit juices is to a large extent a study of water and sugar solutions. The nucleation and growth processes are greatly affected by the presence of additives that affect the surface properties of the crystals. Pectic compounds are present in most fruit juices and therefore the effect of pectins on ice crystallization is of particular importance and worth to study. Previous studies on ice nucleation in the presence of macromolecular compounds such as polymers, dextran and pectins (5,13,14) showed that there is a considerable effect of these solution additives on the ice nucleation rate. It was observed that probably due to an increase in the viscosity of the solution, the nucleation rate decreases with an increase in the macromolecule concentration.

In our study, the amount of liquid entrapped was obtained by measuring the sucrose in the ice layer. The conditions that minimize the liquid entrapment were then identified.

EXPERIMENTAL STUDY

For this study (15) a batch crystallization system was employed. The experimental set-up is shown in figure 1. The basic parts of the apparatus were the following;

1. The Crystallizer Cell Where All The Experiments were Performed. This was a 1 liter pyrex vessel covered by a 1 cm thick plexiglass cover. Holes in the plexiglass cover permitted the introduction of a Beckman thermometer and a specially constructed stirrer capable of accommodating two ice seed crystals on its blades. These seed crystals were attached to the tips of the stirrer blades by a slow freezing process of water in contact with the tips of the stirrer.

2. The Constant Temperature Bath. This was a glass container of a 30 cm diameter and 20 cm high, covered by a 1 cm thick plexiglass cover. The bath was filled with a 50% solution of ethylene glycol in water and it was cooled by continuously circulating ethylene glycol from a refrigerator unit (VWR model 1140) through a 1/2" I.D. copper tubing. The temperature of the bath was controlled by a temperature controller that regulated the power of a 150 Watt, heating element submerged in the cooled water-glycol mixture. The external surfaces of the bath were covered by a 2" thick fiber glass insulation. In this way the temperature of the bath was stable within a 0.005 K interval. A Beckman thermometer was used to measure the relative temperature differences between the crystallizing cell and the constant temperature bath.

The whole set of experiments was conducted using the following solutions:
1. Pure water.
2. 5% sucrose solution.
3. 10% sucrose solution.
4. 5% sucrose solution with 1 g/l pectins.

After the bath attained its equilibrium temperature, the crystallizer was charged with about 400 ml of liquid and was inserted into the bath. After about 30 minutes the system attained a constant temperature and a subcooling (difference of equilibrium temperature and constant temperature before initiation of crystallization) was established. Introduction of the seed crystals (after being allowed to warm for a period of a few seconds) on the specially prepared stirrer initiated crystallization (secondary nucleation) and resulted in a change in the temperature of the crystallizer (figure 2). The temperature of the crystallizer attained an equilibrium value of a few minutes after nucleation occurred. The concentration of the sucrose solutions was measured using a refractometer (.1% accuracy).

RESULTS AND DISCUSSION OF RESULTS

During the experimental procedure the following observations concerning the behavior of the system were made:

When pure water was used we observed that primary nucleation occurred at subcoolings higher than $1^{\circ}K$. For smaller subcooling values no primary nuclei were formed, and the system remained in a metastable state. Introduction of the seed crystals caused the formation of a large number of secondary nuclei and a change in the temperature of the crystallizer was observed that followed a sigmoidal curve (figure 2) and is characteristic of a system moving from a metastable to a stable equilibrium state. Because of this behavior of the system, a temperature difference (driving force for heat transfer) is established between the bulk of the outside constant temperature bath (and as a result of the walls of the crystallizer) and the bulk of the crystallizer (this will be expressed as ΔT from now on). The relatively

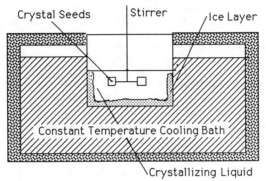

Figure 1. Experimental set-up.

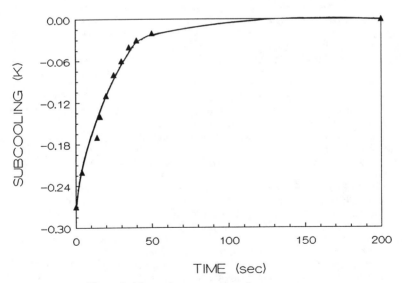

Figure 2. Thermal response curve for pure water.

high stirring rates of our experiments caused a uniform temperature distribution in the bulk of the crystallizer. An ice layer is formed at the cooled walls of the crystallizer as a result of the established temperature difference. ΔT is the driving force for the process that is heat transfer controlled. This is different than the temperature difference T_s-T_0 (where T_s is the temperature at the ice layer-solution interface and T_0 is the equilibrium temperature that is the melting point of ice) which is the driving force for the surface integration process (this is not the controlling step). The ice layer formed is either a large transparent ice monocrystal, for the case of pure water crystallization, or a non-transparent, polycrystalline ice layer where some sucrose is entrapped, for the case of the crystallization of sucrose solutions. Pictures of these two types of layers are shown in figure 3.

If we examine the variation of the sucrose entrapment in the ice layer (for ice layer thickness equal to about 1 cm) the following observations are made:
1. The amount of sugar entrapped in the ice layer varies with the overall temperature difference   between the bulk of the crystallizer and the surrounding cooling medium (figures 4,5).
2. A minimum is observed at a subcooling value smaller than the subcooling required for spontaneous nucleation. For instance in figure 4 spontaneous nucleation in the absence of the seeds occurred at $\Delta T > 1.7^{\circ}K$. The minimum at $\Delta T = 1.2^{\circ}K$ is observed in the region in which only secondary nucleation takes place. This result points out that in processes in which the ice is formed by spontaneous nucleation, as in most of the industrial processes involving scraped ice from cooled surfaces, the entrapment will be always higher than in processes that involve secondary nucleation.

A possible explanation for the variation of the sucrose entrapment in the ice layer as a function of the temperature difference between crystallizer and cooling bath and the existence of the minimum is the following. For high subcoolings the nucleation rate is high. The ice layer consists of a large number of crystals which grow fast and as a result there is a higher entrapment of the liquid which contains not only the initially dissolved sucrose but also that rejected from the growing ice crystals. For low temperature differences the heat removal rate (driving force for the formation of the ice layer) is not adequate and the ice layer does not start forming from the beginning but after some time and after a large amount of secondary crystals has been formed. These secondary crystals, thrown against the wall, find themselves in a region of lower temperature and start growing to form a "layer" of secondary crystals. At low ΔT the growth is rather slow and the layer grows mostly by capturing secondary nuclei, a condition which favors entrapment. As ΔT increases the secondary nuclei grow faster and a layer is formed consisting of larger crystals. At larger ΔT the number of nuclei is higher. This combined with faster growth leads to a more porous layer and increased entrapment. Thus at intermediate ΔT we have a minimum.

The effect of soluble pectins was also examined. Addition of pectins in aqueous solutions affects nucleation and growth rates. A comparative study of the system in the presence and without the presence of pectins (figures 6,7) will give some insight to whether it would be desirable to depectinize fruit juices of high pectin content or to use pectins as additives for juices that contain little or no pectins. In figure 7, where the results for a 5% sucrose solution with and without pectins are compared, a relative displacement of the curve to the higher temperature difference region is observed when pectin addition is taking place. Minimum losses of sucrose occur at much higher heat removal rates (higher temperature differences between the bulk of the crystallizer and the walls) and the concentration of sucrose entrapped at the point of minimum loss, has a slightly higher value. This may not be significant since the concentration difference is within the experimental accuracy. These results can be

Figure 3. Photographs of ice layers. Top, pure water; bottom, 5% sucrose solution.

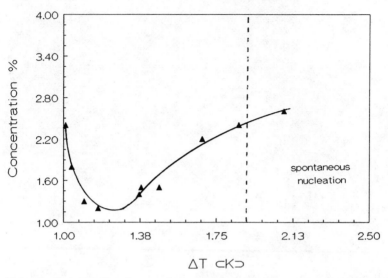

Figure 4. Concentration of sucrose in ice layer versus ΔT for a 5% sucrose solution.

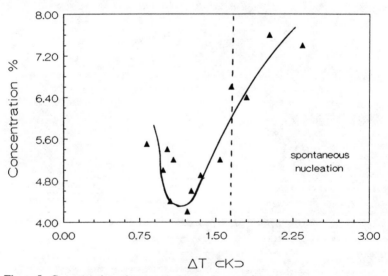

Figure 5. Concentration of sucrose in ice layer versus ΔT for a 10% sucrose solution.

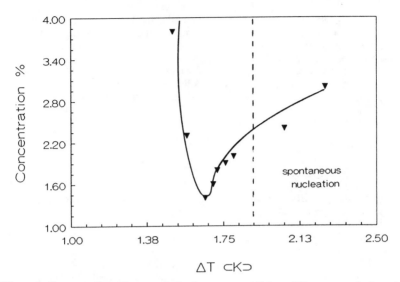

Figure 6. Concentration of sucrose in ice layer versus ΔT for a 5% sucrose solution where pectins have been added.

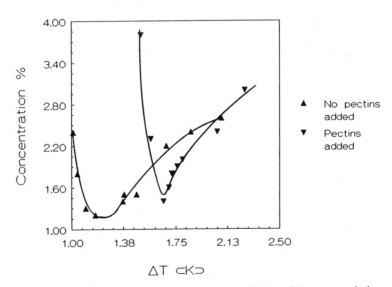

Figure 7. Concentration of sucrose in ice layer versus ΔT for a 5% sucrose solution with and without pectins.

explained as follows. Pectins are high molecular weight compounds. Their addition to a crystallizing system increases its viscosity and causes lower nucleation rates. In order to have nucleation rates comparable to those of the sucrose solutions that have no pectic additives, higher subcooling values are required. As a result, minimum entrapment is observed at higher subcooling values.

These results show that it may be more efficient to depectinize fruit juices before their concentration by freezing because this would give minimum losses at lower heat removal rates and thus at conditions of more economical operation. The implication of these results for the design of a scraped surface crystallizer are currently being examined.

Literature Cited

1. Sulc D. Confructa studien 1984, 3, 258.
2. Thijssen H.A. C.J. Food Technol. 1970, 5, 211.
3. Deshpande S.S.; Bolin H.R.; Salunke D.K. Food Technology 1982, May, 68.
4. Muller J. G. Food Technology 1967, 21, 49.
5. Omran A. M.; King C.J. AIChE J. 1974, 20, 795.
6. Stocking J.H.; King C.J. AIChE J. 1976, 22, 131.
7. Evans T.W.; Margolis G. and Sarofim A. F. AIChE J., 1974, 20, 950.
8. Strickland-Constable R.F. Chem. Eng. Progr. Symp. Ser., 1972, 68, 1.
9. Garabedian H.; Strickland-Constable R.F. J. Crystal Growth, 1974, 22, 188.
10. Estrin J.; Wang W.L. and Youngquist G.R. AIChE J., 1975, 21, 392.
11. Huige N.J.J.; Thijssen H.A.C. J. Crystal Growth, 1972, 13/14, 483.
12. Nyvlt j.; Sohnel O.; Matuchova M.; Broul M. The Kinetics of Industrial Crystallization; Chemical Engineering monographs 19, Elsevier, 1985, 310.
13. Shirai Y.; Nakanishi K.; Matsuno R. and Kamikubo T. AIChE J., 1985, 31, 676.
14. Shirai Y.; Nakanishi K.; Matsuno R. and Kamikubo T. J. Food Science, 1985, 50, 401.
15. Kyprianidou-Leodidou T. C. M.S. Thesis, Tufts University, 1987.

RECEIVED June 5, 1990

Chapter 28

Shape Factor of a Magnesium Sulfate Crystal Growing in Its Aqueous Solution

Yoji Taguchi, Takashi Matsumura, Masayuki Satoh, and Tetsuya Ohsaki

Department of Chemical Engineering, Niigata University, Ikarashi 2-nocho, Niigata 950-21, Japan

A magnesium sulfate heptahydrate ($MgSO_4 \cdot 7H_2O$) crystal particle was grown in supersaturated aqueous solutions of different concentrations, and the surface area and weight of the crystal were measured at regular time intervals in order to understand the variation of crystal shape. The shape factor was calculated from these measured values, and the influence of operating conditions on the shape factor was examined in batchwise experiments. It was found that the shape factor varied with time periodically, and in each case it resembled a single oscillation curve. In addition, the periods of the oscillation curves were influenced by the operating conditions such as degree of supersaturation, agitation speed, and gas flow rat'. The amplitude was not influenced by these conditions, but influenced by the solution temperature.

It is important to use highly accurate surface area values of a crystal particle in growth rate equations, when the equations include a term of the surface area. In practice, however, it is difficult to measure the surface area unless the crystal possesses a simple geometry (e.g., sphere, cube, octahedron, etc.), or unless the B.E.T. method of measuring surface area is applied. Therefore, a characteristic diameter is usually defined, and the area calculated from the diameter is used.

In this study, the surface area of a magnesium sulfate heptahydrate crystal ($MgSO_4 \cdot 7H_2O$) was measured by an adsorption method using DBPC vapor. After a magnesium sulfate crystal was grown in an agitated crystallizer, the surface area and weight of the crystal were measured. The growth process of the same crystal was continued by returning it to the crystallizer; its surface area and weight were repeatedly measured. From these measured values, the shape factor(1) (i.e., Carman's shape factor(2)) of the crystal was calculated. The relationships between the shape factor and the

0097-6156/90/0438-0373$06.00/0
© 1990 American Chemical Society

operating conditions (i.e., supersaturation degree, agitation speed, gas flow rate, and solution temperature) were considered. The calculated shape factors showed a periodic oscillation curve during the crystal growth. This phenomenon might be of significant interest for research into the crystal growth.

Experimental

The crystallizer was an agitated vessel with an inside diameter of 9.0 cm and a volume of about 1 ℓ. It was equippet with four vertical baffles, a water jacket to keep the solution temperature constant, and a nozzle through which nitrogen gas was introduced in several experiments to suspend a speed crystal more effectively in the solution. Agitation was accomplished with a 5.0 cm stainless steel marine propeller having three blades driven by a variable speed motor.

The experimental procedure was as follows. One litre of supersaturated aqueous solution was prepared, of which 0.8 ℓ was poured into the crystallizer. After a constant temperature (of 298.2 K) was attained, the solution concentration was determined, and a seed crystal (0.59-0.71 mm) was put into the solution and was allowed to grow for either 600 or 900 s. The crystal was then taken out of the crystallizer and put into a 50 mℓ beaker in which a small quantity of alcohol was placed. The crystal was removed from the beaker and dried in the air for a few minites, then its weight and surface area were measured. The same crystal was again placed into the crystallizer and allowed to resume growth.

A vapor of DBPC (2,6-di-tert-butyl-p-cresol) was used for measuring the surface area. DBPC adhered to the surface was washed off by a small amount of benzene. The amount of dissolved DBPC in the benzene was determined by gas chromatography.

The shape factor, ϕ_c, was calculated using the following equation.

$$\phi_c = (36\pi)^{1/3} \, v^{2/3} \cdot s^{-1}$$
$$= 4.8359 \, (w/\rho_p)^{2/3} \cdot s^{-1} \qquad (1)$$

in which ρ_p is the density of the magnesium sulfate heptahydrate crystal. The values of ϕ_c can range between 0 and 1.0, where $\phi_c=1.0$ indicates a spherical particle.

In this paper, the concentration, C [kg·m^{-3}], is always represented as C kg of $MgSO_4 \cdot 7H_2O$ in 1 m^3 of solution. The density of a solution which was necessary for determining the concentration of magnesium sulfate was preliminarily measured, and the results were consistent with the data from previous research[3].

Results and Discussion

Variation of Shape Factor. Three typical variations in the shape factor of $MgSO_4 \cdot 7H_2O$ crystal are shown in Figure 1. The factors show the results obtained from three experiments of crystal growth in the solution without gas admittance at a supersaturation degree (ΔC) of 1.78 kg·m^{-3}, and a temperature of 298.2 K. In this solution, secondary nuclei[4] were hardly found. The curves connecting the points in the figure were determined by Fourier series equations to

which the shape factors were fitted. Fourier series equations were
used to fit the data because they were most suitable for the
establishment of a curve satisfying the data points. And the curve
enabled us to predict with ease the positions of the peaks and
troughs in the curve using a calculator.

The $MgSO_4 \cdot 7H_2O$ crystal in an aqueous solution grew along a
periodic oscillation curve of the shape factors. The variations in
the curve suggest that a crystal exhibits an analogous shape almost
periodically within a certain range of shape factors.

If the shape of the $MgSO_4 \cdot 7H_2O$ crystal is assumed to be
ellipsoidal, an increase in the ratio of the long axis to the short
axis of the ellipsoid will yield a decrease in the shape factor. The
two linear growth rates in the directions of the long and short axes
usually differ during the crystal growth. Around the peak of the
oscillation curve, the growth rate in the short-axis direction is
larger than that in the long-axis direction. On the other hand,
around the lower part of the valley in the curve, the growth rate in
the long-axis direction is larger.

Figure 1 also shows that the period of oscillation between two
successive peaks or troughs seems to be a function of agitation
speed. An increase in speed causes a small reduction in period (i.e.,
from 3,900, through 3,780, to 3,720 s, as the speed is increased from
5.0, through 6.7, to 8.3 s^{-1}, respectively), if compared from the
beginning of crystal growth. Those periods were determined as the
intervals of the two first-appearing peaks or troughs in the Fourier
series equations.

To ascertain the period estimated from the Fourier series
equation, the interval from the measurement of weight and surface
area to the next measurement was changed from 900 to 600 s. Three
curves with eight data points each were also obtained for a total
crystal growth time of 4,200 s at $\Delta C = 4.57$ kg·m^{-3}. The periods of
the three curves were 3,060, 2,880 and 2,820 s respectively. They
were slightly smaller than the previous values which appeared in
Figure 1. Accordingly, more narrow superimposed curves with shorter
periods could not exist in the estimated Fourier series equations.
From these curves, it could also be observed that the shape factor
during the crystal growth displayed a periodic oscillation. Also, an
increase in ΔC brought about a decrease in the period.

In general, the crystal growth rate becomes faster as the
agitation speed is increased. In other words, a shorter period or a
higher oscillation frequency may have enhanced the crystal growth
rate. The curve of the shape factor, though oscillating, probably has
a central value (e.g., 0.75) which will be revealed after a certain
time of crystal growth. The time is probably determined by the
operating conditions and the crystal's shape at the beginning of
crystal growth. In each of the oscillation curves, hardly any
difference could be found in the amplitudes between the highest and
lowest values for any time or any agitation speed.

Effects of Supersaturation Degree on Period and Amplitude.   The
relation between the period of the oscillation curve and the
supersaturation degree is shown in Figure 2. An increase in
supersaturation degree brought about a considerable reduction in
period. If a more supersaturated solution is used (there should be an
operation restriction on the crystal growth), the period will be

shorter and the oscillation frequency will be larger, so that the
crystal growth rate will be enhanced. On the other hand, a decrease
in supersaturation degree expanded the period, and the larger period
reduced the crystal growth rate. In Figure 2, an effect of agitation
speed on period is also indicated, but it is smaller than that of the
supersaturation degree.

It is occasionally seen in a crystallization process that a
certain gas is blown to react with some components of the solution
and to yield a crystalline product. The admittance of the gas into
the crystallizer was also expected to promote an agitation effect.
The effect of gas flow rate of nitrogen (non-reactive) on shape
factor is shown in Figure 3. The flow rate did not influence the
amplitude, but a slight decrease in period was observed as the flow
rate increased.

The relation between the amplitude of the oscillation curve and
the supersaturation degree is shown in Figure 4. The effect of the
agitation speed on amplitude may also be seen. The amplitude value is
always about 0.04, and is not influenced by the supersaturation
degree or the agitation speed.

Dependence of Amplitude on Solution Temperature.   In the foregoing
discussions, the constancy of amplitude has been shown for any
supersaturation degree or any agitation method chosen for the
operating conditions. However, the solution temperature did influence
the amplitude, as shown in Figure 5. These experimental values were
obtained by setting n at 5.0 s^{-1}, ΔC at 1.78 to 11.07 $kg \cdot m^{-3}$, and the
gas flow rate at zero. The increase in temperature yielded an almost
linear increase in amplitude. In general, the increase in temperature
raised the crystal growth rate. In other words, hifger temperature
raised the amplitude, and then the higher amplitude increased the
rate of crystal growth. The value of the amplitudes shown in Figure 5
may be considered as a characteristic or inherent value for $MgSO_4 \cdot 7H_2O$, and was dependent only upon the solution temperature.

Convergency of Amplitudes.   Three sets of lines connecting the
highest and lowest values for each of the three oscillation curves
are depicted in Figure 6, in which the difference between the upper
and lower lines shows the amplitude. The curves are for three
different seed crystals (a, b, and c) at a supersaturation degree of
1.78 $kg \cdot m^{-3}$, an agitation speed of 8.3 s^{-1}, and a temperature of
298.2 K. At the beginning of crystal growth the values of the shape
factors showed different values, but at about 12,000 s they were
between 0.73 and 0.77. This suggests that 12,000 s will be necessary
to obtain crystals of similar shape. The attainment of a steady state
of the shape factor, however, does not necessarily mean the
production of exactly the same shaped crystals. The shape factors
could not exceed the shown values (of 0.73-0.77), which were
determined by the operating conditions. The amplitudes were always at
the same value (of 0.04) during the crystal growth.

As is well known, the shape of $MgSO_4 \cdot 7H_2O$ is an orthorhombic
crystal with three different lattice parameters(5), 11.86, 11.99, and
6.858. From these values, the shape factor can be calculated as
0.778. This value closely matches the upper lines of Figure 6. If
crystals having a shape factor of 0.778 are wanted, the lower values
should be raised so that the amplitude becomes more narrow and the

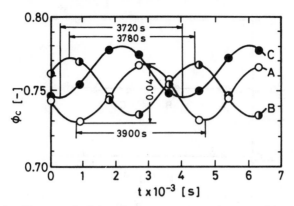

Fig. 1   Change of shape factor at $\Delta C=1.78$ kg·m^{-3}, T=298.2 K,
Qg=0.   n [s^{-1}]: A; 5.0, B; 6.7, C; 8.3

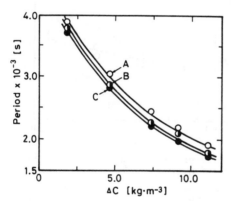

Fig. 2   Dependence of period on supersaturation degree
at T=298.2 K and Qg=0.   n [s^{-1}]: A; 5.0, B; 6.7, C; 8.3

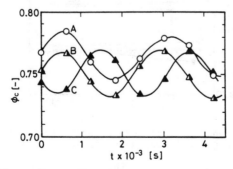

Fig. 3   Effect of gas flow rate on variation of shape factor
obtained at $\Delta C=7.35$ kg·m^{-3}, n=5.0 s^{-1}, T=298.2 K.
Qg [m^3·s^{-1}]: A; 0, B; 8.33, C; 16.7

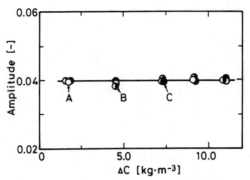

Fig. 4   Constancy of amplitude at T=298.2 K and Qg=0.
n [s⁻¹]: A; 5.0, B; 6.7, C; 8.3

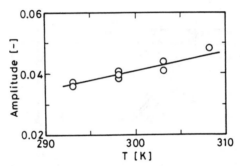

Fig. 5   Dependence of amplitude on solution temperature at
1.78≦ΔC≦11.07 kg·m⁻³, n=5.0 s⁻¹, Qg=0.

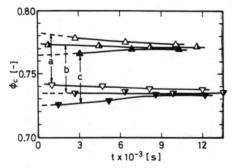

Fig. 6   Convergence of amplitude of shape factor at ΔC=1.78 kg
·m⁻³, n=8.3 s⁻¹, T=298.2 K, Qg=0.

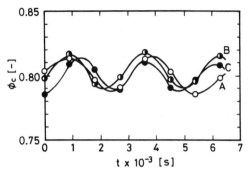

Fig. 7   Shape factor of K_2SO_4 crystal at $\Delta C=1.22$ kg·m^{-3},
T=298.2 K, Qg=0.
n [s^{-1}]: A; 5.0, B; 6.7, C; 8.3

period shorter. This corresponds to the neccesity of having a lower
temperature and a quiescent solution.

That is, a growing crystal takes on various shapes while
displaying a periodic oscillation curve with a constant period, a
constant amplitude, and a fixed axis at the steady state of the shape
factor. The amplitude is altered by the temperature (we are currently
researching the amplitudes of various other crystals, of which one
example follows).

Shape Factors of Potassium Sulfate Crystals.   As expected, if the
chemical substance forming a crystal is changed, the shape factor may
be different from that of $MgSO_4 \cdot 7H_2O$. One example for K_2SO_4 is shown
in Figure 7. The shape factors are between 0.786 and 0.819, which are
slightly larger than those for $MgSO_4 \cdot 7H_2O$. The period of the
oscillation curves was reduced (resembling that for $MgSO_4 \cdot 7H_2O$) from
2,940, through 2,580, to 2,400, as the agitation speed was increased
from 5.0, through 6.7, to 8.3 s^{-1}. The amplitude also showed an
almost constant value, as with $MgSO_4 \cdot 7H_2O$, but at a smaller value of
0.03.

Conclusions

It was found that the shape factor of a magnesium sulfate
heptahydrate crystal growing in its aqueous solution showed periodic
oscillation. The period of oscillation was influenced by the
operating conditions, i.e., the degree of supersaturation, agitation
speed, and gas flow rate, while the amplitude was not influenced by
the above operating conditions. Only an increase in the solution
temperature resulted in an increase in amplitude. The average of the
oscillation curve approached a stable value. The magnesium sulfate
heptahydrate crystal had an inherent amplitude which was determined
by the solution temperature.

Legend of Symbols

C	concentration of solution	$[kg \cdot m^{-3}]$
ΔC	degree of supersaturation	$[kg \cdot m^{-3}]$
n	agitation speed of impeller	$[s^{-1}]$
Qg	gas flow rate	$[m^3 \cdot s^{-1}]$
s	surface area of a crystal	$[m^2]$
T	temperature of solution	$[K]$
t	time	$[s]$
v	volume of a crystal	$[m^3]$
w	weight of a crystal	$[kg]$
ρp	density of crystal	$[kg \cdot m^{-3}]$
ϕc	shape factor	$[-]$

Literature Cited

1.   Wadell, H. J. Geol., 1932, 40, 443-451.
2.   Carman, P. C. Trans. Inst. Chem. Engrs., London, 1937, 15, 150-166.
3.   Kubota, N.; Shimizu, K.; Wakabayashi, S.; Kawakami, T. Kagaku Kogaku Ronbunshu, Tokyo, 1978, 4, 202-205.
4.   Cayey, N. W.; Estrin, J. Ind. Eng. Chem. Fund., 1967, 6, 13-20.
5.   Joint Comittee on Powder Diffraction Standards Powder Diffraction Files, ASTM, New York, 1967, Set 6-10, p 399.

RECEIVED May 30, 1990

Chapter 29

Continuous Crystallization of Calcium Sulfate Phases from Phosphoric Acid Solutions

G. J. Witkamp and G. M. van Rosmalen

Delft University of Technology, Leeghwaterstraat 44, 2628 CA Delft, Netherlands

Calcium sulfate crystals were precipitated in a Continuous Mixed Suspension Mixed Product Removal (CMSMPR) crystallizer by mixing of calcium phosphate and sulfuric acid feed streams. The formed calcium sulfate hydrate (anhydrite, hemihydrate and dihydrate) mainly depends on the temperature and the solution composition. The uptake of cadmium and phosphate ions in these hydrates has been studied as a function of residence time and solution composition. In anhydrite, also the incorporation of other metal ions has been investigated. The uptake was found to be a function of both thermodynamics and kinetics.

Calcium sulfate is formed as a byproduct in industrial processes such as flue gas desulfurization and the production of zinc, fluoride, organic acids and phosphoric acid, in amounts of many million tons per year. In this study the attention is focussed on calcium sulfate from the production of phosphoric acid for fertilizer applications. It is precipitated, from solution after digestion of phosphate ore by addition of sulfuric acid according to [1]:

$$Ca_9(PO_4)_6 \cdot CaF_2 + 10\ H_2SO_4 + 10x\ H_2O \longrightarrow 10\ CaSO_4 \cdot xH_2O\downarrow + \\ 6\ H_3PO_4 + 2\ HF\uparrow$$

With varying temperature and acid concentration either anhydrite (AH, x=0), hemihydrate (HH, x=1/2) or dihydrate (DH, x=2, also called gypsum) is crystallized from the acidic solution (see figure 1). As this so-called phosphogypsum is usually disposed, the quality itself is not

0097–6156/90/0438–0381$06.00/0
© 1990 American Chemical Society

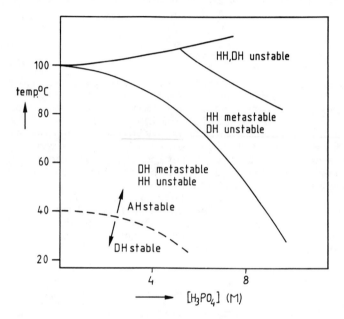

Figure 1. Stability of calcium sulfates in phosphoric acid-water mixtures.

a guiding factor for process design. Since, however, the byproduct is formed in large quantities (more than 2 tons of gypsum per ton phosphate), the separation of liquid product and crystals has to proceed efficiently to maintain a good yield. This implies that a good filterability and washability is required, but also that the uptake of phosphate in the crystals should be minimized. This uptake (typically 1%) not only reduces the process efficiency, but also causes eutrofication of surface waters. Also minimization in uptake of other compounds originating from the phosphate ore is important, especially when hazardous compounds are concerned, which lead to environmental pollution when the gypsum is disposed. Such a compound is cadmium, which is toxic even at its low concentration of a few ppm. The presence of radium hampers the application of the calcium sulfate as plaster in the building industry because its decay product, radon, may cause an enhanced radioactivity level in poorly ventilated rooms. Other impurities which can become incorporated follow from table 1, where the composition of a typical phosphate ore is given.

Table 1. Typical composition of a phosphate ore

compound	%	compound	ppm	compound	ppm
HPO_4^{2-}	43	Zn	200	Al	3000
Ca	37	Ni	50	Fe	1300
SO_4^{2-}	2	Cu	40	V	200
F	4.2	Cd	20	U	150
SiO_2	2.2	Mn	15	As	15
Na	0.7	Pb	3	Sb	3
Ce, La etc.	max. 1 %	(magmatic ores)		Hg	0.01-0.05
	100 ppm	(sedimentary ores)		Ra	0.00004

The level of impurity uptake can be considered to depend on the thermodynamics of the system as well as on the kinetics of crystal growth and incorporation of units in the growing crystal. The kinetics are mainly affected by the residence time which determines the supersaturation, by the stoichiometry (calcium over sulfate concentration ratio) and by growth retarding impurities. The thermodynamics are related to activity coefficients in the solution and the solid phase, complexation constants, solubility products and dimensions of the foreign ions compared to those of the ions of the host lattice [2,3,4].
The impurities can become incorporated in the solid phase in three modes: i) interstitially, i.e. between regular lattice positions, ii) by coprecipitation as a separate insoluble phase or iii) by isomorphous substitution of either Ca^{2+} or SO_4^{2-}. Interstitial uptake (mode i) is difficult to identify, and the dependance of uptake as a function of process conditions is hard to predict. Formation

of sparingly soluble compounds as $PbSO_4$, $RaSO_4$ and $BaSO_4$ explains the very high uptake of these metal ions according to mode ii. In this paper only those ions which are expected to become incorporated by isomorphous substitution (mode iii) will be discussed.

A generally used expression to quantify the uptake is the distribution coefficient K, which for cadmium as an example is defined as: $K(Cd)=[Cd^{2+}](crystal)/[Cd^{2+}](solution)$.
For isomorphous substitution also a partition coefficient D can be defined which takes into account the competition between the impurity and substituted ions. For cadmium D is given by:
$D = [Cd^{2+}]/[Ca^{2+}] (crystal) / [Cd^{2+}]/[Ca^{2+}] (solution)$.
When the D-value is experimentally found to be constant over a wide range of calcium and cadmium concentrations, the incorporation most likely proceeds by isomorphous substitution.
The aim of this work is to study the incorporation of cadmium and phosphate in the three calcium sulfate modifications. The uptake of other metal ions in AH will also be described. Kinetic effects of operating conditions such as the residence time, sulfuric acid and phosphate concentration upon the phosphate and cadmium uptake has been investigated. In addition the influence of a growth retarding impurity, AlF_3, on the cadmium and phosphate uptake will be given.

Experimental

Chemically pure reagents were used. Cadmium was added as its sulfate salt in concentrations of about 50 ppm. Lanthanides were added as nitrates. For the experiments with other metal ions so-called "black acid" from a Nissan-H process was used. In this acid a large number of metal ions were present. To achieve calcium sulfate precipitation two solutions, one consisting of calcium phosphate in phosphoric acid and the other of a phosphoric acid/sulfuric acid mixture, were fed simultaneously in the 1 liter MSMPR crystallizer. The power input by the turbine stirrer was 1 kW/m^3. The solid content was about 10%. Each experiment was conducted for at least 8 residence times to obtain a steady state. During the experiments liquid and solid samples were taken for analysis by ICP (Inductively Coupled Plasma spectrometry, based on atomic emission) and/or INAA (Instrumental Neutron Activation Analysis). The solid samples were washed with saturated gypsum solution (3x) and with acetone (3x), and subsequently dried at 30 °C. The details of the continuous crystallization experiments are given in ref. [5].

Results

General. In order to precipitate either AH, HH or DH the appropriate temperature and phosphate and/or sulfate concentration must be selected. In table 2 the operational conditions are listed for each experiment. In the experiments where DH is crystallized only the sulfuric acid concentration, the temperature and the residence time were varied. In case of HH crystallization various sulfuric acid and phosphoric acid concentrations were applied. AH crystallization has been studied at various residence times, in the presence of lanthanides or AlF_3 and also in black acid from a Nissan-H process.

Phosphate and cadmium uptake in DH. In figure 2 the weight fraction of phosphate in the DH crystals is plotted as a function of the H_2SO_4 concentration in the solution (exp. 1-8). The solution contains 3.5 M H_3PO_4 at a temperature of 60 or 70 °C, while the residence time varies from 1200 to 3600 seconds. The curve for 1200 seconds residence time shows a decrease in phosphate uptake at increasing sulfate concentrations, which can be explained from a competition between phosphate and sulfate ions for a lattice position. At higher sulfate concentrations the relative amount of phosphate ions in the solution becomes smaller. The D-value for the phosphate uptake given by $D=[HPO_4^{2-}]/[SO_4^{2-}]$ (crystal) / $[H_3PO_4]/[H_2SO_4]$ (solution) equals $\approx 2 \cdot 10^{-3}$ as follows from the slope of the plot presented in figure 3. This value is 10 times higher than for HH (section 3.3). This might result from the difference in crystal lattice between the two calcium sulfate modifations or from the difference in solution composition, since DH is crystallized from weaker phosphate solutions (3.5 M) than HH (5.5 M or higher). At higher phosphate concentration, the percentage phosphate dimer ($H_5P_2O_8^-$) increases, while the single phosphate ions are almost completely protonized. This means that at high acid concentrations the incorporation in the calcium sulfate lattice requires to a larger extent a deprotonation or splitting up of dimers. The higher energy required for such a step lowers the uptake. Another explanation is an effect of the temperature, which is about 60 °C for DH and 90 °C for HH. Earlier data obtained from recrystallization experiments [6] for DH indicate a strong decrease in uptake at increasing temperature. From figure 2, however, only a small difference in uptake between 60 and 70 °C can be seen. At longer residence time the fraction phosphate in the crystals and the D-value become smaller. The smaller values can only result from a kinetic effect. The lower uptake at longer residence time is caused by the lower supersaturation or growth rate. Such a kinetic effect must be related to an entrapment of the impurity ions at the crystal surface by propagating steps.
The uptake of cadmium increases at higher sulfate concentration as can be seen from figure 4. This is due to

Table 2. Experimental conditions and results

		solution					crystals	
exp. nr.	phase	$[H_3PO_4]$ M	$[H_2SO_4]$ M	$[Ca^{2+}]$ 10^{-3} M	T °C	res.time 10^3 s	K(Cd)	D(Cd) $\times 10^3$	$[HPO_4^{2-}]$ fraction
1	DH	3.2	0.08	125	60	1.2	0.13	2.8	0.041
2	DH	3.3	0.13	80	60	1.2	0.15	2.1	0.021
3	DH	3.5	0.28	39	60	1.2	0.45	3.0	0.015
4	DH	3.5	0.55	18	60	1.2	0.45	1.4	0.009
5	DH	3.9	1.10	11	60	1.2	0.85	1.6	0.006
6	DH	3.7	0.60	28	60	2.4	0.40	1.9	0.006
7	DH	3.5	0.53	20	60	3.6	0.30	1.03	0.005
8	DH	3.6	0.53	30	70	1.2	0.50	2.6	0.008
9	HH	5.5	0.02	270	92	1.2	0.40	15.7	0.027
10	HH	5.5	0.04	180	92	1.2	0.16	4.1	0.021
11	HH	5.5	0.09	67	92	1.2	0.11	1.0	0.008
12	HH	5.5	0.10	67	92	1.2	0.25	2.9	0.008
13	HH	5.5	0.11	54	92	1.2	0.17	1.3	0.006
14	HH	5.5	0.16	41	92	1.2	0.20	1.2	0.004
15	HH	5.4	0.17	43	92	1.2	0.21	1.3	0.004
16	HH	5.5	0.18	35	92	1.2	0.35	2.1	0.006
17	HH	5.5	0.23	29	92	1.2	0.36	1.5	0.003
18	HH	5.5	0.35	30	92	1.2	0.55	2.8	0.003
19	HH	5.5	0.35	20	92	1.2	0.50	1.4	0.003
20	HH	6.0	0.03	110	92	1.2	0.20	3.1	0.018
21	HH	6.1	0.15	25	92	1.2	0.30	1.1	0.004
22	HH	6.1	0.29	13	92	1.2	0.75	1.4	0.003
23	HH	6.3	0.12	25	92	1.2	0.57	2.1	0.006
24	HH	6.5	0.14	25	92	1.2	0.65	2.3	0.005
25	HH	6.6	0.22	9	92	1.2	1.30	1.7	0.003
26	HH	6.4	0.22	16	92	1.2	1.31	3.1	0.003
27*	AH	5.5	0.70	9	95	3.6	25	30	n.a.
28**	AH	5.5	1.0	4	95	2.4	70	39	0.007
29#	AH	5.5	1.0	4	95	2.4	70	38	0.006
30	AH	5.5	1.0	3	95	2.4	35	16	0.005
31	AH	5.3	0.9	2	95	2.4	50	11	0.006
32	AH	5.5	1.0	2	95	4.8	35	8	0.005
33	AH	5.5	0.9	2	95	7.2	25	6	0.006

*:with 100 ppm lanthanides **:with 0.8% AlF_3 #:in black
acid

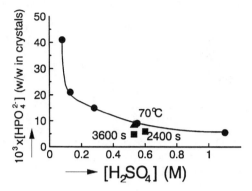

Figure 2. Phosphate uptake in DH, crystallized from 3.5. M H_3PO_4, vs. sulfate content of solution with residence times from 1200 (circles) to 3600 s, at 60 ° C (circles) or 70 ° C.

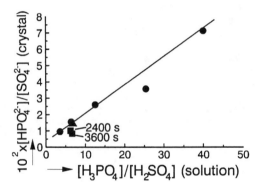

Figure 3. Phosphate over sulfate ratio in the crystals vs. the same ratio in the solution for residence times of 1200 (circles) to 3600 s.

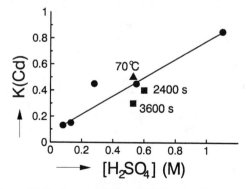

Figure 4. Cadmium distribution coefficient in DH vs. sulfate content of solution for residence times of 1200 (circles) to 3600 s at 60 and 70 ° C.

a competition between calcium and cadmium. At higher sulfate concentrations the calcium concentration is lower because of the constant solubility product of calcium sulfate. In figure 5 $[Cd2+]/[Ca^{2+}]$ in the crystals is plotted versus the same concentration ratio in the solution. The slope D equals about $2 \cdot 10^{-3}$ for the experiments with a residence time of 1200 seconds. This D-value for DH is approximately the same as for HH (section 3.3). To show the effect of kinetics, the residence time in the crystallizer has been changed from 1200 seconds to 2400 or 3600 seconds. As a result, the D-value decreased from about 2 to $1 \cdot 10^{-3}$. Whether this value is close to its thermodynamic value cannot be verified experimentally since at longer residence times, the stable anhydrite phase starts to develop.

Phosphate and cadmium uptake in HH. In the experiments described here the phosphate concentration has been varied between 5.5 and 6.5 M with sulfate concentrations between 0.02 and 0.3 M. In this whole range the incorporation of phosphate can be described by a D-value of $1.5-2 \cdot 10^{-4}$, which is 10 times lower than for DH. As pointed out in section 3.2, this can be caused by differences in crystal structure of HH and DH or in solution composition, or by the higher temperature of HH crystallization.
In figure 6 the resulting Cd-concentration in the crystals is plotted versus the sulfate content of the solution. In 5.5 M phosphate solutions, K(Cd) seems to decrease from 0.02 to 0.5 M sulfate contents, while above 0.1 M sulfate K(Cd) increases. This tendency of higher uptake at higher sulfate concentrations has also been found for DH. The high K values at very low sulfate contents indicates a major influence of kinetics.
There is also an increase in K at higher phosphate concentrations.
To be able to interpret these results and to correct for the lower calcium concentrations at high sulfate and phosphate concentrations, the partition coefficients D have been determined. These values follow from the slopes of the curves in figure 7. For 5.5 and 6.0 M H_3PO_4 a D of about $1.5 \cdot 10^{-3}$ is obtained. A similar D-value for both acid concentrations should indeed be obtained, when the activity coefficients of the ions in solution is not strongly affected by the acid concentration. The D-value for 6.5 M H_3PO_4 lies somewhat higher. This could e.g. be caused by a higher activity coefficient of cadmium compared to calcium at this acid concentration. The thermodynamic D-value cannot be determined by increasing the residence time, because a residence time of 2400 seconds already caused anhydrite formation.

Phosphate and cadmium uptake in AH. The D for phosphate uptake in AH is about 10^{-3}, which is ten times higher than for HH, and equal to DH. This can only be explained from the difference in crystal strucure between HH and AH, since the

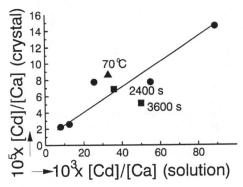

Figure 5. Cadmium over calcium ratio in DH vs. the same ratio in the solution. The slope (D) is about $1.5 \cdot 10^{-3}$. Residence time 1200 (circles) to 3600 s.

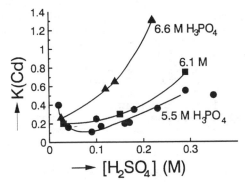

Figure 6. Distribution coefficient K for cadmium in HH as function of sulfate concentration for various phosphate contents.

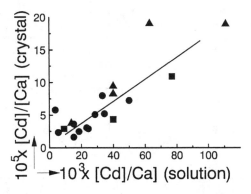

Figure 7. Cadmium over calcium ratio in HH versus the same ratio in the solution for various phosphate concentrations.

solution composition and the temperature are approximately the same. The D-value of cadmium in AH has been determined in solutions containing 5.5 M H_3PO_4 and 1.0 M H_2SO_4 at 95 °C, for residence times between 1200 and 3600 seconds (figure 8). Obviously the uptake is much higher at shorter residence times. This effect also occurred for DH, and leads in case of AH to differences in uptake by a factor of three in this range of residence times.

The supersaturation is too low in all experiments to be measured accurately, but it seems reasonable to assume that the effect of residence time is imposed through the kinetics. Another observation is that the D-value for cadmium uptake in anhydrite is about ten times higher than in HH or DH. An explanation for this higher D seems to be related to the crystal structures of the calcium sulfates. Only the AH structure matches with an anhydrous $CdSO_4$ phase, while no hemi- or dihydrate phase of $CdSO_4$ exists.

Effect of AlF_3 on cadmium and phosphate uptake in AH. From chemically pure 5.5 M H_3PO_4 and 1.0 M H_2SO_4 containing solutions agglomerates of long platelets of AH are formed at 95 °C and a residence time of 2400 seconds. These platelets have lengths up to 100 μm, widths of maximally 10 μm and a thickness of less than one μm (figure 9). In the presence of 0.8 % AlF_3 in the acid solution, however, the growth of the top faces is strongly retarded (figure 9). As a result, the supersaturation increases, causing a faster growth of the slower growing prism faces, which causes formation of block-shaped crystals with dimensions of typically 10x10x10 μm. The uptake of both cadmium and phosphate is higher, due to an apparent kinetic effect. The D(Cd) values are respectively $2 \cdot 10^{-2}$ and $4 \cdot 10^{-2}$, while the fractions of phosphate in the crystals are 5 and $7 \cdot 10^{-3}$ respectively.

The fact that the block-shaped crystals contain more Cd and HPO_4^{2-} than the long platelets indicates that changes in the growth rate of a given crystal face lead to differences in uptake and that the absolute growth rate is not the only determining factor for the uptake. In other words, if one crystal face grows faster than the other, it does not necessarily mean that also the Cd or P uptake through that crystal face is higher. This means that the higher uptake in the presence of AlF_3 is caused by the faster growth of the prism faces. Another explanation for a higher uptake could be that the impurity changes the surface structure or even the growth mechanism, resulting in a different uptake.

Uptake of other metal ions in DH and AH. In this section the results from experiments 27 and 29 are discussed. The partition coefficients of several metal ions in AH are plotted versus their ionic radii in figure 10. These data demonstrate that the radius is an important parameter. The closer the radius is to that of Ca^{2+} (112 pm in an eightfold coordination), the higher the D-value becomes. The phase of

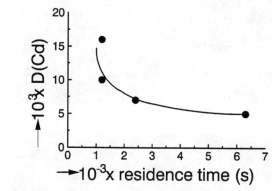

Figure 8. Partition coefficient of cadmium in AH vs. residence time.

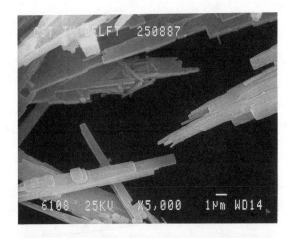

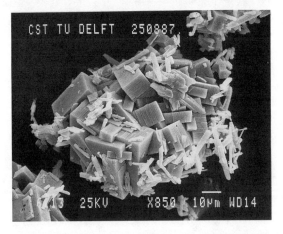

Figure 9. AH crystals grown in pure phosphoric acid (top) and acid containing 0.8% aluminumfluoride (bottom).

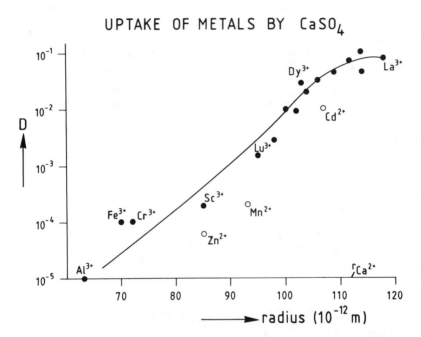

Figure 10. Partition coefficients of several metal ions in AH vs. their ionic radii.

the calcium sulfate is of less influence. Only in case of
cadmium the partition coefficient is ten times higher for
AH than for DH. The uptake of trivalent metal ions occurs
more easily than of divalent ions. This was unexpected,
since a 3+ charged ion, replacing Ca^{2+} needs to be
compensated in the lattice. These trivalent ions are
probably more attracted to the sulfate ions at the crystal
surface. Other parameters, such as dehydration energy or
frequency, are not of importance [4].

Conclusions

-In the investigated range of conditions the uptake of
 cadmium and phosphate in calcium sulfate by isomorphous
 substitution can be described by a D-value.

-In AH the D(Cd) is ten times higher than in HH or DH. D for
 phosphate is for AH and DH ten times higher than for HH.

-D increases with the growth rate or supersaturation i.e.
 at shorter residence times and the in presence of
 impurities.

-The D for metal ions in AH is mainly determined by the ion
 radii of the impurities.

-For uptake proceeding by isomorphous substitution, the
 partition coefficient D depends on thermodynamic parameters
 such as ionic radius of the impurity ions and the phases
 of the calcium sulfate as well as on kinetics.

References

1. Becker, P. Phosphates and Phosphoric acid. Fertilizer
   Science and Technology Series, Vol. 6,
   Dekker, New York (1989).
2. Balarew, C. Proc. 13th gen. meeting Int. Mineral. Ass.,
   Varma (1982)   Publ. House Bulg. Acad. Sci. p. 287.
3. Witkamp, G.J. Thesis, Delft University of Technology,
   1989.
4. Heavy Metal in the Hydrological cycle, M. Astruc, Ed.,
   Selper Ltd,   London 1988, p. 108.
5. van der Sluis, S., Witkamp, G.J. and van Rosmalen, G.M.,
   J. Crystal Growth 79 (1986) 620-29.
6. Witkamp, G.J. and van Rosmalen, G.M., Proc. Sec. Int.
   Symp. Phosphogypsum, Florida Inst. Phosphate Res., 1988,
   377-405.

RECEIVED May 17, 1990

Chapter 30

Acceleration Phenomenon on Solidification of CaCl$_2$ • 6H$_2$O with a Nucleation Agent

Yasuyuki Watanabe[1], Tomonari Saito[1], and Tasuku Nakai[2]

[1]Nok Corporation, 4–3–1 Tsujido-shinmachi, Fujisawa, Kanagawa 251, Japan
[2]Hiroshima Institute of Technology, 2–1–1 Miyake, Saeki-ku, Hiroshima 731–51, Japan

The effects of thermal history on nucleation characteristics of CaCl$_2$·6H$_2$O melt without nucleation agent and the same melt with nucleation agent, which makes to accelerate the nucleus formation of CaCl$_2$·6H$_2$O melt, were studied experimentally using thirteen kinds of agent by the cooling method at constant rate of 1K/min. CaCl$_2$·6H$_2$O melt was supercooled till the temperature of 20~30K lower than its melting point for a superheating sample obtained beyond 5K. On the other hand, CaCl$_2$·6H$_2$O melt including nucleation agent occurred nucleation at the temperature of 2~4K below its melting point. However, when the superheating temperature elected a value higher than an upper limiting temperature, the supercooling temperature for solidification began an increase to some large value. The magnitude of this upper limiting temperature varied 21 ~37K higher than melting point of CaCl$_2$·6H$_2$O by the sort of nucleation agent. The magnitude of supercooling temperature also varied 4~25K lower than melting point.

Through the investigation on such nucleation characteristics, it became clear that the adhesion force of CaCl$_2$·6H$_2$O clusters, which remained on surfaces of precipitated nucleation agent in melt, and the acceleration ability of agent surfaces for heterogeneous nucleation were closely related each other on the acceleration mechanism for the solidification of CaCl$_2$·6H$_2$O melt.

The method of latent heat storage based on liquid-solid phase transition is available to make smaller the volume of heat storage tank, because of its higher thermal density than that of sensible heat storage. Therefore, a substance which has a large amount of latent heat of fusion is more profitable as a heat storage material.

$CaCl_2 \cdot 6H_2O$ (1) has superior thermal characteristics for latent
heat storage, because of its melting point of 302K and latent heat
of fusion of 176kJ/kg. But, at cooling stage $CaCl_2 \cdot 6H_2O$ melt
exhibits large degree of supercooling without solidification (2).
Such a fact results in lowering of running efficiency for heat
storage system. A large number of experimental studies on nucleation agents (3) forcused effectiveness for the depression of supercooling have been actively carried out in recent years. At present,
about sixty kinds of agent have been proposed and nevertheless the
depression mechanism of each nucleation agent for solidification
have not yet been clarified.

In our laboratory, as one of approaches to make clear the
function of these nucleation agents, the effects of thermal history
on the solidification of $CaCl_2 \cdot 6H_2O$ melt including agent have been
studied in detail. Through a number of experiments on the nucleation characteristic of the melt with agent, the following facts
were found out and reported already (4). Firstly, many kinds of
nucleation agent added to $CaCl_2 \cdot 6H_2O$ melt became active for the
depression of such supercooling only after experience the solidification of $CaCl_2 \cdot 6H_2O$ melt. Secondly, the activities of these agents
for solidification had each peculiar upper limiting temperature at
heating and gradually changed to lower value by heating over their
limiting temperatures.

On the industrial application of heat storage system, for the
case of higher heat storaging temperature the rate of storaging and
also the thermal density become larger. In order to use $CaCl_2 \cdot 6H_2O$
for heat storage, the clarification of acceleration mechanism for
the nucleation of $CaCl_2 \cdot 6H_2O$ melt with nucleation agent is one of
greatly important subjects.

In this paper, the nucleation characteristics of the melt
including agent were investigated experimentally in more detail for
thirteen kinds of agent.

Experimental

Sample Preparation. $CaCl_2 \cdot 6H_2O$ melt was obtained in a 20ml volume
cylindrical glass bottle by dissolving $CaCl_2 \cdot 2H_2O$ crystal powder of
6.71g in distilled water of 3.29g at 333K. The composition of this
solution corresponds to that of $CaCl_2 \cdot 6H_2O$ melt.

The preparation procedure of samples including nucleation
agents is shown in Figure 1. Thirteen kinds of material used as
nucleation agents are listed in Table I together with their particle sizes. All crystal powders as commercial reagents of special
grade were used without pretreatment. The amount of nucleation
agent added to the melt was fixed at 3mol% being enough quantity to
occur the precipitation in the melt. All samples thus prepared were
solidified one time by cooling to 243K at constant rate of 1K/min
from 333K before using for nucleation experiments. This preparation
process through melt solidification was called as activation treatment, in this paper.

Experimental Procedure.   A schematic diagram of experimental apparatus is shown in Figure 2. About ten of cylindrical glass bottles
enclosed samples of same composition were set in a thermostatic air
chamber. The samples melted perfectly by holding for about 2h at

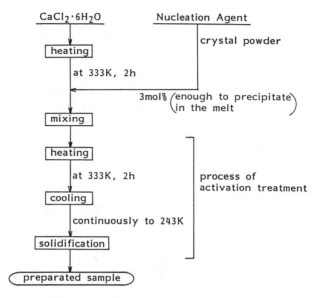

Figure 1. Process scheme for preparation of CaCl$_2$ · 6H$_2$O melt including nucleation agent.

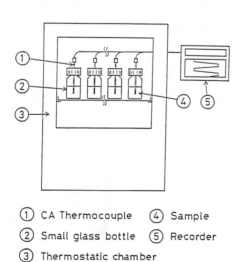

① CA Thermocouple ④ Sample

② Small glass bottle ⑤ Recorder

③ Thermostatic chamber

Figure 2. Schematic diagram of experimental apparatus used for investigation on nucleation characteristic under thermal history.

Table I. List of Nucleation Agents

Numbers	Chemical Formula	Particle Diameter (μm)	
		Average	Standard Deviation
1	$Ba(CH_3COO)_2$	140.0	65.0
2	$BaCO_3$	3.1	1.8
3	$BaCl_2 \cdot 2H_2O$	160.0	120.0
4	$BaS_2O_3 \cdot H_2O$	21.0	8.0
5	$BaSiF_6$	4.2	1.7
6	$BaTiO_3$	1.6	0.3
7	Na_2HPO_4	82.0	35.0
8	$NaKCO_3$	210.0	120.0
9	$Na_3PO_4 \cdot 12H_2O$	290.0	210.0
10	$Na_4P_2O_7 \cdot 10H_2O$	290.0	200.0
11	$Pb(CH_3COO)_2 \cdot 3H_2O$	190.0	170.0
12	$PbCl_2$	16.0	8.0
13	$SrSO_4$	5.5	2.1

prescribed heating temperature Tu were cooled to 238K at constant rate of 1K/min. The sample temperature was measured by CA thermocouple inserted into the center of sample. One of cooling curves of melted samples is shown in Figure 3. During continuous cooling, because a large amount of the heat of solidification was generated by the occurrence of nucleation of melted sample, the sample temperature began to rise rapidly. The temperature of this point just starting elevation was named as nucleation temperature Tsc of the sample.

The nucleation characteristics of samples were examined by the relation between superheating temperature ΔTu ($=Tu-Tm$) and supercooling temperature ΔTsc ($=Tm-Tsc$). Where, Tm is the melting point of $CaCl_2 \cdot 6H_2O$ sample, that is, 302K.

Result and Discussion

Nucleation Characteristic of $CaCl_2 \cdot 6H_2O$ Melt without Nucleation Agent. The effect of thermal history on ΔTsc of $CaCl_2 \cdot 6H_2O$ melt is shown in Figure 4. At $\Delta Tu = 2K$, ΔTsc converge to about 3K. At ΔTu range of 5~20K, the number of samples for large ΔTsc increase gradually according to the increase in ΔTu. At ΔTu range larger than 20K, ΔTsc diverge in the range of 20~30K.

The nucleation mechanism of $CaCl_2 \cdot 6H_2O$ melt may be considered similarly to the usual nucleation mechanism of solute from aqueous solution. That is, $CaCl_2 \cdot 6H_2O$ clusters are generated firstly in the supercooled $CaCl_2 \cdot 6H_2O$ melt. Thereafter, the generated clusters gradually grow up by mutual agglomeration and after the appearance of clusters grown beyond critical diameter, the nucleation of the melt occurs irreversibly. Therefore, if $CaCl_2 \cdot 6H_2O$ clusters are remained in $CaCl_2 \cdot 6H_2O$ melt even beyond its melting point, the melt will occur easily the nucleation even if small degree of supercooling (5).

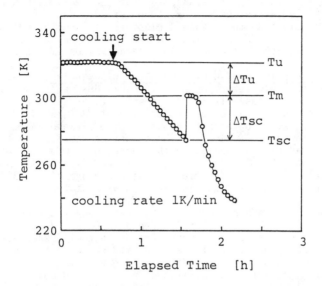

Figure 3. Representative cooling curve of CaCl$_2$ · 6H$_2$O melt without nucleation agent.

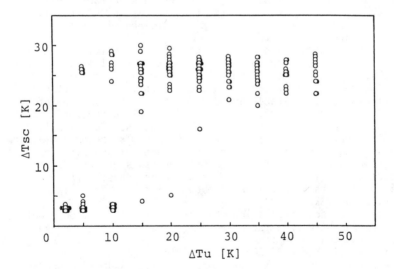

Figure 4. Effect of thermal history on supercooling temperature ΔTsc of CaCl$_2$ · 6H$_2$O melt without nucleation agent.

As shown in Figure 4, it is assumed that the small value of
ΔTsc at ΔTu$=$2K is suggesting the existence of such CaCl$_2\cdot$6H$_2$O
clusters and at ΔTu$=$5K, the remained clusters will begin the
thermal dissociation. As this dissociation progresses completely at
ΔTu beyond 20K, all samples come to show large values of ΔTsc.

Nucleation Characteristic of CaCl$_2\cdot$6H$_2$O Melt with Nucleation Agent.
The effects of thermal history on ΔTsc of CaCl$_2\cdot$6H$_2$O melt including
PbCl$_2$ and BaS$_2$O$_3\cdot$H$_2$O respectively are shown in Figures 5 and 6.
Figure 5 shows nucleation characteristic of CaCl$_2\cdot$6H$_2$O melt
including PbCl$_2$. ΔTsc converge to about 4K at ΔTu range smaller
than 22K. At ΔTu range of 24$\sim$28K, the number of samples for large
ΔTsc increase gradually according to the increase in ΔTu. At ΔTu
range larger than 28K, ΔTsc diverge in the range of 11$\sim$19K. Such
transition behavior of ΔTsc distribution at ΔTu range of 24$\sim$28K is
similar to the behavior of ΔTsc distribution without nucleation
agent at ΔTu range of 5$\sim$20K, as shown in Figure 4. Therefore, this
transition behavior is also seemed to be suggesting the thermal
dissociation of CaCl$_2\cdot$6H$_2$O clusters remained in CaCl$_2\cdot$6H$_2$O melt.
However, in this case starting temperature of thermal dissociation
is about 20K higher than that of the case without nucleation agent.
The reason of cluster preservation in superheating is supposed
as follows. PbCl$_2$ crystals precipitated in CaCl$_2\cdot$6H$_2$O melt are
wrapped into CaCl$_2\cdot$6H$_2$O crystals as a solid inclusion through the
activation treatment as mentioned above. At the consecutive heating
process, CaCl$_2\cdot$6H$_2$O crystals are melted and however PbCl$_2$ crystals
included in melt are remained as precipitating crystals. At this
moment, some tiny parts of CaCl$_2\cdot$6H$_2$O crystals contacted with PbCl$_2$
crystals are remained as clusters on surfaces of PbCl$_2$ crystals.
These clusters can be supposed to be maintained stably on sufaces
even till large ΔTu of 22K.
The value of critical ΔTu, at which the transition of ΔTsc
distribution is just beginning, is named as upper limiting temperature ΔTl, in this paper. Then, ΔTl corresponds to a temperature at
which the remained clusters begin the desorption from surfaces of
precipitated PbCl$_2$ crystals and dissociate gradually. Therefore,
the value of ΔTsc at ΔTu range smaller than ΔTl corresponds to the
magnitude of activity of such remained clusters. That is, for the
case of higher activity of remained clusters ΔTsc becomes smaller.
And also, the value of ΔTl means the strength of adhesion force of
remained clusters to surfaces of PbCl$_2$ crystals against Brownian
movement. That is, for the case of stronger adhesion force of
remained clusters this ΔTl value becomes larger. From Figure 5, ΔTl
is decided as 23K.
Figure 6 shows nucleation characteristic of CaCl$_2\cdot$6H$_2$O melt
including BaS$_2$O$_3\cdot$H$_2$O. ΔTsc converge to about 3K at ΔTu range
smaller than 28K. The transition of ΔTsc distribution is recognized
slightly at ΔTu range of 30$\sim$32K. At ΔTu range larger than 34K,
ΔTsc converge to about 8K. ΔTl for this case is decided as 29K.
Since the tendency of this transition behavior of ΔTsc distribution
is similar to that of the behavior shown in Figure 5, the nucleation
mechanism of this case is considered to be similar to the case with
PbCl$_2$. However, ΔTsc of CaCl$_2\cdot$6H$_2$O melt including BaS$_2$O$_3\cdot$H$_2$O is
similar than that of the melt including PbCl$_2$ at each of ΔTu range

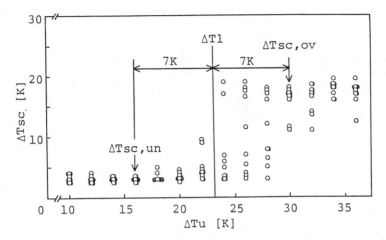

Figure 5. Effect of thermal history on supercooling temperature ΔTsc of $CaCl_2 \cdot 6H_2O$ melt with $PbCl_2$ as nucleation agent.

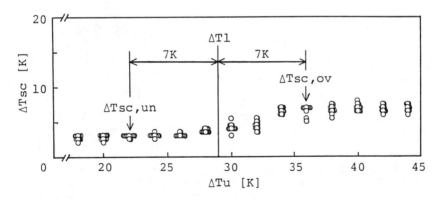

Figure 6. Effect of thermal history on supercooling temperature ΔTsc of $CaCl_2 \cdot 6H_2O$ melt with $BaS_2O_3 \cdot H_2O$ as nucleation agent.

beyond ΔTl. Because the particles of $PbCl_2$ and $BaS_2O_3 \cdot H_2O$ have
almost the same average diameter and standard deviation as shown in
Table I, it is difficult to consider that the difference between
such nucleation characteristics is influenced by the particle size
of nucleation agent.

The nucleation of $CaCl_2 \cdot 6H_2O$ melt including $BaS_2O_3 \cdot H_2O$ occurred
at cooling from ΔTu beyond ΔTl is considered to be caused by heterogeneous nucleation on surfaces of $BaS_2O_3 \cdot H_2O$ crystals which are
precipitated in melt. It can be considered that two experimental
facts mentioned below support this interpretation. One is a fact
that the occurrence of $CaCl_2 \cdot 6H_2O$ nuclei is observed during cooling
in the vicinity of surfaces of precipitated agent crystals using
optical microscope. The other fact is that the nucleation agent
lost activity by the heating beyond ΔTl shows the same acceleration
ability with that recognized for nucleation agent before activation
treatment ($\underline{4}$).

Consequently, $BaS_2O_3 \cdot H_2O$ crystals seem to have strong acceleration ability for the heterogeneous nucleation of $CaCl_2 \cdot 6H_2O$ melt.
On the other hand, the acceleration ability of $PbCl_2$ crystals seem
to be weak.

Mutual Contribution of Nucleation Mechanisms. In all cases of
nucleation characteristics of $CaCl_2 \cdot 6H_2O$ melt added thirteen kinds
of nucleation agent, the remained $CaCl_2 \cdot 6H_2O$ clusters can be assumed
to finish completely their thermal dissociation at ΔTu being 7K
higher than ΔTl. Then, ΔTsc, ov, which is an average value of ΔTsc
distribution at cooling from ΔTu being 7K higher than ΔTl, is
selected as a value representing the acceleration ability for
heterogeneous nucleation of $CaCl_2 \cdot 6H_2O$ melt on surfaces of precipitated agent crystals. On the other hand, ΔTsc, un, which is an
average value of ΔTsc distribution at cooling from ΔTu being 7K
lower than ΔTl, is selected as a value representing the activity of
remained $CaCl_2 \cdot 6H_2O$ clusters, because any thermal dissociation is
not confirmed in this heating condition.

The relationships between ΔTl, ΔTsc, un and ΔTsc, ov are respectively shown in Figures 7, 8 and 9. The number of each symbol is
corresponding to the number of nucleation agent listed in Table I.
A solid line added to each symbol shows the width of its ΔTsc
distribution.

Figure 7 shows a relationship between ΔTl and ΔTsc, un for all
salts. In the low region of ΔTl, ΔTsc, un is ranging in 3~4K and
the width of ΔTsc distribution is ranging in 2~11K. According to
the increase in ΔTl, ΔTsc, un decreases gradually and the width also
becomes smaller. As discussed above, the inversely proportional
relationship is recognized slightly between ΔTsc, un and ΔTl. The
tendency of such relationship is seemed to be suggesting that for
the case of stronger adhesion force of remained $CaCl_2 \cdot 6H_2O$ clusters
the activity of those clusters becomes higher, that is, the size of
those clusters becomes larger.

Figure 8 shows a relationship between ΔTsc, ov and ΔTsc, un for
all salts. According to the increase in ΔTsc, ov, ΔTsc, un increases
slightly and finally reaches to about 4K. This tendency probably
means that when the acceleration ability of nucleation agent for
heterogeneous nucleation is remarkably strong, the remained clusters
become more active. This phenomenon can be explained from the
reason that these acceleration functions for nucleation appear

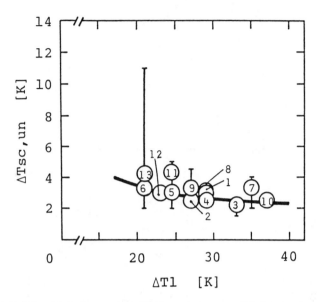

Figure 7. Relationship between upper limiting temperature ΔTl for superheating at which increase in supercooling temperature will begin and average value of supercooling temperature $\Delta Tsc,un$ at cooling from a temperature of 7 K lower than ΔTl.

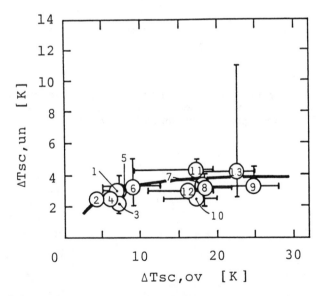

Figure 8. Relationship between average value of supercooling temperature $\Delta Tsc,ov$ at cooling from a temperature of 7 K higher than upper limiting temperature ΔTl and that of supercooling temperature $\Delta Tsc,un$ at cooling from a temperature of 7 K lower than ΔTl.

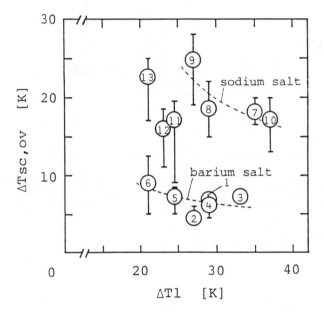

Figure 9. Relationship between upper limiting temperature ΔTl for superheating at which increase in supercooling temperature will begin and average value of supercooling temperature $\Delta Tsc,ov$ at cooling from a temperature of 7 K higher than ΔTl.

through both interfacial phenomena between precipitated nucleation agent and $CaCl_2 \cdot 6H_2O$ melt. Furthermore, a noteworthy point in Figure 8 is that $\Delta Tsc, ov$ of barium salt group distribute in a different range from those of other salt groups.

Figure 9 shows a relationship between ΔTl and $\Delta Tsc, ov$ for all salts. Any relationship between them is not clear on the whole. However, inversely proportional relationships are slightly observed for each group of barium salts and sodium salts. Considering such experimental result, it may be supposed that the adhesion force of remained clusters is controlled by more strong factors, such as a crack of the surface of nucleation agent occurred by the solid inclusion during the activation treatment.

Finally, it was found that $BaCO_3$ and $Na_4P_2O_7 \cdot 10H_2O$ have superior acceleration ability for the nucleation of $CaCl_2 \cdot 6H_2O$ melt in the comparison with the acceleration ability of nucleation agents shown in Figures 7, 8 and 9.

Conclusion

The effects of thermal history on the nucleation characteristics of $CaCl_2 \cdot 6H_2O$ melt without nucleation agent and the same melt with nucleation agent were studied experimentally by the cooling method at constant rate of 1K/min.

Two kinds of different acceleration mechanism for the nucleation of melt with agent were cleared. One of mechanisms was the nucleation accelerated by $CaCl_2 \cdot 6H_2O$ clusters which were remained without thermal dissociation on surfaces of precipitated agent even in a high temperature beyond its melting point. Another mechanism was the nucleation accelerated by heterogeneous nucleation from the supercooled melt to surfaces of agent. It was also considered that the nucleation behavior accelerated by heterogeneous nucleation appeared instead of the behavior accelerated by remained clusters and also according to the heating temperature rising, because such remained clusters began dissociation by heating beyond upper limiting temperature.

Furthermore, it was cleared that the adhesion force of the remained clusters to the agent and the acceleration ability of the agent for heterogeneous nucleation were closely related each other on the acceleration mechanism for the solidification of $CaCl_2 \cdot 6H_2O$ melt.

Literature Cited

1. Kosaka, M. Kogyozairyo 1978, 26, 54.
2. Malatidis, N. A.; Abhat, A. Forsch. Ing.-Wes. 1982, 48, 15.
3. Kimura, H.; Kai, J. Solar Energy 1984, 33, 557.
4. Watanabe, Y.; Saito, T. Proc. Intern, Symp. PFMIC '89 Osaka, 1989, P 52.
5. Nakai, T. Bull. Chem. Soc. Japan 1969, 42, 2143.

RECEIVED May 30, 1990

INDEXES

Author Index

Affiliation Index

Tokyo University of Agriculture and
　Technology, 251
Tufts University, 364
Università di Ferrara, 72
Università di Torino, 72
Universität Bremen, 43,210

University of Arizona, 115
University of Queensland, 198,292
The University of Texas at Austin, 102
University of Wisconsin, 316
Villanova University, 230
Waseda University, 271,281

Subject Index

Production: Paula M. Befard
Indexing: Deborah H. Steiner
Acquisition: Cheryl Shanks

Books printed and bound by Maple Press, York, PA
Dust jackets printed by Sheridan Press, Hanover, PA

Paper meets minimum requirements of American National Standard
for Information Sciences—Permanence of Paper for Printed Library
Materials, ANSI Z39.48–1984   ∞

Other ACS Books

Chemical Structure Software for Personal Computers
Edited by Daniel E. Meyer, Wendy A. Warr, and Richard A. Love
ACS Professional Reference Book; 107 pp;
clothbound, ISBN 0–8412–1538–3; paperback, ISBN 0–8412–1539–1

Personal Computers for Scientists: A Byte at a Time
By Glenn I. Ouchi
276 pp; clothbound, ISBN 0–8412–1000–4; paperback, ISBN 0–8412–1001–2

Biotechnology and Materials Science: Chemistry for the Future
Edited by Mary L. Good
160 pp; clothbound, ISBN 0–8412–1472–7; paperback, ISBN 0–8412–1473–5

Polymeric Materials: Chemistry for the Future
By Joseph Alper and Gordon L. Nelson
110 pp; clothbound, ISBN 0–8412–1622–3; paperback, ISBN 0–8412–1613–4

The Language of Biotechnology: A Dictionary of Terms
By John M. Walker and Michael Cox
ACS Professional Reference Book; 256 pp;
clothbound, ISBN 0–8412–1489–1; paperback, ISBN 0–8412–1490–5

Cancer: The Outlaw Cell, Second Edition
Edited by Richard E. LaFond
274 pp; clothbound, ISBN 0–8412–1419–0; paperback, ISBN 0–8412–1420–4

Practical Statistics for the Physical Sciences
By Larry L. Havlicek
ACS Professional Reference Book; 198 pp; clothbound; ISBN 0–8412–1453–0

The Basics of Technical Communicating
By B. Edward Cain
ACS Professional Reference Book; 198 pp;
clothbound, ISBN 0–8412–1451–4; paperback, ISBN 0–8412–1452–2

The ACS Style Guide: A Manual for Authors and Editors
Edited by Janet S. Dodd
264 pp; clothbound, ISBN 0–8412–0917–0; paperback, ISBN 0–8412–0943–X

Chemistry and Crime: From Sherlock Holmes to Today's Courtroom
Edited by Samuel M. Gerber
135 pp; clothbound, ISBN 0–8412–0784–4; paperback, ISBN 0–8412–0785–2

For further information and a free catalog of ACS books, contact:
American Chemical Society
Distribution Office, Department 225
1155 16th Street, NW, Washington, DC 20036
Telephone 800–227–5558